APPLIED STATISTICS
FOR ENGINEERS

APPLIED STATISTICS FOR ENGINEERS

second edition

WILLIAM VOLK

McGRAW-HILL BOOK COMPANY

New York   St. Louis   San Francisco   London
Sydney   Toronto   Mexico   Panama

APPLIED STATISTICS FOR ENGINEERS

Copyright © 1969, 1958 by McGraw-Hill, Inc. All Rights Reserved. Printed in the United States of America. No part of this publication may be reproduced, stored in a retrieval system, or transmitted, in any form or by any means, electronic, mechanical, photocopying, recording, or otherwise, without the prior written permission of the publisher. *Library of Congress Catalog Card Number* 69-11710

67551

1234567890 MAMM 754321069

For DOROTHY
whose confidence is without limits

PREFACE TO THE SECOND EDITION

Many advances have been made in the application of statistics to chemical engineering data handling since the first edition of this book was written. The principal change has been the increased availability of small computer facilities to engineering laboratories and colleges so that statistical analysis can be more readily applied. There has been a noticeable increase in the use of statistical criteria in the reporting of data in the chemical engineering literature.

This revised edition of "Applied Statistics" has four principal differences from the first edition.

1. It reflects its use as a text in applied statistics courses for engineers taught both by the author and by many others who have kindly made suggestions and pointed out errors in the first edition. To all who wrote commenting on the original edition, the author expresses his thanks.

2. Chapter 8 on fractional factorials is completely new in this edition. It presents a discussion of the use of the full and fractional 2^n factorial for the design and analysis of experimental data.

3. A short section on optimization in experimental design is included in Chapter 9 dealing with correlation and regression analysis.

4. A new chapter, Chapter 12, dealing with the use of computers for

statistical analysis, is included. This chapter discusses the use of computers for relatively small, routine statistical problems of analysis and correlation. Although there may be need of a book that deals solely with statistical analysis by computer, covering both the statistics and programming, Chapter 12 assumes some programming knowledge and discusses the use and interpretation of the programs that are available to engineers in the computer systems that are most readily available.

William Volk

PREFACE TO THE FIRST EDITION

This book was written with three specific classes of reader in mind: the engineer who has finished his formal training and who wishes to learn about the application of statistics to industrial data; the engineer or technician who is interested in applying statistics to some specific problems; and the instructor in the engineering college who wants a text dealing with statistical techniques rather than statistics theory.

In the last two decades the engineering profession has vastly expanded the scope of its operations. Low temperatures, new alloys and metals, fluid-solid processes, nuclear energy, jet engines, electronic computers—to name only a few of the new fields—are now all part of the engineer's domain. And the horizon is just as distant as ever. In the same way that an engineer will employ all the latest processes without a complete mastery of the basic sciences involved, so, too, many are desirous of applying statistics without necessarily relearning all the mathematics behind the formulas. This book is for them.

An attempt has been made to make each chapter an integral unit. Statistics is, of course, all interrelated, and it is not possible, for example, to discuss analysis of variance without a mention of means and the t test; nor is it possible to discuss correlation without mentioning variances. The reader wishing to use only some specific method from the text is advised to read Chapter 4 dealing with definitions and fundamental equations and then to turn to the specific chapter for the problem at hand. Each statistical method is illustrated with a numerical problem to serve as an example of the procedure discussed.

This book is the outgrowth of a course entitled Application of Statistics to Chemical Engineering Data, inaugurated at the Polytechnic Institute of Brooklyn in 1951 in the Graduate Division and under the Chemical Engineering Department. The book might well be titled, to use Professor Wilks' phrase, "The Ad Hoc Statistics Book for Chemical Engineers." The material presented here, plus the usual digressions for specific problems facing graduate students, covers a full year—two semesters—of instruction. The chapters dealing with probability and distributions are brief summaries of these subjects and are meant as refreshening sources of this information. Chapters 10 and 11 dealing with sequential analysis and nonparametric statistics are included to acquaint the engineer with these subjects and to supply some of the formulas for their application to data. The principal offerings to engineers are Chapters 4 to 9 dealing with the application of statistical tests to samples of data taken from the literature.

"No man is an island entire of itself." This statement is especially applicable to the author of this book. None of the statistical developments presented is original, although some of the handling may differ from the stylized forms to facilitate the application. My debt to earlier authors is therefore profound, and the creditors too numerous to list. Specific references are given at the end of each chapter. I am obliged, however, to state that I am indebted to Professor Sir Ronald A. Fisher, Cambridge, to Dr. Frank Yates, Rothamsted, and to Oliver & Boyd, Ltd., Edinburgh and London, for permission to reprint tables from their book "Statistical Tables for Biological, Agricultural and Medical Research"; and to Dr. George W. Snedecor and the Iowa State College Press for permission to reprint tables from their book "Statistical Methods," fifth edition. I am also indebted to Miss Dorothy Kichinko and Mrs. Katherine Ellington for graciously typing the manuscript; to Nicholas Capone for the illustrations; and to the Friden Calculating Machine Co. for the use of one of their desk calculators for speeding the arithmetic involved in this book. My brother, Victor Volk, and my friend Dr. A. N. Peiser read parts of the manuscript and offered criticisms, for which I am grateful.

Although I am indebted to many, named and unnamed, for this book, I must accept responsibility for any errors still remaining. I shall be happy to hear from readers and users comments favorable and unfavorable which might improve the book either in substance or presentation.

William Volk

CONTENTS

Preface to the second edition   vii
Preface to the first edition   ix

one	PROBABILITY	1
two	PERMUTATIONS AND COMBINATIONS	13
three	DISTRIBUTIONS	26
	Binomial 27	
	Poisson 46	
	Normal 53	
four	MEASURES OF VARIABILITY	63
five	X^2, CHI SQUARE	82
six	THE t TEST	109
seven	ANALYSIS OF VARIANCE	149
eight	THE 2^n FACTORIAL	235
nine	CORRELATION—REGRESSION	260
ten	SEQUENTIAL ANALYSIS	342
eleven	NONPARAMETRIC STATISTICS	364

| twelve | COMPUTER CALCULATIONS | 389 |

TABLES 401

Area under the Normal Curve, $F(z)$ 403
X^2 404
t table 405
$F_{0.05}$ 406
$F_{0.01}$ 408

Index 411

**APPLIED STATISTICS
FOR ENGINEERS**

chapter one
PROBABILITY

If a weight balance is reported as 98.6 per cent, might the true value be 98.5? 98.3? 95.0? If one process gives an octane number of 87 and another an octane number of 85, is the first really different from the second? When you draw a line through some data on graph paper, is there really a correlation, or could you get just as good results by scattering grass seed over the paper? It is to the solution of problems of this type that the statistics of this book is applied.

There are two meanings to the word *statistics*. As a plural, statistics refer to the accumulation of data, as by the Bureau of Labor Statistics, for example. There are all sorts of organizations collecting all sorts of statistics: football, baseball, horse racing, life insurance, automobile production, movie attendance, etc. If statistics of some particular nature, no matter how esoteric, are desired, it is usually possible to find them. If you want to know the number of delicatessen stores in the United States—9,909 in 1939 and 8,212 in 1948.[1] If you want to know the number of one-hundred-dollar bills in circulation in the United States—9.19 million in 1939 and 50.74 million in 1948.[2] More one-hundred-dollar

[1] Statistical Abstract of the United States, U.S. Department of Commerce, Washington, 1955.
[2] The World Almanac, New York World Telegram, New York, 1951.

bills: fewer delicatessen stores. (Before you draw any conclusions from these few statistics, please read the rest of this book.)

The singular usage of the word statistics refers to the application of mathematical laws of probability to samples of data so that statements made from them about the source from which the samples were taken can be set within definite bounds. If all the information about a material or process is available, then statistics is not necessary to evaluate the information, although it may help in simplifying the presentation of the data. If information about only a sample of the material or process is available, then statements about the sample may not necessarily be true for the whole system. Statistics provides a mathematical basis for setting probability limits to statements made about the total system from a study of the sample.

In this book we deal only with the second meaning of the word statistics.

Although the laws of probability are involved in the background of all statistical statements, it is not necessary for an engineer to be thoroughly familiar with all the mathematics of the various probability distributions involved in order to apply statistics to his data. However, the engineer who does apply statistics usually will find that he is expected to be able to solve problems dealing with probability, sometimes outside of the field of engineering—at the bridge table for example. Besides, the concept of probability is basic to the statistical analysis of data, so that an understanding of probability is essential to the correct interpretation of statistical conclusions.

This chapter deals with the elementary probability theorems so that the student who is new to the subject can become acquainted with the fundamental ideas. For the engineer who has studied probability, perhaps with freshman algebra, this chapter will provide a brief review.

1.1 Definitions

Probability in a numerical sense is a quantitative measure of chance or likelihood in a manner similar to that in which temperature is a quantitative measure of heat level. The engineer in New York, at the Arctic Circle, and at the Equator might mean different things by hot, warm, cold; but they all would agree on what is meant by 212°F. Likewise, two men might not mean the same thing when they say an event has a "fair chance" of occurring; but they would agree that if an event had a 0.6 probability of occurring it would be expected to happen on the average 6 times out of 10.

Probability is expressed as a fraction between 0 and 1 and is usually designated p. The lower case p indicates the general probability, while the upper case P, that is, $P(A)$ indicates the probability of a specific event, A. An impossible event has a probability of 0, and an event that is certain to occur has a probability of 1. A probability value between 0 and 1 indicates that in the long run an event will occur that fraction of the total number of times. If the probability of a failure in a process is 0.02, we expect a failure to occur on the average twice out of every hundred trials.

In mathematical terms, if N events are possible and equally likely and n of them have an attribute A, and if we designate the probability of A as $P(A)$, then the probability of A occurring in any single event is

$$P(A) = \frac{n}{N} \tag{1.1}$$

If a plant has a fleet of 27 trucks, 3 of which are defective, and if the likelihood of being assigned any one truck is the same as that of being assigned any other truck, then the probability of getting a defective vehicle is $3/27 = 0.111$. Or perhaps a more common example: the probability of picking an ace from a deck of cards is $4/52 = 0.0769$. The probability of picking a black ace is $2/52 = 0.0385$. And the probability of picking one particular ace, say diamonds, is $1/52 = 0.0192$.

Odds is a word that is misused when the meaning intended is probability. Odds does not have as precise a mathematical definition as probability, but its common usage is to supply a multiplier for purposes of a wager in the form of a ratio of the amount to be gained to the amount to be lost. With "true" odds, the amount to be gained times the probability of winning will equal in absolute value the amount to be lost times the probability of losing. For example, if the probability of event A, $P(A)$, is 0.25, so that the probability of event A not occurring, $P(\tilde{A})$, is 0.75, the odds for event A are 3/1. If A occurs, 3 units are gained, and if A does not occur, 1 unit is lost. $P(A) \times 3 = P(\tilde{A}) \times 1$; $(0.25)(3) = (0.75)(1)$. In general, if the probability is p and $1 - p = q$, then the odds are q/p, usually stated as the ratio of two whole numbers. If the product of the probability of an event and the numerator of the odds is greater than the product of the probability of the event not occurring and the denominator of the odds, then the odds are "favorable." If the reverse is true, and the former product is less than the latter product, then the odds are "unfavorable." If the products are equal, the odds are correct, or fair, or true.

Mathematical expectation is a method of evaluating events in terms of their probabilities and numerical value, particularly for multiple choice situations. The mathematical expectation, ME, is the algebraic sum of the products of the dollar value and the probability of events. If the dollar value of event A is symbolized $\$(A)$, the mathematical expectation can be expressed as

$$ME = \sum_i \$(A_i) \cdot P(A_i)$$

where the summation is taken over all possible events.

If the mathematical expectation is positive, the situation is advantageous. If it is negative, the situation is disadvantageous. If the mathematical expectation is exactly zero, the situation is completely fair or indifferent.

For example, I can invent a game with a deck of 52 cards in which I say that I will select a card at random and then you select a card at random. If my card is higher than yours, you pay me two dollars, and if your card is higher than mine, I will pay you one dollar. If they are both the same, I will pay you five dollars. What is your mathematical expectation? Is the game advantageous to you? How much should I pay you when the cards are the same to make the game completely fair?

As shown later in this chapter in Example 1.5, the probability of your card being higher than mine is 24/51, of being lower than mine is 24/51, and of being exactly the same as mine is 3/51. Your mathematical expectation therefore is

$$\$1.00(24/51) + (-\$2.00)(24/51) + \$5.00(3/51) = -\$9/51$$
$$= -\$0.176$$

The negative value indicates the game is not in your favor. To have the game completely fair, ME should equal zero. The payment for matching cards to make the ME zero can be solved for by rearranging the equation:

$$ME = \$1.00(24/51) + (-\$2.00)(24/51) + \$X(3/51) = 0.00$$
$$X = \$8.00$$

Another use of mathematical expectation is to have a quantitative method for comparing alternative choices. The one with the higher ME is mathematically the better choice. For example if a company building motels wants to choose between two alternative locations for its next units, it might proceed as follows.

In the first location there is a 5/8 chance of being successful and making $40,000 a year, a 1/4 chance of only being moderately successful and making $5,000 a year, and a 1/8 chance of having a failure and losing $15,000 a year. In the second location, there is only a 1/2 chance of being successful, but if it is successful it will profit $80,000 a year. There

PROBABILITY

is a 1/4 chance of losing $5,000 a year and a 1/4 chance of losing $15,000 a year. (Note that in both cases the total probability has to add up to 1.0.) The ME for the two alternatives are calculated as follows:

$$ME_1 = \$40,000(5/8) + \$5,000(1/4) + (-\$15,000)(1/8)$$
$$= \$24,375$$
$$ME_2 = \$80,000(1/2) + (-\$5000)(1/4) + (-\$15,000)(1/4)$$
$$= \$35,000$$

The second alternative is mathematically better, even though it has less chance of making a profit.

In comparing mathematical expectations, the alternatives are, of course, not limited to two.

In defining probability in the form of Eq. (1.1), it is customary to stipulate that the events be mutually exclusive and equally likely. These restrictions are implicit in the definition.

When the probability of an event is set purely by a definition of the system, this is known as an *a priori* probability. The a priori probability of drawing an ace from a deck of cards is $4/52 = 0.0769$. This probability is strictly a mathematical probability and does not necessarily have to apply (although it may) to some particular deck of cards. It is based on the idealized situation that there are 4 aces and 48 cards which are not aces, that each card has an equal likelihood of being drawn, and that only one card will be drawn. In a real situation other factors, known or unknown to the card drawer, often enter to change the actual effective probabilities.

When the numerical value of the probability is arrived at from accumulated data or experience, it is referred to as an *empirical* probability. Insurance rates are the most common example of empirical probabilities; actually they are odds based on empirical probabilities. Generally, engineering statistics is based on a combination of both types of probabilities; inspection of processes and data which leads to empirical probabilities, and comparison with statistical distributions which involves a priori probabilities.

1.2 Some Laws of Probability Relationships

1.2.1 Multiplication Law.
If A and B are independent events and the occurrence of one does not affect the chance of the occurrence of the other, then the probability of both events is equal to the product of the probabilities of the separate events.

$$P(A \text{ and } B) = P(A,B) = P(A)P(B) \tag{1.2}$$

example 1.1. The probability of drawing two aces, one each from two separate decks of cards, is equal to $4/52 \times 4/52 = 1/169 = 0.00592$.

If more than two events are involved and their individual chances of occurrence do not affect each other, then the probability of them all occurring is equal to the product of all the separate probabilities.

$$P(B_1, B_2, B_3, \ldots, B_r) = \prod_{i=1}^{r} P(B_i) \tag{1.3}$$

1.2.2 Addition Laws. If A and B are independent events *and are mutually exclusive*, then the probability of A or B equals the sum of the individual probabilities.

$$P(A \text{ or } B) = P(A) + P(B) \tag{1.4}$$

By mutually exclusive we mean that both events cannot occur; that is, $P(A,B) = 0$. Getting both heads and tails in a single toss of a coin is an example of mutually exclusive events.

example 1.2. The probability of drawing either an ace or a king in one draw from a deck of cards is $4/52 + 4/52 = 8/52 = 0.154$.

If more than two independent and mutually exclusive events are involved, the probability of any one of them is equal to the sum of all the individual probabilities.

$$P(B_1 \text{ or } B_2 \text{ or } B_3 \text{ or } \cdots B_r) = \sum_{i=1}^{r} P(B_i) \tag{1.5}$$

If A and B are independent events and *are not mutually exclusive*, the probability of at least one of these events is equal to the sum of the individual probabilities less the probability of both events.

$$P(A \text{ and/or } B) = P(A) + P(B) - P(A,B) \tag{1.6}$$

example 1.3. The probability of drawing at least one ace in two separate draws, one from each of two separate decks of cards, is $4/52 + 4/52 - 1/169 = 25/169 = 0.148$.

The generalization of this law for more than two categories of events is obtained by adding all the odd-number combinations and subtracting all the even-number combinations. For three categories, this is

$$P(A \text{ and/or } B \text{ and/or } C) = P(A) + P(B) + P(C) \\ - P(A,B) - P(A,C) - P(B,C) + P(A,B,C) \quad (1.7)$$

The correctness of including the negative terms in Eqs. (1.6) and (1.7) may not be obvious. Consider the case of tossing two coins. What is the probability of obtaining at least one head? The probability of obtaining one head with one coin is 0.5. If we were simply to add the probabilities for heads for the two coins, we should obtain $0.5 + 0.5 = 1.0$, indicating that the probability of obtaining at least one head was 1.0, or a certainty. But this is not true since the event of two tails can occur. The true probability of obtaining at least one head is $0.5 + 0.5$ minus the probability of two heads, $(0.5)(0.5) = (0.25)$, giving a correct answer of 0.75. Consider the case of tossing three coins. What is the probability of obtaining at least one head? If the events were mutually exclusive, the probability would be $0.5 + 0.5 + 0.5 = 1.5$, an answer without meaning. There are eight ways the three coins may be tossed: HHH, HHT, HTH, THH, HTT, THT, TTH, TTT. Seven of these results have at least one head, so that the probability is $7/8 = 0.875$. Using Eq. (1.7) the same result is obtained as follows: $P = 0.5 + 0.5 + 0.5 - (0.5)(0.5) - (0.5)(0.5) - (0.5)(0.5) + (0.5)(0.5)(0.5) = 0.875$; this is the sum of the probabilities of heads for each coin separately, less the probability of heads for each possible pair of coins, plus the probability of heads for the three coins simultaneously.

1.2.3 Law of Conditional Probability. If an event B is contingent upon the prior occurrence of event A, then the probability of B is termed a conditional probability, i.e., on the condition that A has occurred, and is designated $P(B|A)$. The probability of a compound event A and B is equal to the probability of one event times the conditional probability of the other event under the condition that the first event has occurred.

$$P(A,B) = P(A)P(B|A)$$
$$= P(B)P(A|B) \quad (1.8)$$

If the events are independent and neither is contingent on the other, then $P(A|B) = P(A)$, and $P(B|A) = P(B)$, and Eq. (1.8) reduces to Eq. (1.2).

example 1.4. The probability of drawing two aces from one deck of cards in two consecutive draws without replacing the first card is $(4/52)(3/51) = 12/2{,}652 = 0.00452$.
    $P(\text{first card is ace}) = 4/52$

$P(\text{second ace}|\text{first ace}) = 3/51$ since if first card is an ace there are 3 aces left in 51 cards

$P(2 \text{ aces}) = P(\text{first ace}) \times P(\text{second ace}|\text{first ace}) = (4/52)(3/51)$.

Table 1.1 will illustrate the law of conditional probability. The extension of this table to include more than two categories will be obvious from the explanation which follows. The occurrence of event A is designated A, and the nonoccurrence of event A is designated $\tilde{A}$, that is, not A. The number of occurrences of event A is n_A, of event B is n_B; the number of occurrences of events A and B occurring together is n_{AB}, and the number of occurrences of events which include neither A nor B is n. The total number of events is $n + n_A + n_B + n_{AB} = N$.

TABLE 1.1 Frequency Table of Events A and B

Category	A	$\tilde{A}$	Sum
B	n_{AB}	n_B	$n_{AB} + n_B$
$\tilde{B}$	n_A	n	$n_A + n$
Sum........	$n_{AB} + n_A$	$n_B + n$	N

From Eq. (1.1), in which the probability of an event was defined as the frequency of that event over the frequency of all events, the following probabilities can be established:

$$\text{Probability of } A, \; P(A) = \frac{n_{AB} + n_A}{N} \qquad (1.9)$$

$$\text{Probability of } B, \; P(B) = \frac{n_{AB} + n_B}{N} \qquad (1.10)$$

$$\text{Probability of } A \text{ and } B, \; P(A,B) = \frac{n_{AB}}{N} \qquad (1.11)$$

Conditional probability of A if B has occurred:

$$P(A|B) = \frac{n_{AB}}{n_{AB} + n_B} \qquad (1.12)$$

Conditional probability of B if A has occurred:

$$P(B|A) = \frac{n_{AB}}{n_{AB} + n_A} \qquad (1.13)$$

Multiplying Eq. (1.9) by (1.13) or (1.10) by (1.12), we get

$$P(A,B) = P(A)P(B|A) = P(B)P(A|B) \qquad (1.8)$$

Table 1.1 may also be used to illustrate the second addition law of Eq. (1.6). The frequency of events A or B or both is $n_A + n_B + n_{AB}$. Therefore the probability of A and/or B is

$$P(A \text{ and/or } B) = \frac{n_A + n_B + n_{AB}}{N} \qquad (1.14)$$

If Eqs. (1.9), (1.10), and (1.11) are substituted into (1.14), we get

$$P(A \text{ and/or } B) = P(A) + P(B) - P(A,B) \qquad (1.6)$$

The probability relationships are summarized in the following tabulation:

TABLE 1.2

Event relationship	Probability of both A and B $P(A$ and $B), P(A,B)$	Probability of A or B or both $P(A$ and/or $B)$
Mutually exclusive	0	$P(A) + P(B)$
Independent	$P(A) \cdot P(B)$	$P(A) + P(B) - P(A,B)$
Not independent	$P(A) \cdot P(B|A)$	$P(A) + P(B) - P(A,B)$

The probability relationships given in Table 1.2 and the definition of probability given in Eq. (1.1) are relatively simple to employ. Difficulties arise in enumerating the number of events with the given attribute, enumerating the total possible events, and determining when events are independent and conditional. A few examples will illustrate these remarks. No general rules can be given, and each situation must be analyzed individually.

example 1.5. Previously in the chapter we said that if you picked a card from a deck of 52 cards, the probability of the next card picked being higher than yours was 24/51. Before we illustrate the solution we will define the problem more completely. The 52-card deck consists of 4 each of 13 denominations ranging from 1 to 13. The 1 is the lowest value and the 13 is the highest value. Any one of the 52 cards is equally likely to be picked. The card picked first is not returned to the deck before the second card is picked.

The probability of a higher second card, $P(H)$, is the sum of the probabilities of thirteen mutually exclusive compound probabilities, one for each of the thirteen different denominations of cards that can be drawn first and the probabilities that the next card is higher.

$$P(H) = \sum_i P(H,X_i) \tag{1.15}$$

Each individual compound probability, $P(H,X_i)$ is the product of the probability of drawing card X_i on the first draw and the conditional probability of drawing a higher card on the second draw given X_i on the first draw.

$$P(H,X_i) = P(X_i) \cdot P(H|X_i) \tag{1.16}$$

The total probability of a higher second card is therefore the sum of the thirteen individual probabilities.

$$P(H) = \sum_i P(H,X_i) = \sum_i P(X_i) \cdot P(H|X_i) \tag{1.17}$$

The value of $P(X_i)$ is the same for all values of i from 1 to 13, that is, for all possible denominations of cards drawn first. There are four cards of each denomination out of a possible 52 cards, so that all $P(X_i) = 4/52$.

The value of the conditional probability, $P(H|X_i)$ is different for each value of i. If the first card drawn is a 1, there are 12 denominations or 48 cards higher. Since one card has been drawn from the deck, there are 51 remaining and the probability of drawing a higher card than a 1 is 48/51. If the first card is a 2 there are 44 cards higher and the probability of drawing one of these is 44/51. In general, the number of cards higher than the card drawn first can be seen to be $4(13 - i)$ and the probability corresponding to each value of i is $4(13 - i)/51$. Equation (1.17) therefore can be rewritten

$$P(H) = \sum_i^{13} (4/52)(4/51)(13 - i) = 24/51 \tag{1.18}$$

The sum of the probabilities of all the possible and impossible (with probabilities equal to zero) outcomes of any given situation must, by definition, equal 1.0. In a situation similar to example 1.5, it is possible to check the calculated probabilities by calculating the probabilities for all events and determining whether their sum equals 1.0. In example 1.5 only three possibilities exist: the second card is either higher, equal to, or lower than the first. We have illustrated the calculation for the first possibility with a result of 24/51. The interested reader can calculate the other two possibilities, which are 3/51 and 24/51 to give a total of 51/51 or 1.0.

Example 1.5 was intended to illustrate the necessity of enumerating all of the possible events in order to calculate the probability of any specific combination. The next example is intended to show the necessity of determining the dependence of events on each other in order to calculate their probabilities.

example 1.6. Suppose you have a vessel containing 4,000 red and 6,000 white chips. You put your hand in and draw one chip. What is the probability it is red? The answer appears to be 0.400.

But suppose the vessel is divided into three compartments of equal size so that you are equally likely to put your hand into one compartment as into another. The first compartment contains 2,000 red and 3,000 white chips. The second contains 1,000 red and 2,000 white chips, and the third compartment contains 1,000 of each. What is the probability in this case of drawing a red chip? There are still 4,000 red and 6,000 white chips, but now the probability of drawing a red one is a conditional probability, conditional on which compartment your hand has entered.

$$P(\text{red}) = \sum_i P(\text{compartment}_i) \cdot P(\text{red}|\text{compartment}_i)$$

In each case the probability for the first term is 1/3, but the probability of the second term changes for each compartment. The total probability of obtaining a red chip is

$$P(\text{red}) = 1/3 \times 2/5 + 1/3 \times 1/3 + 1/3 \times 1/2 = 37/90 = 0.411$$

Suppose there is twice as much chance of putting your hand into the first compartment (the opening is twice as large as each of the other openings) as into either of the other compartments. What is the probability in that case of getting a red chip? $P(\text{red}) = 49/120 = 0.408$.

These rather simple examples are merely to introduce the concept of probability calculations and to show the necessity of completely defining the situation before calculating the probability. Again, as in the previous example, since either a red or a white (according to the definition of the problem) chip must be drawn, the sum of the two probabilities will equal 1.0. The reader can calculate the probability of getting a white chip for each case as an exercise.

1.3 Bayes' Theorem of Inverse Probability

An extension of the law of conditional probability can be applied to establish the probability that an event which has already occurred might have occurred in a particular way. For example, in the situation of example 1.6, if you have drawn a red chip, what is the probability it came from the second compartment?

If an event A is dependent on one of several mutually exclusive events $B_1, B_2, \ldots, B_n$, and if event A has occurred, then the probability that event B_i has also occurred is defined by Bayes' theorem as

$$P(B_i|A) = \frac{P(B_i)P(A|B_i)}{P(A)} \tag{1.19}$$

example 1.7. Applying Bayes' theorem to the data of the last part of example 1.6 (probability of first compartment 1/2, probability of each of the other compartments 1/4; red chips in first compartment 2,000/5,000, in second compartment 1,000/3,000, and in the third compartment 1,000/2,000), we have obtained one chip and it is red. What is the probability it came from the second compartment?

$P(B_i) = P(\text{compartment 2}) = 1/4$

$P(A|B_i) = P(\text{red chip}|\text{compartment 2}) = 1/3$

$P(A) = P(\text{red chip}) = 49/120$ (from example 6.1)

$P(B_i|A) = P(\text{compartment 2}|\text{red chip})$

$\qquad = (1/4)(1/3)/(49/120) = 10/49 = 0.204.$

chapter two
PERMUTATIONS AND COMBINATIONS

The solution of problems of a priori probability involves the determination of the number of events that have the attribute whose probability is to be found and the total number of events possible within the situation studied. The calculation of the number of events can often be facilitated by recourse to certain rules of permutations and combinations. These rules are also sometimes useful in experimental design in determining the number of experiments required to cover all levels of all the variables under consideration. In sampling of discrete items from a large source of such items, the sample is a combination of the items drawn, while the order of drawing is a permutation of the items.

The introduction of the concept and formulas for the systematic study of the arrangement of numbered events is usually made in a course of algebra. This chapter presents the subject briefly as a matter of review for the sake of completeness.

2.1 Permutations

A permutation is an arrangement. AB and BA are two different permutations of the letters A and B taken together. The word *permutation* is usually restricted to arrangements of things taken without replacement.

Selecting five cards from a deck of 52 without replacement and noting the order in which the cards are obtained is an example of obtaining a permutation of 52 things taken five at a time. It is also an example of sampling without replacement. If the 52 cards were mounted on a 52-sided polygon, (a dopentadecagon if there were such a thing), this polygon rolled five times, and the upturned face recorded for each roll, the resulting order of the exposures would be an example of an arrangement of 52 things taken five at a time with replacement. It is also an example of sampling with replacement. This latter arrangement is sometimes referred to as a permutation with replacement.

Words are permutations of letters. "Eat" and "tea" are two different permutations of the same three letters and designate two different words. There is a word game which involves making the most words out of some large word; for example, how many words can be made from the single word "chemistry"? "Mystic," "chest," "hysteric," "itchy," "miser," etc. These are all simply different permutations of the letters selected from the original word, with the added restriction in this case that the permutation must be an intelligible word. The examples of the various laws of permutations will be illustrated in terms of letters, but the application to any system of categories will be the same.

Law 1. The number of arrangements of n different things taken n at a time from a supply in which the different items are present in a quantity at least as large as n, that is, with replacement, is n^n.

example 2.1. The number of arrangements of the letters of the alphabet taken 26 at a time with replacement is $(26)^{26}$. The first letter can be any one of 26, the second letter likewise can be any one of 26, and so on, so that the total number of ways the 26 letters can be taken is $(26)(26)(26) \cdots$ for 26 successive terms.

Law 2. The number of arrangements of n different things taken r at a time with replacement is n^r.

example 2.2. The number of arrangements of five-letter code words that can be made from the letters of the alphabet is $(26)^5$. This illustration follows directly from the first. With a five-letter code word, the first letter may be selected in 26 ways, the second letter in 26 ways, and so on for all five letters.

Law 3. The number of permutations (arrangements) of n different things taken n at a time without replacement is $n!$. The symbol for this permutation is $P_n{}^n$, therefore

$$P_n{}^n = n! \tag{2.1}$$

example 2.3. The number of arrangements of the 26 letters of the alphabet taken all at one time without duplicating any letter is 26!. The first letter may be taken 26 ways, leaving 25 ways for the next letter, 24 for the third, and so on, so that the total number of ways is $(26)(25)(24) \cdots (2)(1) = 26!$.

Law 4. The number of permutations of n different things taken r at a time without replacement is $n!/(n-r)!$. The symbol for this permutation is $P_r{}^n$; therefore

$$P_r{}^n = \frac{n!}{(n-r)!} \tag{2.2}$$

example 2.4. The number of arrangements containing five letters each that can be made from the 26 letters of the alphabet without duplication is $26!/21!$. The first letter can be selected in 26 ways, the second in 25, and so on, so that all five letters can be selected in $(26)(25)(24)(23)(22)$ ways. This product is equivalent to $26!/21!$ or $26!/(26-5)!$.

Law 5. The number of permutations of n things, not all different, of which there are n_1 of one kind, n_2 of another, n_3 of a third, up to n_k, wherein $\sum_{i=1}^{k} n_i = n$, taken all at one time, is equal to $n!/n_1!n_2!n_3! \cdots n_k!$.

example 2.5. The number of distinguishably different arrangements of the letters in the word "statistician" is $12!/(2!)(3!)(2!)(3!)(1!)(1!)$; there being 12 letters made up of two s's, three t's, two a's, three i's, one c, and one n. The number of permutations of n_1 things taken all at once is, by Eq. (2.1), $n_1!$. If all these things are alike, then the different permutations will be indistinguishable from each other. Therefore if n_1 is part of a larger n, then the total permutations of $n!$ will be decreased proportionally to the number of permutations of these like terms. In the example with the word "statistician," the total permutations of the 12 letters, 12!, include, for example, various arrangements of the three t's. The number of arrangements (permutations) of these three t's is 3!. Since none of these arrangements is distinguishable from another, the total number of permutations is decreased by a factor equal to 3!. Likewise for all the other replicated letters.

2.2 Combinations

A combination of things is a group of these things without regard to the order in which they are taken. A sample of 10 items, made up of eight acceptable and two defective, is considered one combination regardless of

TABLE 2.1 Combinations of Five Things Taken in Different Group Sizes

Group size	Combinations	No. of combinations
0	0	1
1	A, B, C, D, E	5
2	AB, AC, AD, AE, BC, BD, BE, CD, CE, DE	10
3	ABC, ABD, ABE, ACD, ACE, ADE, BCD, BCE, BDE, CDE	10
4	ABCD, ABCE, ABDE, ACDE, BCDE	5
5	ABCDE	1

whether the two defectives are the first items, the last items, or whether they are scattered throughout the sample.

If there are five different things, A, B, C, D, and E, the number of combinations of these taken in different group sizes is as shown in Table 2.1.

Law 6. In general, the number of combinations of n things taken r at a time, designated $\binom{n}{r}$, is

$$\binom{n}{r} = \frac{n!}{(r!)(n-r)!} \tag{2.3}$$

example 2.6. The number of five-card hands that can be made from a 52-card deck is

$$\binom{52}{5} = \frac{52!}{(5!)(47!)} = 2{,}598{,}960$$

PERMUTATIONS AND COMBINATIONS

The number of five-card hands all of one suit (flushes) that can be made from a 52-card deck consisting of 13 cards each of four different suits is

$$4\binom{13}{5} = \frac{(4)(13!)}{(5!)(8!)} = 5{,}148$$

The probability, from Eq. (1.1), of getting a flush on a five-card draw from a deck of 52 cards is

$$P(\text{flush}) = \frac{5{,}148}{2{,}598{,}960} = 0.00198$$

about 2 chances in 1,000.

The number of combinations of n things taken n at a time, $\binom{n}{n}$, from Eq. (2.3), is $n!/(n!)(n-n)!$. Since $0!$ is by definition equal to 1, $\binom{n}{n} = n!/n! = 1.0$.

2.3 Relation between Permutations and Combinations

The number of permutations of n things taken r at a time, by Law 4, is $n!/(n-r)!$. Included in the total number of permutations is the number of permutations of each different group of r things taken r at a time. Therefore there is a factor corresponding to $r!$ which represents the different permutations of the same combination of r things. The number of different combinations, then, of n things taken r at a time is the number of permutations of n things taken r at a time divided by the number of permutations of r things taken r at a time. $\binom{n}{r} = P_r^n/P_r^r$.

$$\binom{n}{r} = \frac{n!/(n-r)!}{r!} = \frac{n!}{(r!)(n-r)!} \tag{2.3}$$

which is identical with the equation of Law 6.

Another way of looking at this relation is by reference to Law 5, in which the number of permutations of n things, some of which were identical, taken n at a time, is defined. If the n things are of two kinds, r of one kind and $n-r$ of another, by Law 5, the number of permutations of these n things taken n at a time is $n!/r!(n-r)!$. This is identical with the number of combinations of n things taken r at a time. Therefore the number of combinations can be looked upon as the number of ways the n things can be split into two groups, one containing r and the other $n-r$ things.

2.4 Combinations of Events

Law 7. If the number of ways event A can occur is k, and the number of ways event B can occur is m, and A and B can not occur simultaneously, then the number of ways events A or B can occur is $k + m$.

example 2.7. If you can lose at dice on the first roll in four ways, and if you can win in eight ways, you can either win or lose on the first roll in $8 + 4 = 12$ ways. And if you can roll two dice in 36 ways, the probability of winning or losing on the first roll, the value of Eq. (1.1), is 12/36 or 0.333.

Law 8. If the number of ways event A can occur is k, and the number of ways event B can occur is m, the compound event (A,B) can occur in $k \cdot m$ ways.

example 2.8. If a laboratory has four analysts, and if an analysis can be done in three different ways, then if an analytical result is left entirely to chance, it can be done in $3 \times 4 = 12$ different ways.

An event may be a single observation such as a failure in a reaction vessel or the completion of a successful experiment; or an event might be a combination of a number of observations such as a sample of 20 items with two defectives, or a week's operation with 98 per cent successful runs. We refer to the single events as *observations* or *specimens* and to events made up of several observations as *samples*. Of course, samples may actually be single observations, but as we hope to point out, samples made up of only one observation are of very limited value.

Suppose you have a batch of material (a population) consisting of N items. Suppose that x of these are defective. What is the probability that in a sample of n items you will find r defectives? The number of ways you can draw r defectives when there are x available is $\binom{x}{r}$. Call this event A. The number of ways you can draw $(n - r)$ good items from the $(N - x)$ good items in the population is $\binom{N - x}{n - r}$. Call this event B. Then the number of ways you can get r defectives and $(n - r)$ good items in a sample size n from the specified population, i.e., the compound event (A,B), is $\binom{x}{r}\binom{N - x}{n - r}$, by Law 8. The number of ways you can draw a sample of size n from the population of N items is $\binom{N}{n}$.

PERMUTATIONS AND COMBINATIONS

Therefore the probability of getting exactly r defectives in a sample size n is

$$P(r)_n = \frac{\binom{x}{r}\binom{N-x}{n-r}}{\binom{N}{n}}$$

$$= \frac{n!(N-x)!n!(N-n)!}{r!(n-r)!(n-r)!(N-x-n+r)!N!}$$

This probability, the hypergeometric probability, involves four variables and is cumbersome to tabulate. The calculation by the use of factorials, however, is relatively simple.

2.5 Factorials

The determination of the values of permutations and combinations involves the evaluation of the factorials $n!$, $r!$, etc. For relatively small values of n, the factorials can usually be as quickly obtained on a desk calculator as they can be looked up in a table. One of the popular makes of desk calculators permits the progressive multiplication involved in computing factorials by the automatic reentry of the product into the machine as multiplicand without the actual punching of the keyboard. Since the ultimate calculation usually involves the quotient of at least two factorials, its answer can be obtained by canceling the common terms in the numerator and denominator and then calculating the remainder on a desk calculator.

For example, the number of five-card hands from a deck of 52 cards in example 2.6 is the combination $\binom{52}{5} = 52!/(5!)(47!)$. This may be written

$$\frac{52 \times \overset{17}{\cancel{51}} \times \overset{5}{\cancel{50}} \times 49 \times \overset{12}{\cancel{48}} \times \cancel{47!}}{1 \times \cancel{2} \times \cancel{3} \times \cancel{4} \times \cancel{5} \times \cancel{47!}} = 52 \times 17 \times 49 \times 60 = 2{,}598{,}960 \quad (2.4)$$

For more cumbersome calculations, or if a desk calculator is not available, the quotient of factorials can be obtained by means of logarithms. The equation of example 2.6 could be written

$$\log\binom{52}{5} = \log 52! - \log 5! - \log 47! \quad (2.5)$$

The solution to Eq. (2.5) by means of logarithms is:

$\log 52! = 67.90665$

$\log 5! = 2.07918$

$\log 47! = 59.41267$

$\log \binom{52}{5} = \log 52! - \log 5! - \log 47!$

$\qquad = 67.90665 - 2.07918 - 59.41267$

$\qquad = 6.41480$

$\binom{52}{5} = 2.599 \times 10^6$

Tables of $n!$ and $\log n!$ are readily available in handbooks used by engineers. The "Handbook of Chemistry and Physics"[1] gives values of $n!$ and $\log n!$ for values of n up to 100. Lange's "Handbook of Chemistry"[2] gives values of $n!$ for n up to 20 and values of $\log n!$ for n up to 199. Grant's "Statistical Quality Control"[3] gives the values of $\log n!$ for n up to 1,009. For convenience the values of $n!$ and $\log n!$ for values of n up to 52 are given here in Table 2.2. An extensive table giving $n!$ to 16 significant figures for values of n up to 1,000 is published by the U.S. Bureau of Standards.[4]

The distribution corresponding to the probability equation in example 3.8 is known as the hypergeometric distribution and represents the fundamental probability in sampling problems and in various games of chance. The probability of each successive event is affected by the occurrence of the events that have already taken place—in drawing samples (cards from a deck or items from a shipment) the quality of the universe changes with successive drawings. The mathematical distributions discussed in the next chapter are all based on the assumption of some fixed probability p applying to each event, independent of previous events. In tossing a true coin, the probability of "heads" on any toss is 0.5. Even if each of twelve previous tosses turned up a head, the probability on the next toss, providing the coin was not biased, would still be 0.5. On the other hand, if you had a shipment of mixed fittings, half of one size and half of another, the probability of drawing one item from the original shipment of either

---

[1] "Handbook of Chemistry and Physics," 36th ed., Chemical Rubber Publishing Co., Cleveland, Ohio, 1955.

[2] N. A. Lange, "Handbook of Chemistry," 5th ed., McGraw-Hill Book Company, New York, 1944.

[3] E. L. Grant, "Statistical Quality Control," 2d ed., McGraw-Hill Book Company, New York, 1952.

[4] Herbert E. Salzer, Tables of $n!$ and $\Gamma(n + \frac{1}{2})$ for the First 1,000 Values of n, *Nat. Bur. Standards (U.S.) Appl. Math. Ser.* 16, 1951.

TABLE 2.2 Factorials

n	$n!$	$\log_{10} n!$	n	$n!$	$\log_{10} n!$
1	1	0.00000	27	$1.0888869450 \times 10^{28}$	28.03698
2	2	0.30103	28	$3.0488834461 \times 10^{29}$	29.48414
3	6	0.77815	29	$8.8417619937 \times 10^{30}$	30.94654
4	2.4×10	1.38021	30	$2.6525285981 \times 10^{32}$	32.42366
5	1.20×10^2	2.07918	31	$8.2228386541 \times 10^{33}$	33.91502
6	7.20×10^2	2.85733	32	$2.6313083693 \times 10^{35}$	35.42017
7	5.040×10^3	3.70243	33	$8.6833176188 \times 10^{36}$	36.93869
8	4.0320×10^4	4.60552	34	$2.9523279903 \times 10^{38}$	38.47016
9	3.62880×10^5	5.55976	35	$1.0333147966 \times 10^{40}$	40.01423
10	3.628800×10^6	6.55976	36	$3.7199332679 \times 10^{41}$	41.57054
11	3.9916800×10^7	7.60116	37	$1.3763753091 \times 10^{43}$	43.13874
12	4.79001600×10^8	8.68034	38	$5.2302261746 \times 10^{44}$	44.71852
13	6.227020800×10^9	9.79428	39	$2.0397882081 \times 10^{46}$	46.30959
14	$8.7178291200 \times 10^{10}$	10.94041	40	$8.1591528324 \times 10^{47}$	47.91165
15	$1.3076743680 \times 10^{12}$	12.11650	41	$3.3452526613 \times 10^{49}$	49.52443
16	$2.0922789888 \times 10^{13}$	13.32062	42	$1.4050061178 \times 10^{51}$	51.14768
17	$3.5568742809 \times 10^{14}$	14.55107	43	$6.0415263063 \times 10^{52}$	52.78115
18	$6.4023737057 \times 10^{15}$	15.80634	44	$2.6582715748 \times 10^{54}$	54.42460
19	$1.2164510041 \times 10^{17}$	17.08509	45	$1.1962222086 \times 10^{56}$	56.07781
20	$2.4329020082 \times 10^{18}$	18.38612	46	$5.5026221598 \times 10^{57}$	57.74057
21	$5.1090942172 \times 10^{19}$	19.70834	47	$2.5862324151 \times 10^{59}$	59.41267
22	$1.1240007278 \times 10^{21}$	21.05077	48	$1.2413915592 \times 10^{61}$	61.09391
23	$2.5852016739 \times 10^{22}$	22.41249	49	$6.0828186403 \times 10^{62}$	62.78410
24	$6.2044840173 \times 10^{23}$	23.79271	50	$3.0414093201 \times 10^{64}$	64.48307
25	$1.5511210043 \times 10^{25}$	25.19065	51	$1.5511187533 \times 10^{66}$	66.19064
26	$4.0329146113 \times 10^{26}$	26.60562	52	$8.0658175171 \times 10^{67}$	67.90665

size would be 0.5. But after you had withdrawn twelve of one size, the probability would no longer be one half for either size. It should be obvious that the change in the probability from drawing to drawing will depend on the size of the original shipment or universe and on the initial probability.

For example, if a bridge player and his partner hold eight cards of one suit, what is the probability the other five cards are split three and two among the opponents? You might reason that the probability of one of the five cards being in either of the opponents' hands is 0.5 and therefore the probability of three of the five cards being in one hand would be $(0.5)^3$, the probability of the other two not being in this hand would be $(0.5)^2$, and since you are considering any three of the five cards, these could be arranged $\binom{5}{3}$ ways; and the probability would be doubled for the alternate cases of either opponent with three cards, making the total probability (2) $\binom{5}{3}(0.5)^3(0.5)^2 = 0.625$. However, the probabilities are not exactly 0.5 for each card, as they are affected by the distribution of the other cards. The exact probability would be the number of ways of arranging three out of the specified five cards times the number of ways of arranging ten of the remaining 21 cards divided by the number of ways of arranging 13 cards out of 26, multiplied by two for the two hands, or (2) $\binom{5}{3}\binom{21}{10} \Big/ \binom{26}{13} = 0.678$.

2.6 Poker

Poker is a card game of many variations. One simple and popular variation is played with a deck made up of 52 cards, 13 each of four different suits. Each suit consists of 13 cards numbered consecutively from 1 to 13. Each player gets five cards, and after certain formalities the best hand wins. As an example of the operation of the various laws of permutations and combinations, the number of possible different five-card hands is listed in Table 2.3 together with the formulas for the derivation of each result. The following is an explanation of each formula in terms of some of the laws enumerated in this chapter.

Straight Flushes. These are sequences of numbered cards all of the same suit. There are four ways the suits can be taken, and there are 10 ways the sequences can be taken; an acceptable sequence may start with any number from 1 to 10. By Law 8 the dual event of sequence and suit can occur in 4×10 ways.

Four of a Kind. There are 13 different denominations of cards, so there are 13 different fours of a kind. The four of a kind is a combination of four things taken four at a time and equals 1.0. With the four of a kind, there are 48 ways of getting the fifth card. Again by Law 8, the combined event of a particular denomination, four of a kind, and a fifth card can occur in $13 \times 1 \times 48$ ways.

Full House. A full house is made of three cards of one denomination and two of another; three 10's and two 6's for instance. There are 13 ways the three-card group can be named. The three cards are a combination of four things taken three at a time. After the three-card group is named, there are 12 ways the two-card group can be named. The two cards are a combination of four things taken two at a time. The total number of ways the full house may occur is the product of these four quantities—Laws 6 and 8.

PERMUTATIONS AND COMBINATIONS

TABLE 2.3   Number of Five-card Poker Hands

Card hand	Formula	Number	Probability
Straight flush............	$(4)(10)$	40	0.000015
Four of a kind..........	$(13)\binom{4}{4}(48)$	624	0.00024
Full house...............	$(13)\binom{4}{3}(12)\binom{4}{2}$	3,744	0.00144
Flush...................	$(4)\binom{13}{5}$ − straight flushes	5,108	0.0020
Straight................	$(10)(4)^5$ − straight flushes	10,200	0.0039
Three of a kind.........	$(13)\binom{4}{3}\left[\binom{48}{2} - 12\binom{4}{2}\right]$	54,912	0.0211
Two pair................	$\binom{13}{2}\binom{4}{2}^2 44$	123,552	0.0475
One pair................	$13\binom{4}{2}\binom{12}{3}(4)^3$	1,098,240	0.4226
No pair.................	$\binom{13}{5}(4)^5$ − straights, flushes, straight flushes	1,302,540	0.5012
			0.999995
Total number of hands....	$\binom{52}{5}$	2,598,960	1.0

Flush. A flush is five cards all of the same suit. The suit may occur in four ways. The five cards of the suit can be any combination of the 13 cards of the suit taken five at a time. The product of these two quantities gives the total number of flushes possible—Laws 6 and 8. However, this total includes the straight flushes, so the quantity of those hands is deducted in Table 2.3.

Straight. A straight is a sequence of cards regardless of the suit. A sequence may start with any number from 1 to 10, so there are 10 sequences. Each of the five cards of the sequence can be any one of the four suits. This is equivalent to four different suits taken five at a time with replacement—Law 2; or four different events occurring five times—Law 8. The product gives the total number of possible straights, which includes straight flushes, so this quantity is subtracted in Table 2.3.

Three of a Kind. There are 13 denominations of cards that can be made into three of a kind, and the three of a kind can be made up from the four of that denomination in $\binom{4}{3}$ ways. The other two cards in the hand can be made up from the 48 cards which are not of the denomination in question in $\binom{48}{2}$ ways. $12\binom{4}{2}$ of these ways would result in a full house and have already been counted in that category and so must be subtracted. The total number of ways is the product of these three independent events.

Two Pair. The two pair can be made of any combination of two of the 13 denominations, $\binom{13}{2}$ ways. Each of the pair can be made in $\binom{4}{2}$ ways. The last card of the hand can be any of the 44 that are of a different denomination than either of the two pair. The total number of ways is the product of these four independent events.

One Pair. A pair can be of any of the 13 denominations. The two cards of the pair can be had from the four of that denomination in $\binom{4}{2}$ ways. The remaining three cards can be any three of the other 12 denominations, and so can be taken in $\binom{12}{3}$ ways. Since there are four suits of each of the three cards taken, each one can be taken in four ways, or a total of $(4)^3$ ways. The product of these four terms gives the total number of ways one pair may be taken.

No Pair. No pair may be made by taking any combination of the 13 denominations, five at a time, $\binom{13}{5}$ ways. Each of these five cards can be taken from four different suits, $(4)^5$ ways. Since these combinations would include all straights and flushes, the number of these possibilities is subtracted to give the number of ways of getting five cards of no distinction.

The total number of hands is the 52 cards taken five at a time, $\binom{52}{5}$ and is equal to the sum of all the separate number of individual hands.

The probability of any particular type of hand is the ratio of the number of hands of that type over the total number of possible hands [Eq. (1.1)].

Note: There is more to the game than this.

2.7 Problems

1. An urn contains three white and seven black balls. What is the probability that one drawn at random will be white? Two drawn without replacement will be black? Two drawn with replacement will be black and white, not necessarily in that order? 0.3; 0.467; 0.42.

2. How many license plates can be made using three letters and three digits? 17,576,000.

3. If 2 million license plates are issued and they are numbered sequentially starting with number 1, and if they are issued strictly at random, what is the probability of getting a license starting with the digit 1? The digit 2? 0.555; 0.0555.

4. How many bridge hands are there (13 cards)? How many possible bridge dealings are there (four hands, of 13 cards each, not considering the seating arrangement)? $\binom{52}{13}$, $\binom{52}{13}\binom{39}{13}\binom{26}{13}\binom{13}{13}$.

5. What is the probability of getting a perfect bridge hand (all of one suit)?

6. What is the probability of two or more aces in a five-card poker hand? 0.042.

7. What is the probability of winning at dice (two cubes, with the faces numbered 1 to 6, are tossed; you win if you show a 7 or an 11, you lose if you show a 2, 3, or 12, and you toss again if you show a 4, 5, 6, 8, 9, 10; you win if you repeat your first toss before you toss a 7, and you lose if you toss a 7 before you repeat your first toss)? 0.493.

8. Three urns contain colored balls as listed in the table below. Two balls are drawn from one urn at random, and they are both red. What is the probability they came from urn 3? 0.070.

	Number of balls in:		
	Urn 1	Urn 2	Urn 3
White.........	2	3	4
Black.........	1	2	3
Red..........	3	4	2

9. A coin is tossed. If the outcome is a head, a black ball is placed in an urn; if the outcome is a tail, two white balls are placed in the same urn. After two tosses of the coin, one ball is withdrawn from the urn. What is the probability that it is black? 0.416.

10. If one-third of the class fail each year, what is the strictly mathematical probability of successfully completing a four-year course?

chapter three
DISTRIBUTIONS

The major portion of the statistics in this book deals with the calculation of a number from a sample of data. This number will, in general, be called a *statistic*. The mean, the slope of a line through the data, a measure of the scatter of the data, the fraction of the data with some characteristic, e.g., fraction defective—are all examples of statistics calculated from data. These statistics are estimates of the corresponding term that applies to the entire population from which the sample was taken. The population value is called a *parameter*. The sample statistic is an estimate of the population parameter. If a large number of samples are taken from a population, statistics calculated from the individual samples will differ and there will be a certain distribution of deviations between the statistics and the population parameter. The distribution of these deviations has been calculated for various types of populations and for various parameters. It is the distribution of these deviations that is the basis for statistical judgments.

A statistical test involves the measure of the discrepance between the sample statistic and an assumed value of the population parameter. If the deviation is one that can be expected to occur with fairly high frequency, then there is no basis for rejecting the assumed value of the population parameter. If, on the other hand, the deviation corresponds

DISTRIBUTIONS

to a value that occurs very infrequently, the probability that the statistic was calculated from a sample drawn from a population with the assumed parameter is assumed to be small. The deviation is said to be *significant*. The probability for a deviation larger than some specified value is said to be the significance level of the specified value. The specified value is called the *critical* value, and the range of deviations beyond the critical value is called the critical range. In other words, if the deviation between the statistic and the assumed population parameter is beyond some specified critical value, the observed statistic is said to be significantly different from the assumed value at some preset significance level of probability.

The idea of comparing a measured quantity with a known distribution is introduced here so that the subject of distributions will have more than academic interest.

We shall discuss in some detail three distributions: the binomial, the Poisson, and the normal. The binomial distribution may have been encountered before by the engineer in the expansion of a binomial to a power, $(p + q)^n$, where n is a whole number. The binomial distribution is probably the most frequently used discrete distribution in statistics. The Poisson distribution is an approximation to the binomial which simplifies the operations when the probability is very small and the sample very large. The Poisson is also used when the expected value is known but the probability is unknown. The normal distribution can be briefly defined in several ways: it is the limit of the binomial distribution as the value of n increases to infinity; it is the distribution of the value of a variable when the variable is influenced by a large number of small effects acting in a random manner. Even though the population may not be normally distributed, means calculated from samples will in general have a normal distribution. Thus the normal distribution is fundamental for many statistical tests.

While it is not essential for the application of statistical tests to have a thorough knowledge of the distributions involved, familiarity with these distributions will contribute to a better understanding of the statistical method.

3.1 Binomial Distribution

3.1.1 Binomial Expansion.
The binomial expansion should be familiar to all engineers. It is

$$(p + q)^n = p^n + nqp^{n-1} + \frac{n(n-1) q^2 p^{n-2}}{2!}$$
$$+ \frac{(n)(n-1)(n-2)}{3!} q^3 p^{n-3} + \cdots + q^n \quad (3.1)$$

The coefficients of the terms of Eq. (3.1) correspond to a combination expression for n things taken progressively 0, 1, 2, 3, . . . at a time. The binomial expansion can therefore be written:

$$(p + q)^n = \binom{n}{0} q^0 p^{n-0} + \binom{n}{1} q p^{n-1} + \binom{n}{2} q^2 p^{n-2} + \cdots$$
$$+ \binom{n}{n} q^n p^{n-n} \quad (3.2)$$

or

$$(p + q)^n = \sum_{r=0}^{n} \binom{n}{r} q^r p^{n-r} \quad (3.3)$$

3.1.2 Pascal's Triangle. The coefficients of the binomial-expansion terms, the combination factors, may be evaluated by calculating the corresponding factorial ratios as explained in the previous chapter. An interesting and elegant way of evaluating the coefficients is by means of Pascal's triangle, which illustrates the systematic arrangement of these terms. The triangle is illustrated in Table 3.1 and is self-explanatory. Each term is equal to the sum of the term directly above and the term above and to the left. The sum of all the terms for each value of n is equal to the value of 2^n, which affords a check. Because of the symmetry of the arrangement the terms beyond $r = 8$ have been omitted from the table.

example 3.1. In the expansion of $(x + y)^{11}$, what is the term containing x^6? In the form of Eq. (3.3), $n = 11$ and $r = 5$. From Table 3.1, $\binom{11}{5} = 462$. Therefore the term in question is $462 x^6 y^5$.

3.1.3 Inductive Derivation. Assume we have a drum of ⅛-in. catalyst pellets, about 3 million pellets in the drum, and that these pellets are of two kinds indistinguishable except by chemical analysis. Assume there are 60 per cent of one kind, say iron (Fe), and 40 per cent of the other kind, say cobalt (Co). If the drum is well mixed and we reach in and take one pellet at random, the probability that it will be an Fe is, by Eq. (1.1), the number of ways we can select an Fe pellet, $0.60N$, divided by the total number of pellets we can select, N, since there are 60 per cent Fe pellets. This probability $P(\text{Fe}) = 0.60N/N = 0.60$. And the probability of selecting a Co pellet $P(\text{Co}) = 0.40N/N = 0.40$. The probability of selecting two Fe pellets is, by the multiplication law of probabilities, $(0.60)(0.60) = 0.36$. (This probability is only strictly true if the first event does not affect the probability of the second event. In our

DISTRIBUTIONS

TABLE 3.1 Pascal's Triangle of Binomial Coefficients, $\binom{n}{r}$

n \ r	0	1	2	3	4	5	6	7	8	Σ
1	1	1								2
2	1	2	1							4
3	1	3	3	1						8
4	1	4	6	4	1					16
5	1	5	10	10	5	1				32
6	1	6	15	20	15	6	1			64
7	1	7	21	35	35	21	7	1		128
8	1	8	28	56	70	56	28	8	1	256
9	1	9	36	84	126	126	84	36	9	
10	1	10	45	120	210	252	210	120	45	
11	1	11	55	165	330	462	462	330	165	
12	1	12	66	220	495	792	924	792	495	
13	1	13	78	286	715	1,287	1,716	1,716	1,287	
14	1	14	91	364	1,001	2,002	3,003	3,432	3,003	
15	1	15	105	455	1,365	3,003	5,005	6,435	6,435	
16	1	16	120	560	1,820	4,368	8,008	11,440	12,870	

assumed case, the ratio of the pellets in the drum would not be materially affected by the withdrawal of one pellet. To be absolutely accurate, we would have to assume that the first pellet was returned and the drum well mixed before the drawing of the second pellet. But since we stated that the pellet could not be identified except by analysis, we cannot assume the pellet could be returned to the drum in its pristine state.)

Suppose we selected five pellets. What would be the probability that three of them would be Fe? The number of ways of selecting the five pellets so that three of them would be one kind and two of them another is equivalent to the number of permutations of five things, three of which are of one kind and two of another, taken five at a time, which is equivalent to the number of combinations of five things taken either three at a time or two at a time. The total probability of the combined event is the product of the number of ways the event can occur times the product of the probabilities of the individual occurrences making up the total event. Since multiplication is a commutative operation, i.e., the order of the multiplication does not affect the product, the probability of three Fe pellets out of five, $P(\text{Fe})_{3/5}$, is $\binom{5}{3}(0.60)^3(0.40)^2 = 0.3456$.

Table 3.2 lists the various combinations of probabilities for selecting one to five pellets, which by extension develops into the general binomial probability distribution. If p the probability of a single event $A = P(A)$, and $1 - p = q$ is the probability of that event not occurring, $P(\tilde{A})$, then the probability that A will occur r times in n trials is

$$P(A)_{r/n} = \binom{n}{r} p^r q^{n-r} \tag{3.4}$$

This is the general term of the binomial expansion.

example 3.2. If you toss eight coins, what is the probability of getting exactly four heads? The probability of getting a head on one coin, assuming no bias, is 0.5; $P(H) = 0.5$; and the probability of not getting a head, $P(\tilde{H})$, is also 0.5. The total number of trials n is 8, and the number of events whose probability is desired r is 4; therefore

$$P(H)_{4/8} = \binom{8}{4}(0.5)^4(0.5)^4 = 0.273$$

Note that the probability of getting four heads in eight tosses of a coin is not a half, but just slightly more than a quarter. The probability of getting exactly eight heads in 16 tosses is only 0.196.

The calculation of a binomial probability may of course be made using Eq. (3.4) as illustrated in example 3.2. For cases where n and r are small, this is usually the easiest way. For cases where n is large, it is customary to refer to tables of the binomial probability distribution for values of the function. Extensive tables are available from the Bureau of Standards[1] for values of n up to 49 and for values of p from 0 to 0.5 in increments of 0.01. The "Handbook of Probability and Statistics"[2] has a table of the binomial distribution function for values of n up to 20 and for values of p in increments of 0.05.

3.1.4 Cumulative Binomial Probability. The binomial probability equation (3.4) gives the probability of the occurrence of a single event under conditions specified. What is usually of more interest is the probability of an event occurring less than or more than some specified number of times under the conditions involved. For example, with our hypothetical drum of catalyst pellets, we might be interested in the probability of getting less than three Fe pellets in a sample of 10, if the mixture

---

[1] Tables of the Binomial Probability Distribution, *Nat. Bur. Standards (U.S.) Appl. Math. Ser.* 6, 1950.

[2] R. S. Burington and D. C. May, "Handbook of Probability and Statistics with Tables," Handbook Publishers, Inc., Ohio, 1953.

TABLE 3.2 Probability of Observing Fe r Times in n Trials if $P(\text{Fe}) = 0.60 = p$ and $P(\widetilde{\text{Fe}}) = 0.40 = q$

$P(\text{Fe})_{r/n}$

n \ r	0	1	2	3	4
1	0.4	0.6			
2	$(0.4)(0.4)$ $= (0.4)^2$	$(0.6)(0.4)$ $+ (0.4)(0.6)$ $= \binom{2}{1}(0.6)(0.4)$	$(0.6)(0.6)$ $= (0.6)^2$		
3	$(0.4)(0.4)(0.4)$ $= (0.4)^3$	$(0.6)(0.4)(0.4)$ $+ (0.4)(0.6)(0.4)$ $+ (0.4)(0.4)(0.6)$ $= \binom{3}{1}(0.6)(0.4)^2$	$(0.6)(0.6)(0.4)$ $+ (0.6)(0.4)(0.6)$ $+ (0.4)(0.6)(0.6)$ $= \binom{3}{2}(0.6)^2(0.4)$	$(0.6)(0.6)(0.6)$ $= (0.6)^3$	
4	$(0.4)^4$	$\binom{4}{1}(0.6)(0.4)^3$	$\binom{4}{2}(0.6)^2(0.4)^2$	$\binom{4}{3}(0.6)^3(0.4)$	$(0.6)^4$
n	q^n	$\cdots$	$\binom{n}{r}p^r q^{n-r}$	$\cdots$	p^n

in the drum were actually 60 per cent Fe pellets. This probability would be the cumulative probability of $P(\text{Fe})_{0/10} + P(\text{Fe})_{1/10} + P(\text{Fe})_{2/10}$.

If you are sampling a shipment which the vendor claims to be not more than 5 per cent defective, and if in a sample of 100 you find seven defectives, the probability that you would be interested in is not that of getting exactly seven defectives, but the probability of getting seven or more defectives if the shipment was actually only 5 per cent defective. You might reason something like this: The sample you take from the batch will not have the identical fraction defectives that the lot has, and two samples from the same batch will probably be different. You cannot afford to examine the entire shipment, so you will have to be governed by the sample. If you are willing to be about 95 per cent correct in your judgment about refusing to accept shipments based on your sampling, you will want to know the number of defectives to expect in your samples 95 per cent of the time when the shipments meet specification of 5 per cent defective. In terms of our symbols, to what value of r does the cumulative binomial probability extend to reach the 95 per cent level?

TABLE 3.3 Individual and Cumulative Binomial Probabilities

$n = 100$

$p = 0.05$

r	$P(r)_n$	$\sum_{x=0}^{r} P(x \leq r)$
0	0.0059	0.0059
1	0.0312	0.0371
2	0.0812	0.1183
3	0.1396	0.2578
4	0.1781	0.4360
5	0.1800	0.6160
6	0.1500	0.7660
7	0.1060	0.8720
8	0.0649	0.9369
9	0.0349	0.9718

Table 3.3 gives the individual binomial probabilities, $P(r)_n = \binom{n}{r}(0.05)^r(0.95)^{n-r}$, and the cumulative binomial probabilities, $P(x \leq r) = \sum_{x=0}^{r} P(x) = \sum_{x=0}^{r} \binom{n}{x}(0.05)^x(0.95)^{n-x}$, for $p = 0.05$ and $n = 100$ up to values of $r = 9$. The probability values corresponding to $r = 6$ mean

DISTRIBUTIONS

that there is a probability of 0.1500 of finding exactly six defectives in a sample of 100 when the probability of selecting a single defective item is 0.05. And the probability of finding any number of defectives from zero to six in the sample of 100 is 0.7660. The corollary to this last statement is that there is a probability of 0.2340 of finding more than six defectives in a sample of 100 when the per cent defectives in the batch is only 5. If, then, we reject our shipment on the basis of the original sample with seven or more defectives, we shall be rejecting acceptable batches about 23 per cent of the time. In order to be about 95 per cent certain of not rejecting shipments with only 5 per cent defectives, Table 3.3 indicates that we should accept batches with samples of size 100 showing eight defectives. About 94 per cent of the samples of 100 will have up to eight defectives. Six per cent of the time a sample of 100 with 9 or more defectives will be drawn from an acceptable shipment having 5 per cent defectives. However, in order to cut down the chance of accepting shipments with greater than 5 per cent defectives we are willing to reject 6 per cent of the good shipments. The difference between the probability of accepting bad shipments as opposed to the probability of rejecting good ones is the difference between statistical errors of type II and type I and is discussed in more detail in Chap. 6.

3.1.5 Evaluating Cumulative Binomials. The evaluation of a cumulative binomial probability can of course be obtained by calculating the individual terms and summing them. With the value of one probability, it is possible to obtain the value of each succeeding term from the relations

$$P(r+1) = \frac{n-r}{r+1}\frac{p}{q} P(r) \tag{3.5}$$

or

$$P(r-1) = \frac{r}{n-r+1}\frac{q}{p} P(r) \tag{3.6}$$

We leave the algebraic proof of Eqs. (3.5) and (3.6) to the interested student. The proof follows from Eq. (3.4).

Tables of the cumulative binomial distribution are available and offer a much easier solution, especially for large values of r and n, than the actual arithmetic computation. Perhaps the most extensive of the tables readily available are those published by the Harvard University Press,[1] which cover values of n to 1,000, p to 0.50 in increments of 0.01 and also include values of p in increments of 1/16 and 1/12. The U.S. Army Ordnance[2] has issued a table of cumulative binomial probabilities for

---

[1] Tables of the Cumulative Binomial Probability Distribution, vol. 35, Harvard University Press, Cambridge, Mass., 1955.

[2] Tables of Cumulative Binomial Probabilities, Ordnance Corps, U.S. Army ORDP20-1, 1952.

values of n to 150 with p in increments of 0.01. With a table of cumulative binomial distributions it is possible to obtain the value of any single binomial within the range of the table by taking the difference between two successive cumulative values. For example, for $n = 100$, $p = 0.05$,

$$P(r = 7) = \binom{100}{7}(0.05)^7(0.95)^{93}$$

$$= \sum_{r=7}^{100} \binom{100}{r}(0.05)^r(0.95)^{100-r}$$

$$- \sum_{r=8}^{100} \binom{100}{r}(0.05)^r(0.95)^{100-r}$$

$$= 0.23399 - 0.12796 = 0.10603$$

example 3.3. There is a gambling game played with dice in which 10 regular dice are rolled 13 times. The player bets he either rolls 13 or fewer of some particular point or 26 or more of that same point. (A "point" is one of the numbers from 1 to 6 on the dice.) The "house" bets him 3 to 1 that he rolls the point he selected between 14 and 25 times. What is the player's probability of winning? What are the mathematical odds?

Since each roll of a single die is an independent event and is not affected by the outcome of a roll of another die, the rolling of 10 dice 13 times constitutes 130 independent events; $n = 130$. The probability of rolling any specific point, since there are six faces on each die, is $1/6$; $p = 1/6$. To solve our problem, we need the cumulative probability from $r = 0$ to $r = 13$, and from $r = 26$ to $r = 130$; or conversely, 1 minus the probability from $r = 14$ to $r = 25$.

$$P(\text{win}) = 1 - \sum_{r=14}^{25} \binom{130}{r}\left(\frac{1}{6}\right)^r\left(\frac{5}{6}\right)^{130-r} \tag{3.7}$$

From the Harvard University Press tables:

$P(r = 14 \text{ to } 130) = 0.97780$

$P(r = 26 \text{ to } 130) = 0.18205$

$P(r = 14 \text{ to } 25) = 0.79575$

$P(\text{win}) = 0.20425$

The mathematical odds against winning is the probability of not winning over the probability of winning:

$$\frac{P(\tilde{W})}{P(W)} = \frac{0.79575}{0.20425} = 3.89 \text{ to } 1.0$$

DISTRIBUTIONS

3.1.6 Confidence Range of Binomial Distribution. In the examples of the use of the binomial distribution illustrated up to now, we have accepted an a priori value of the probability p and determined the expected frequency of some observation. The more common practice in industrial processes is to start with the observation of a sample and then to set limits to the universe from which it might have come. If we take a sample of 100 and find seven defectives, what is the probable range of defectives in the shipment from which the sample was taken? If in a questionnaire returned by 175 customers, 20 per cent indicate a preference for a change, what is the probable range in the percentage of total customers who would prefer this change?

The cumulative binomial probability distribution can be shown as a stepwise function plotted against the number of observations r having a particular characteristic, the steepness of the plot being a function of the sample size, n and the probability p of observing the characteristic.

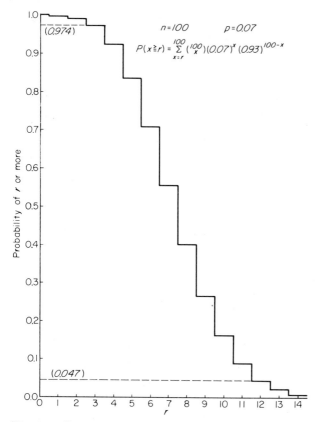

Fig. 3.1 Cumulative binomial probability.

in question. Figure 3.1 shows the cumulative probability plot for $n = 100$ and $p = 0.07$ for values of r from 0 to 14. The "steps" in the curve are drawn at the mid-points between the successive r values inasmuch as r can occur only in discrete integers. From this curve it can be seen that there is a somewhat better than 95 per cent chance ($P = 0.974$) of r being 3 or more; i.e., there is about a 3 per cent chance of r being less than three. Also from the curve, there is about a 5 per cent chance ($P = 0.047$) of r being 12 or more. Therefore there is an approximately symmetrical 0.927 probability range that r is from 3 to 11. A 0.95 probability range for r would be from 3 to 12 $[P(r \geq 13) = 0.022]$. In other words, from a shipment of material containing 7 per cent defective items, 95 per cent of samples of 100 will have between 3 and 12 defective items. If, now, we start with the sample of 100 and find seven defectives, what is the 95 per cent probability range for the per cent defectives in the lot? Or,

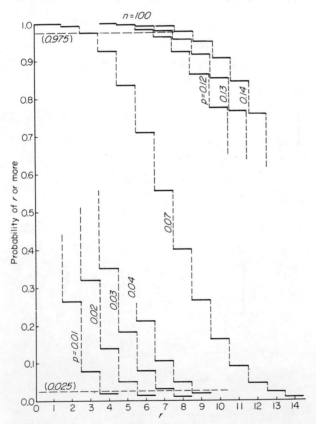

Fig. 3.2 Cumulative binomial probabilities for different values of p.

DISTRIBUTIONS

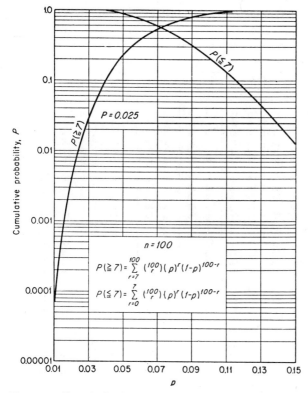

Fig. 3.3 Cumulative binomial probability distribution as a function of p.

what values of p in binomial distributions with $n = 100$ will include $r = 7$ in a 0.95 probability range?

In Fig. 3.2 is plotted a series of cumulative binomial probability curves for $n = 100$ and different values of p. If we are interested in which curves include the value of $r = 7$ in their 0.95 probability range, we would include all curves which have at least 0.025 probability of $r \geq 7$ and exclude all curves which have more than 0.975 probability of $r \geq 8$. This is a range of $p = 0.03$ to $p = 0.14$. The actual values from the binomial tables are $p = 0.03$, $P(r \geq 7) = 0.031$, and at $p = 0.14$, $P(r \geq 8) = 0.977$.

An alternative way of showing this relation is as in Fig. 3.3 where the cumulative probability of $P(r \geq 7)$ and $P(r \leq 7)$, that is, $1 - P(r \geq 8)$, are plotted for $n = 100$ against different values of p. The exact 0.95 range is from about $p = 0.029$ to $p = 0.138$. However, since in our example p can only be in units of 0.01, we say that the 0.95 range of per cent defectives in the shipment, with 7 per cent defectives in the sample of 100, is from 3 to 14 per cent.

TABLE 3.4 95 Per Cent Confidence Interval (Per Cent) for Binomial Distribution†

Number observed r	Size of sample, N						Fraction observed r/N	Size of sample	
	10	15	20	30	50	100		250	1,000
0	0 31	0 22	0 17	0 12	0 7	0 4	.00	0 1	0 0
1	0 45	0 32	0 25	0 17	0 11	0 5	.01	0 4	0 2
2	3 56	0 40	1 31	1 22	0 14	0 7	.02	1 5	1 3
3	7 65	4 48	3 38	2 27	1 17	1 8	.03	1 6	2 4
4	12 74	8 55	6 44	4 31	2 19	1 10	.04	2 7	3 5
5	19 81	12 62	9 49	6 35	3 22	2 11	.05	3 9	4 7
6	26 88	16 68	12 54	8 39	5 24	2 12	.06	3 10	5 8
7	35 93	21 73	15 59	10 43	6 27	3 14	.07	4 11	6 9
8	44 97	27 79	19 64	12 46	7 29	4 15	.08	5 12	6 10
9	55 100	32 84	23 68	15 50	9 31	4 16	.09	6 13	7 11
10	69 100	38 88	27 73	17 53	10 34	5 18	.10	7 14	8 12
11		45 92	32 77	20 56	12 36	5 19	.11	7 16	9 13
12		52 96	36 81	23 60	13 38	6 20	.12	8 17	10 14
13		60 98	41 85	25 63	15 41	7 21	.13	9 18	11 15
14		68 100	46 88	28 66	16 43	8 22	.14	10 19	12 16
15		78 100	51 91	31 69	18 44	9 24	.15	10 20	13 17
16			56 94	34 72	20 46	9 25	.16	11 21	14 18
17			62 97	37 75	21 48	10 26	.17	12 22	15 19
18			69 99	40 77	23 50	11 27	.18	13 23	16 21
19			75 100	44 80	25 53	12 28	.19	14 24	17 22
20			83 100	47 83	27 55	13 29	.20	15 26	18 23
21				50 85	28 57	14 30	.21	16 27	19 24
22				54 88	30 59	14 31	.22	17 28	19 25
23				57 90	32 61	15 32	.23	18 29	20 26
24				61 92	34 63	16 33	.24	19 30	21 27
25				65 94	36 64	17 35	.25	20 31	22 28
26				69 96	37 66	18 36	.26	20 32	23 29
27				73 98	39 68	19 37	.27	21 33	24 30
28				78 99	41 70	19 38	.28	22 34	25 31
29				83 100	43 72	20 39	.29	23 35	26 32
30				88 100	45 73	21 40	.30	24 36	27 33
31					47 75	22 41	.31	25 37	28 34
32					50 77	23 42	.32	26 38	29 35
33					52 79	24 43	.33	27 39	30 36
34					54 80	25 44	.34	28 40	31 37
35					56 82	26 45	.35	29 41	32 38
36					57 84	27 46	.36	30 42	33 39
37					59 85	28 47	.37	31 43	34 40
38					62 87	28 48	.38	32 44	35 41
39					64 88	29 49	.39	33 45	36 42
40					66 90	30 50	.40	34 46	37 43
41					69 91	31 51	.41	35 47	38 44
42					71 93	32 52	.42	36 48	39 45
43					73 94	33 53	.43	37 49	40 46
44					76 95	34 54	.44	38 50	41 47
45					78 97	35 55	.45	39 51	42 48
46					81 98	36 56	.46	40 52	43 49
47					83 99	37 57	.47	41 53	44 50
48					86 100	38 58	.48	42 54	45 51
49					89 100	39 59	.49	43 55	46 52
50					93 100	40 60	.50	44 56	47 53

† This table is Table 1.3.1 from George W. Snedecor, "Statistical Methods," 5th ed., 1956, reprinted with the permission of the author and the publisher, Iowa State College Press, Ames, Iowa.

DISTRIBUTIONS

TABLE 3.5 99 Per Cent Confidence Interval (Per Cent) for Binomial Distribution†

Number observed r	Size of sample, N						Fraction observed r/N	Size of sample	
	10	15	20	30	50	100		250	1,000
0	0 41	0 30	0 23	0 16	0 10	0 5	.00	0 2	0 1
1	0 54	0 40	0 32	0 22	0 14	0 7	.01	0 5	0 2
2	1 65	1 49	1 39	0 28	0 17	0 9	.02	1 6	1 3
3	4 74	2 56	2 45	1 32	1 20	0 10	.03	1 7	2 4
4	8 81	5 63	4 51	3 36	1 23	1 12	.04	2 9	3 6
5	13 87	8 69	6 56	4 40	2 26	1 13	.05	2 10	3 7
6	19 92	12 74	8 61	6 44	3 29	2 14	.06	3 11	4 8
7	26 96	16 79	11 66	8 48	4 31	2 16	.07	3 13	5 9
8	35 99	21 84	15 70	10 52	6 33	3 17	.08	4 14	6 10
9	46 100	26 88	18 74	12 55	7 36	3 18	.09	5 15	7 12
10	59 100	31 92	22 78	14 58	8 38	4 19	.10	6 16	8 13
11		37 95	26 82	16 62	10 40	4 20	.11	6 17	9 14
12		44 98	30 85	18 65	11 43	5 21	.12	7 18	9 15
13		51 99	34 89	21 68	12 45	6 23	.13	8 19	10 16
14		60 100	39 92	24 71	14 47	6 24	.14	9 20	11 17
15		70 100	44 94	26 74	15 49	7 26	.15	9 22	12 18
16			49 96	29 76	17 51	8 27	.16	10 23	13 19
17			55 98	32 79	18 53	9 29	.17	11 24	14 20
18			61 99	35 82	20 55	9 30	.18	12 25	15 21
19			68 100	38 84	21 57	10 31	.19	13 26	16 22
20			77 100	42 86	23 59	11 32	.20	14 27	17 23
21				45 88	24 61	12 33	.21	15 28	18 24
22				48 90	26 63	12 34	.22	16 30	19 26
23				52 92	28 65	13 35	.23	17 31	20 27
24				56 94	29 67	14 36	.24	18 32	21 28
25				60 96	31 69	15 38	.25	18 33	22 29
26				64 97	33 71	16 39	.26	19 34	22 30
27				68 99	35 72	16 40	.27	20 35	23 31
28				72 100	37 74	17 41	.28	21 36	24 32
29				78 100	39 76	18 42	.29	22 37	25 33
30				84 100	41 77	19 43	.30	23 38	26 34
31					43 79	20 44	.31	24 39	27 35
32					45 80	21 45	.32	25 40	28 36
33					47 82	21 46	.33	26 41	29 37
34					49 83	22 47	.34	26 42	30 38
35					51 85	23 48	.35	27 43	31 39
36					53 86	24 49	.36	28 44	32 40
37					55 88	25 50	.37	29 45	33 41
38					57 89	26 51	.38	30 46	34 42
39					60 90	27 52	.39	31 47	35 43
40					62 92	28 53	.40	32 48	36 44
41					64 93	29 54	.41	33 50	37 45
42					67 94	29 55	.42	34 51	38 46
43					69 96	30 56	.43	35 52	39 47
44					71 97	31 57	.44	36 53	40 48
45					74 98	32 58	.45	37 54	41 49
46					77 99	33 59	.46	38 55	42 50
47					80 99	34 60	.47	39 55	43 51
48					83 100	35 61	.48	40 56	44 52
49					86 100	36 62	.49	41 57	45 53
50					90 100	37 63	.50	42 58	46 54

† This table is Table 1.3.1 from George W. Snedecor, "Statistical Methods," 5th ed., 1956, reprinted with the permission of the author and the publisher, Iowa State College Press, Ames, Iowa.

It is not necessary to go through the actual calculation in each case if the symmetrical 0.95 or the 0.99 probability range is desired. Snedecor[1] has published tables for various sample sizes for both confidence ranges in terms of per cent range of the population. These tables are reproduced here as Tables 3.4 and 3.5. The tables are used in two ways, one for samples of 100 or less and another for larger samples. With samples of 100 or less, the number in the sample corresponding to some classification, i.e., number defective, is located in the left column. The two numbers in the same row under the heading for the particular sample size give the 95 per cent range for the percentage of the population in that category. A linear interpolation may be used for samples intermediate of the table headings.

For samples larger than 100, the table gives the fraction of the sample falling within one category and, under the sample size, the 95 per cent range for the percentage of the population coming within that category. For observed fractions larger than 0.50, the confidence range for the alternative category can be found, and the desired confidence range will be 100 minus the tabulated results.

example 3.4. The following samples (Table 3.6) and the corresponding number defective were taken from a large shipment of material. Note that all the samples have the same percentage defectives. What is the 0.95 probability range of per cent defectives in the shipment?

TABLE 3.6

Sample size	Number defective
10	2
20	4
50	10
100	20
500	100
1,000	200

The answers are read directly from Table 3.4 and are given in Table 3.7.

The narrowing of the probability range with the increase in sample size indicates the relative confidence with which statements can be made from samples of different sizes. A quick estimation of the sample required can be made from Tables 3.4 and 3.5 if the desired spread of the confidence range is established in advance.

[1] George W. Snedecor, "Statistical Methods," 5th ed., Iowa State College Press, Ames, Iowa, 1956.

DISTRIBUTIONS

TABLE 3.7

Sample size	0.95 range of per cent defectives in shipment
10	3–56
20	6–44
50	10–34
100	13–29
500	16–25 (by interpolation)
1,000	18–23

example 3.5. In respose to a questionnaire we find that 87 per cent of 250 replies show a certain attitude toward our product. What is the minimum per cent of the population that we should expect with 0.99 probability to show this attitude assuming we have a representative sample and that the binomial distribution applies?

The fraction observed, r/N, is 0.87—beyond our scale. The fraction not showing the attitude in question is 0.13, and the 0.99 probability range is 8 to 19 per cent; therefore the 0.99 probability range of the positive attitude is 81 to 92 per cent. The minimum response we should therefore expect is 81 per cent.

The probabilities employed in Tables 3.4 and 3.5 are symmetrical probabilities, meaning that for the 0.95 level, 0.025 is the probability of falling below the indicated range and 0.025 is the probability of falling above it; and for the 0.99 level, 0.005 probability of being below and 0.005 of being above. If the only value that is of interest is one of the extremes —for example, you may not care how good a batch of material is, but only how poor it might be—then one extreme of the indicated range would be used at a one-sided probability equal to $0.5 + P/2$, where P is the table probability value; i.e., for the 0.99 table, $0.5 + 99/2 = 0.995$.

3.1.7 Binomial-probability Paper. Mosteller and Tukey[1] have developed a graph paper for solving problems involving the binomial probability. The theory and detailed discussion of the use of the paper is presented in the original publication by the authors. We shall merely outline a few of the uses of this paper with examples.

The use of the binomial paper involves a concept that we have not yet discussed, namely, the standard deviation as a measure of distribution. For a normal distribution, the standard deviation defines the distribution of the data around the mean. If we designate the standard deviation by σ and the mean by $\bar{x}$, then with a normal distribution, 68.26 per cent of the data will occur within the range $\bar{x} \pm \sigma$, and 95.44 per cent of the data will fall within $\bar{x} \pm 2\sigma$, and so on for any range around $\bar{x}$ in terms of σ.

---

[1] F. Mosteller and J. W. Tukey, "The Use and Usefulness of Binomial Probability Paper," *J. Am. Statis. Assoc.*, **44**:174 (June, 1949).

The exact factor for 95 per cent of the data is $\pm 1.96\sigma$. The normal distribution is discussed in detail in Sec. 3.3 of this chapter. The binomial distribution approximates the normal, the approximation becoming better as n increases and p approaches 0.5. The standard deviation of the binomial distribution, in all cases, is equal to $(npq)^{1/2}$, and as the binomial approaches the normal, the relative frequency in any range around the mean, np, can be reasonably well defined in terms of the standard deviation $(npq)^{1/2}$. In the example of a sample of 100 items and a fraction defective of 0.07, $n = 100$, $p = 0.07$, and $q = 1 - p = 0.93$. Therefore the standard deviation is $\sqrt{(100)(0.07)(0.93)} = 2.5514$. Using the factor of 1.96 for the 95 per cent range around the mean, this range is $np \pm (1.96)(npq)^{1/2} = 7 \pm (1.96)(2.5514) = 2.0$ to 12.0 compared with the 3 to 14 obtained from the binomial-distribution tables. The difference between these answers is due to a large extent to the low value of p, which results in a skewed binomial distribution.

The binomial distribution is used as a model for counted data that fall into two categories: good or bad, favorable or unfavorable, heads or tails, etc. In its numerical application, the pair of counted data is used as the basis of calculating (or selecting from a table) the binomial probability of such a count if the probability of one category is p and the probability

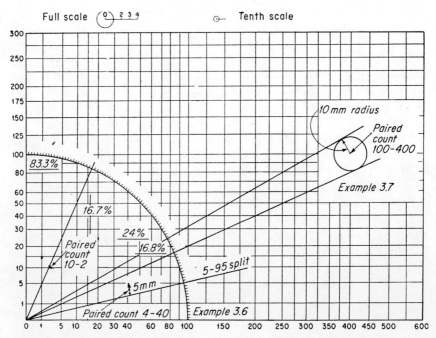

Fig. 3.4 Binomial-probability paper.

DISTRIBUTIONS 43

of the other category is $1 - p$. We can call the count of the data into two categories the *paired count*, and we can call the expected count on the basis of the probabilities p and $q = 1 - p$ the *split*. For simple problems, the application of the binomial-probability paper involves plotting the split and the paired count and measuring the distance between them in units of the standard deviation.

The binomial-probability paper is shown in Fig. 3.4,[1] which also illustrates the examples described in this section. The ordinates of the paper are divided in a square-root scale, and the numbers represent actual counts. The upper left scale, labeled "full scale," is in units of standard deviation to be used with the ordinates of the paper as given. On this basis paired counts up to 600–300 can be plotted directly. If the observations exceed this number, the scales can be multiplied by 10, extending the range to 6,000–3,000, in which case the upper right scale, labeled "tenth scale," gives the units of standard deviation. A split is plotted as a straight line through the origin and the coordinates of the split. Any line drawn through the origin goes through all splits of the same ratio. The line through the origin and the 20–30 split goes through the 2–3 split, the 40–60 split, the 100–150 split, the 150–225 split, etc. The split, therefore, can be plotted as an expected ratio or as the expected distribution of the actual count. The coordinates of the quarter circle add to 100, so that the percentage distribution of the split can be located on the quarter circle and plotted in that manner.

A paired count is plotted as a triangle the sides of which extend one unit in both positive directions from the right angle which has the paired count for its coordinates. Thus the paired count 10–2 is plotted as a triangle with vertices at 11–2, 10–2, and 10–3. This paired count is shown in Fig. 3.4. To find the percentage corresponding to a paired count, draw the line through the right angle and the origin. The percentages can be read from the coordinates of the intersection of the line and the quarter circle. If either value of a paired count is more than 100, it is sufficient to plot a line, or if both values exceed 100, a point.

The standard-deviation unit, when applied to the full scale, is 5.080 mm (on the basis of the paper as printed; the paper reproduced in Fig. 3.4 has been reduced in size so that the measurements cannot be checked from the figure). The number of standard-deviation units corresponding to different probabilities or significance levels is given in Table 3.8 together with the distances for use with the binomial-probability paper.

These values are used in the following manner. A sample is counted, and the distribution into two categories determined. This information is plotted as a paired-count triangle on the binomial-probability paper.

[1] Paper 31298, Codex Book Company, Norwood, Mass.

The split corresponding to the test or theoretical distribution of the population is also plotted on the paper. The shortest distance from the split to the paired-count triangle is measured, and the significance level corresponding to the number of standard-deviation units is the probability of obtaining the paired count from a population with the test split. For instance, if the distance exceeds 9.96 mm (1.96 standard deviations),

TABLE 3.8

Significance level, %	No. of standard deviations	Distance, mm
31.74	1.0	5.08
20.0	1.28	6.50
10.0	1.65	8.38
5.0	1.96	9.96
4.56	2.00	10.16
2.5	2.24	11.38
2.0	2.33	11.84
1.0	2.57	13.06
0.26	3.00	15.24

there is less than 5 per cent probability, and if the distance exceeds 13.06 mm (2.57 standard deviations), there is less than 1 per cent probability of the paired count having come from a population with a distribution represented by the split.

example 3.6. From a shipment of material supposed to be 5 per cent defective, we draw a sample of 44 and find four defectives. We wish to determine whether our sample is significantly different from the assumed population, i.e., whether the probability of observing four defectives in a sample of 44 from a population with 5 per cent defectives is less than 0.05.

We draw the paired count, 4–40. This is the triangle 5–40, 4–40, 4–41. We draw the split, 5–95 corresponding to 5 per cent. The smallest distance from the split to the paired count is almost exactly 5 mm, corresponding to one standard deviation, or a probability of 0.317. The observed paired count is not significantly different from the expected split.

Note that the probability values listed in Table 3.8 give the probability of values falling in the range of the mean plus *or* minus multiples of the standard deviation. The probability tabulated is the total probability;

half applies to values greater than the mean and half to values less than the mean. Therefore, in a problem like that of example 3.6 where we would probably be interested only in samples that had more than the average number of defectives and not in those that had less, we would use probabilities equal to half those listed in Table 3.8. If the distance from the split to the paired count had been 8.38 mm, equal to 1.65 standard deviations and a probability of 0.10, we could say this corresponded to a one-sided deviation of 5 per cent and therefore there was only a 0.05 probability that the paired count came from a population with a distribution represented by the split.

In applying probability analysis to counted data it is important to keep in mind that there may be relatively large differences in probabilities associated with successive counts. For example, if $p = 0.05$, that is, an expected 5 per cent defectives, with a sample of nine, two defectives would not be less than the 5 per cent probability level, but three would be less than the 1 per cent level. This spread in probability levels between two successive counts is called a *significance zone* and is indicated on the binomial-probability paper by the difference between the shortest and longest distance from the split to the paired-count triangle. This zone is obviously much larger for small counts than for large where the data can be represented by a point.

The confidence range on the split of the population from which a sample was drawn can be obtained by finding the splits which have distances to the paired count corresponding to the probability level desired. For the 95 per cent confidence range, from Table 3.8, this distance is 9.96 mm. For the 99 per cent confidence range it is 13.06 mm. (For problems in which the graphical solution offers sufficient accuracy, 10- and 13-mm distances are close enough.) With a paired count represented by a point, the simplest procedure is to draw a circle of the desired radius and to draw two tangents to this circle through the origin. The percentage splits corresponding to the confidence range can be read from the coordinates of the quarter circle where it is intersected by the two tangent lines. For a paired count represented by a triangle, the arcs are drawn from the apexes of the two acute angles of the triangle. The tangents to these arcs are drawn through the origin as described.

example 3.7. In example 3.4 a sample of 500 items was found to have 100 defectives. We wish to establish the 95 per cent confidence range of the per cent defectives in the shipment from which the sample was drawn.

See Fig. 3.4. The paired count 100–400 is plotted, and a circle of 10-mm radius, corresponding to 95 per cent confidence range, or 5 per cent significance level, is drawn around this point. The tangents to this circle drawn through the origin intersect the quarter circle at the splits 24–76 and 16.8–83.2. The 95 per cent confidence range of the per cent defective is therefore 16.8 to 24.0 per cent.

3.2 The Poisson Distribution

3.2.1 Definition. A limiting approximation to the binomial distribution, as p gets smaller and n increases, is the Poisson distribution. As n approaches ∞ and p approaches 0, but in such a manner that np is kept constant, the individual terms of the binomial distribution approach the value of the corresponding terms of the Poisson distribution and give the probability of an event occurring exactly the specified number of times. The expression for the probability of an event occurring exactly r times in the Poisson formula is

$$P(r) = e^{-m}\frac{m^r}{r!} \tag{3.8}$$

e is the base of natural logarithms and equals 2.718282 . . . . The logarithm of e to the base 10, $\log_{10}e$, equals 0.434294. m is the expected number of occurrences and is equal to np, where n is the total number of events and p is the probability of occurrence of the event in question.

The Poisson distribution is used a great deal in the solution of industrial probability problems when the expected frequency is small and total number of events is large. Problems of hourly traffic loadings whether automobile-highway traffic, telephone-circuit traffic, or hospital-emergency traffic, where the total traffic is large but the expected traffic for any single hour is relatively small, respond to the Poisson probability distribution. In sampling items of small size, such as agricultural seed or small machine parts, where a random sample might contain several thousand items and where the expected impurities or defects are small in number, the Poisson distribution offers an easier method of evaluating the observed frequencies. The frequency of radioactive disintegration as measured by Geiger counts per minute, where the number of particles available for disintegration is large and the expected number disintegrating in unit time is small, occurs in a Poisson distribution pattern.

3.2.2 Evaluating Individual Poisson Terms. The Poisson distribution is made up of the series of terms

$$\frac{e^{-m}m^0}{0!} + \frac{e^{-m}m^1}{1!} + \frac{e^{-m}m^2}{2!} + \frac{e^{-m}m^3}{3!} + \cdots \tag{3.9}$$

Since e is defined as

$$e = \frac{1}{0!} + \frac{1}{1!} + \frac{1}{2!} + \frac{1}{3!} + \cdots \tag{3.10}$$

and e^m can be expressed in a similar series

$$e^m = \frac{m^0}{0!} + \frac{m^1}{1!} + \frac{m^2}{2!} + \frac{m^3}{3!} + \cdots \qquad (3.11)$$

it follows that the sum of the terms making up the Poisson series is equal to unity:

$$\sum_{r=0}^{\infty} \frac{e^{-m} m^r}{r!} = 1.0 \qquad (3.12)$$

The individual terms of the Poisson series represent the limit of the corresponding terms in the binomial series, that is, $e^{-m} m^r / r! = \binom{n}{r} p^r (1-p)^{n-r}$, as n becomes infinite. m is equivalent to pn, the expected value. Demonstration of this equivalence can be found in several texts on the theory of probability.[1] Each term, therefore, represents the probability of the event occurring r times in n trials when the mean number of occurrences is m.

example 3.8. The probability for the total population of the United States[2] of dying within the year for persons age 35 to 36 is 0.0039. In a group of 1,000 such persons, what is the probability that exactly 10 will die within the year?

$$p = 0.0039$$
$$n = 1{,}000$$
$$m = pn = 3.9$$
$$r = 10$$
$$P(10) = \frac{(3.9)^{10}}{10! e^{3.9}}$$

$\log P(10) = 10 \log 3.9 - \log 10! - 3.9 \log e$
$\log P(10) = 5.910650 - 6.55976 - 1.69375$
$\log P(10) = -2.342857$
$\log P(10) = 7.657143 - 10$
$P(10) = 0.004541$

Tables of the Poisson distribution function are available, so that the actual calculation of the probability value is usually not necessary. The

---

[1] See, for example, A. M. Mood, "Introduction to the Theory of Statistics," McGraw-Hill Book Company, New York, 1950.
[2] United States Life Tables, U.S. Department of Commerce, 1946.

best known of these tables are Molina's,[1] which cover a range of m from 0.001 to 100. Burington and May[2] have somewhat less extensive tables covering values of m from 0.1 to 20.

An important use of the Poisson distribution is in cases where it is not possible to assign a value to either p or n but the product of pn does have a value. For example, if you note the number of lightning flashes for every 5-min period during a storm, or the number of accidents per month for a busy intersection, you will get a number corresponding to the mean number of incidents under observation; call this m. The probability of a lightning flash, or of an accident, is m/n, but the value of n corresponding to the total of lightning flashes and no lightning flashes, or accidents and no accidents, has no meaning. In testing insulated wire, you can note the number of insulation failures per 10,000 ft, but the number of no failures in the same 10,000 ft has no meaning. However, with the observed number of occurrences per unit time or per unit length of wire, we can use the Poisson distribution to calculate the probability of any number of occurrences. If the mean number of occurrences is m and we desire the probability of r occurrences, then Eq. (3.8) applies, and

$$P(r) = e^{-m}\frac{m^r}{r!} \tag{3.8}$$

example 3.9. In 168 years that the U.S. Supreme Court was in existence, 93 justices were appointed. What is the probability that three justices will be appointed in 1 year?

m = mean number of justices appointed per year

$$= \frac{93}{168} = 0.5535$$

$$r = 3$$

$$P(3) = \frac{e^{-0.5535}(0.5535)^3}{3!} = 0.0162$$

i.e., about 16 times every thousand years three justices will be appointed in 1 year.

To illustrate the application of the Poisson distribution to the Supreme Court appointment data, the actual frequency of the number of appointments is compared with the frequency predicted from the Poisson distribution in Table 3.9 which follows:

[1] E. C. Molina, "Poisson's Exponential Binomial Limit," D. Van Nostrand Company, Inc., Princeton, N.J., 1942.
[2] R. S. Burington and D. C. May, "Handbook of Probability and Statistics with Tables," McGraw-Hill Book Company, New York, 1953.

TABLE 3.9

(1) No. of justices appointed per year r	(2) Actual no. of years in which r justices were appointed f	(3) Justices appointed per year times the number of years rf	(4) Probability of r justices calculated from Poisson distribution $P(r) = \dfrac{e^{-m} \cdot m^r}{r!}$	(5) Poisson prediction of no. of years for r appointments $P(r) \cdot \Sigma f$
0	100	0	0.575	96.6
1	50	50	0.318	53.4
2	14	28	0.0881	14.8
3	3	9	0.0162	2.7
4	0	0	0.00224	⎫
5	0	0	0.00024	⎬ 0.42
6	1	6	0.000023	⎭
	168	93	0.999803	167.92
	Σf	$\Sigma(r \cdot f)$	$m = \Sigma(r \cdot f)/\Sigma f$ $= 93/168$ $= 0.5535$	

The agreement of the observed frequencies (column 2) with the frequencies predicted by the Poisson distribution (column 5) indicates that the same probability was in effect for all of the data, which is the basis for the use of the Poisson distribution, and that $m\ (= 0.5535)$ was the expected number of justices to be appointed in any year.

This criterion is applied to industrial accidents and other similar phenomena to determine whether accidents, for example, are occurring more frequently than would be expected if the probability of an accident was the same for all men, shifts, machines, or whatever units are under observation.

3.2.3 The Cumulative Poisson Distribution. As with the binomial distribution, the probability of obtaining more than or less than some specific term in the Poisson series is often of more interest than the probability of an individual term. The evaluation of the individual terms is necessary to demonstrate whether a distribution conforms to the Poisson, but it is usually a cumulative distribution that is necessary to demonstrate statistical significance, especially with larger values of m. For example, if m is larger than 15, no frequency can be expected to occur more than 10 per cent of the time, and if m is larger than 63, no frequency can be expected to occur more than 5 per cent of the time. If, as is usual in industrial practice, we are interested in providing for the events that occur 95 or 99 per cent of the time, or in taking some action when the events occur which can only be expected 5 or 1 per cent of the time, we need to know the 0.95 or the 0.99 cumulative probability range. If experience has demonstrated that on the average 2 per cent of employees taking a promotional examination fail, and if with a new examination 2 out of 40 fail the first time it is used, is there reason (less than a 0.05 probability) to believe the new examination is more difficult than the previous ones? If 5 out of 100 fail? If 10 out of 200?

The cumulative Poisson probabilities can most easily be obtained by reference tables. Molina's[1] tables also include the cumulative values of the Poisson distribution and cover a range of m from 0.001 to 100. Grant[2]

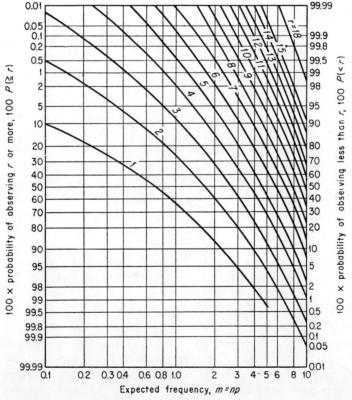

Fig. 3.5 Cumulative Poisson probabilities.

[1] *Molina, op. cit.*
[2] E. L. Grant, "Statistical Quality Control," 2d ed., McGraw-Hill Book Company, New York, 1952.

DISTRIBUTIONS

has tables of cumulative probabilities for values of m from 0.02 to 25, and Burington and May[1] have them for m from 0.1 to 20. Figure 3.5 is a plot drawn from Molina's tables and gives the cumulative probability for observing r or more events when the expected frequency m is in the range 0.1 to 10.

example 3.10. Using the data from the problem posed earlier, we expect 2 per cent failures in a certain examination. Of 40, 100, and 200 who take the test, 2, 5, and 10 fail in each case. The question to be answered is whether a number as large as the observed has more than a 0.05 probability of occurring if the expected frequency were effective. If there is less than 0.05 probability, we can say, with less than 5 per cent chance of being wrong, that the conditions giving the usual frequency do not apply; i.e., in this case the examination is more difficult than usual for the employees involved.

Table 3.10 gives the expected frequencies, the observed frequencies, the probabilities read from Fig. 3.5, and the probabilities taken from Molina's tables.

Only in the last case would we judge the examination to be significantly different from the earlier ones.

TABLE 3.10

Examinees	Expected failures, m	Observed failures, r	$P(\geq r)$	
			Fig. 3.5	Molina
40	0.8	2	0.195	0.191208
100	2.0	5	0.055	0.052653
200	4.0	10	0.008	0.008132

The cumulative value of the Poisson distribution is very readily computed from any one term. A study of Eq. (3.9) for the Poisson series will show that each term exceeds the previous term by the factor m/r.

$$P(0) = e^{-m} \frac{m^0}{0!}$$

$$P(1) = e^{-m} \frac{m^1}{1!} = P(0) \frac{m}{1}$$

[1] Burington and May, *Op. cit.*

$$P(2) = e^{-m}\frac{m^2}{2!} = P(1)\frac{m}{2}$$

.
.
.

$$P(r-1) = e^{-m}\frac{m^{r-1}}{(r-1)!}$$

$$P(r) = e^{-m}\frac{m^r}{r!} = P(r-1)\frac{m}{r}$$

Since $P(0)$ in every case is equal to e^{-m}, m^0 and $0!$ both being equal to unity, it is a relatively simple procedure if tables are not available to evaluate $P(0)$ and multiply successively by $m/1$, $m/2$, etc., until the desired value of r is obtained. The values of $m/1$, $m/2$ decrease rapidly, so that not many terms are required usually before the value of the probability reaches insignificance.

If the problem is to determine the probability of a value larger than r, rather than the probability of a value as large as r, the former can be obtained from the relation

$$P(>r) = 1 - P(\leq r) \tag{3.13}$$

example 3.11. If a long history of experience has demonstrated that the mean number of personal-injury accidents in a particular section of a plant is 0.450 accident per shift in a 6-month period, how many accidents in a 6-month period would prompt the safety department to take more than ordinary action? To answer this question we need to determine the probability of experiencing one or more, two or more, three or more, etc., accidents under the conditions specified. We can then decide to take some action if the number of accidents occur for which the probability is 0.05 or less. Without the use of tables, the problem can be solved as follows:

$$m = 0.450$$
$$P(0) = e^{-m} = e^{-0.450}$$
$$\log P(0) = -0.450 \log e = (-0.450)(0.434294)$$
$$= -0.195432$$
$$= 9.804568 - 10$$
$$P(0) = 0.63763$$
$$P(1) = (0.450)(0.63763) = 0.28693$$
$$P(2) = (0.450/2)(0.28693) = 0.06456$$
$$P(3) = (0.450/3)(0.06456) = 0.009684$$

$$P(\leq 0) = P(0) = 0.63763$$
$$P(\leq 1) = P(0) + P(1) = 0.92456$$
$$P(\leq 2) = P(0) + P(1) + P(2) = 0.98912$$
$$P(\leq 3) = P(0) + P(1) + P(2) + P(3) = 0.99880$$
$$P(\geq 0) = 1.00000$$
$$P(\geq 1) = 1 - P(=0) = 0.36237$$
$$P(\geq 2) = 1 - P(\leq 1) = 0.07544$$
$$P(\geq 3) = 1 - P(\leq 2) = 0.01088$$
$$P(\geq 4) = 1 - P(\leq 3) = 0.00120$$

From the last series of probabilities we see that the probability of having two or more accidents is 0.075, that is, about 1 chance in 14. So while two or more accidents are not very likely, still they might be expected to occur with sufficient frequency so that drastic action is not called for. The probability of three or more accidents is only 0.01, 1 chance in 100, and definitely worth investigating.

The Molina tables do not have the probabilities for $m = 0.45$, as used in example 3.11, but a linear interpolation from the tables between the values for $m = 0.40$ and $m = 0.50$ gives a very close approximation to the calculated values, indicating that linear interpolation from tabulated values gives accuracy to about the third decimal place. A comparison between the calculated results of example 3.11 and the interpolated values from Molina's tables is given in Table 3.11.

TABLE 3.11

	Calculated probabilities for $m = 0.45$	Linear interpolation between $m = 0.4$ and $m = 0.5$
$P(\geq 1)$	0.36237	0.36157
$P(\geq 2)$	0.07544	0.07538
$P(\geq 3)$	0.01088	0.01116
$P(\geq 4)$	0.00120	0.00126

3.3 The Normal Distribution

The normal distribution occupies a central position in the application of statistics. In general practice the engineer will not deal directly with the normal distribution but with the distribution of some statistical function which is directly related to the normal. However, since the normal distribution is basic to statistical analysis, and since the mathematics of the normal distribution have been highly developed and are comparatively simple, a brief description of this distribution with relation to mathematical probability will be given here.

A point about the word *normal*. When this word is used in the phrase "normal distribution," it is strictly a mathematical term and does not have its usual clinical meaning. Distributions which are not normal should not be thought of as abnormal. While many distributions encountered in practice do approach the normal, no distribution of actual measurements is identical with the normal, as the distribution of finite measurements can only approximate the continuous function of the normal distribution. Some functions are definitely different from the normal, as the Poisson distribution for small values of m. The exceptions are merely pointed out to emphasize that the word normal as applied to the distribution is only descriptive and not definitive. It might be better if the distribution were always referred to as the Gaussian or the Demoivre distribution, after its originators.

3.3.1 Definition. The equation of the normal curve in terms of the frequency y of observations of various values of x, as a function of the

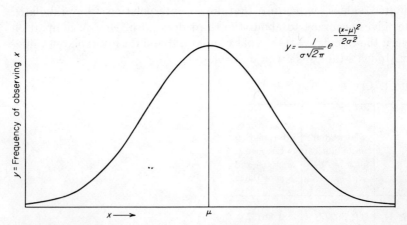

Fig. 3.6 Normal curve.

variable x, is

$$y = \frac{1}{\sigma\sqrt{2\pi}} e^{-\frac{(x-\mu)^2}{2\sigma^2}} \quad (3.14)$$

This equation is illustrated in Fig. 3.6.

DISTRIBUTIONS

μ is the mean value of the variable, and σ is the standard deviation—the square root of the mean-squared deviation from the mean. The total area under the normal curve, i.e., the integral of Eq. (3.14) from $-\infty$ to $+\infty$, is equal to 1.0. The expected frequency of occurrence of the variable between any two values of x, say x_1 and x_2, is measured by the fraction of area under the curve between these values times the total number of observations N. Since the total area is equal to 1.0, the area between any two values of x can be taken as the probability of a normally distributed variable falling within this range.

$$P(x_1 \leq x \leq x_2) = \frac{1}{\sigma\sqrt{2\pi}} \int_{x_1}^{x_2} e^{-\frac{(x-\mu)^2}{2\sigma^2}} dx \qquad (3.15)$$

Likewise, the probability of observing a value of a normally distributed variable equal to or less than some value of x, say x_1, is the area under the curve from $-\infty$ to x_1.

$$P(x \leq x_1) = \frac{1}{\sigma\sqrt{2\pi}} \int_{-\infty}^{x_1} e^{-\frac{(x-\mu)^2}{2\sigma^2}} dx \qquad (3.16)$$

The area under the normal curve has been carefully calculated, and tables of the area for various increments are available in many handbooks used by engineers. An abridged set of values is given here in Table 3.12.

3.3.2 The Standard Deviate. We can express each value of x in units of the number of standard deviations it deviates from the mean: $x = \mu + z\sigma$. z is the standard measure or the standard deviate, i.e., the number of standard deviations the variable deviates from the mean.

$$z = \frac{|x - \mu|}{\sigma} \qquad (3.17)$$

If we rewrite the equations for the normal curve, both the probability density equation (3.14) and the total probability equation (3.16) in terms of the standard deviate, noting from Eq. (3.17) that $dx = \sigma\, dz$, we get

$$y\, dx = \frac{1}{\sqrt{2\pi}} e^{-z^2/2}\, dz \qquad (3.18)$$

and

$$P\left(\frac{x-\mu}{\sigma} \leq z_1\right) = \frac{1}{\sqrt{2\pi}} \int_{-\infty}^{z_1} e^{-z^2/2}\, dz = F(z_1) \qquad (3.19)$$

To state Eq. (3.19) verbally: Any variable that is normally distributed, regardless of the magnitude of the variable or its standard deviation, if expressed in terms of the deviation from the mean divided by the standard deviation, will have a distribution identical with any other variable

TABLE 3.12 Area under the Normal Curve, $F(z)$

$$F(z) = \int_{-\infty}^{z} \frac{1}{\sqrt{2\pi}} e^{-z^2/2}\, dz$$

z	.00	.01	.02	.03	.04	.05	.06	.07	.08	.09
.0	.5000	.5040	.5080	.5120	.5160	.5199	.5239	.5279	.5319	.5359
.1	.5398	.5438	.5478	.5517	.5557	.5596	.5636	.5675	.5714	.5753
.2	.5793	.5832	.5871	.5910	.5948	.5987	.6026	.6064	.6103	.6141
.3	.6179	.6217	.6255	.6293	.6331	.6368	.6406	.6443	.6480	.6517
.4	.6554	.6591	.6628	.6664	.6700	.6736	.6772	.6808	.6844	.6879
.5	.6915	.6950	.6985	.7019	.7054	.7088	.7123	.7157	.7190	.7224
.6	.7257	.7291	.7324	.7357	.7389	.7422	.7454	.7486	.7517	.7549
.7	.7580	.7611	.7642	.7673	.7704	.7734	.7764	.7794	.7823	.7852
.8	.7881	.7910	.7939	.7967	.7995	.8023	.8051	.8078	.8106	.8133
.9	.8159	.8186	.8212	.8238	.8264	.8289	.8315	.8340	.8365	.8389
1.0	.8413	.8438	.8461	.8485	.8508	.8531	.8554	.8577	.8599	.8621
1.1	.8643	.8665	.8686	.8708	.8729	.8749	.8770	.8790	.8810	.8830
1.2	.8849	.8869	.8888	.8907	.8925	.8944	.8962	.8980	.8997	.9015
1.3	.9032	.9049	.9066	.9082	.9099	.9115	.9131	.9147	.9162	.9177
1.4	.9192	.9207	.9222	.9236	.9251	.9265	.9279	.9292	.9306	.9319
1.5	.9332	.9345	.9357	.9370	.9382	.9394	.9406	.9418	.9429	.9441
1.6	.9452	.9463	.9474	.9484	.9495	.9505	.9515	.9525	.9535	.9545
1.7	.9554	.9564	.9573	.9582	.9591	.9599	.9608	.9616	.9625	.9633
1.8	.9641	.9649	.9656	.9664	.9671	.9678	.9686	.9693	.9699	.9706
1.9	.9713	.9719	.9726	.9732	.9738	.9744	.9750	.9756	.9761	.9767
2.0	.9772	.9778	.9783	.9788	.9793	.9798	.9803	.9808	.9812	.9817
2.1	.9821	.9826	.9830	.9834	.9838	.9842	.9846	.9850	.9854	.9857
2.2	.9861	.9864	.9868	.9871	.9875	.9878	.9881	.9884	.9887	.9890
2.3	.9893	.9896	.9898	.9901	.9904	.9906	.9909	.9911	.9913	.9916
2.4	.9918	.9920	.9922	.9925	.9927	.9929	.9931	.9932	.9934	.9936
2.5	.9938	.9940	.9941	.9943	.9945	.9946	.9948	.9949	.9951	.9952
2.6	.9953	.9955	.9956	.9957	.9959	.9960	.9961	.9962	.9963	.9964
2.7	.9965	.9966	.9967	.9968	.9969	.9970	.9971	.9972	.9973	.9974
2.8	.9974	.9975	.9976	.9977	.9977	.9978	.9979	.9979	.9980	.9981
2.9	.9981	.9982	.9982	.9983	.9984	.9984	.9985	.9985	.9986	.9986
3.0	.9987	.9987	.9987	.9988	.9988	.9989	.9989	.9989	.9990	.9990
3.1	.9990	.9991	.9991	.9991	.9992	.9992	.9992	.9992	.9993	.9993
3.2	.9993	.9993	.9994	.9994	.9994	.9994	.9994	.9995	.9995	.9995
3.3	.9995	.9995	.9995	.9996	.9996	.9996	.9996	.9996	.9996	.9997
3.4	.9997	.9997	.9997	.9997	.9997	.9997	.9997	.9997	.9997	.9998

Even percentage points of the normal distribution

$F(z)$	.75	.90	.95	.975	.99	.995	.999	.9995	.99995	.999995
$\alpha = 2[1 - F(z)]$	.50	.20	.10	.05	.02	.01	.002	.001	.0001	.00001
z	0.674	1.282	1.645	1.960	2.326	2.576	3.090	3.291	3.891	4.417

similarly expressed. The probability of observing a deviation from the mean as large as z (in standard deviation units) is the same for any normally distributed variable and is here expressed as $F(z)$.

Table 3.12 gives the area under the normal curve for different values of z. These areas represent the probabilities for normally distributed variables of the occurrence of values deviating from the mean from $-\infty$ to z times the standard deviation. For example, 94.18 per cent of the area under the normal curve is included in the range $-\infty$ to $\mu + 1.57\sigma$, or 94.18 per cent of the total number of observations lie within this range. In terms of probability, there is a 0.9418 probability that a randomly selected measurement from this population will be in the range $-\infty$ to $\mu + 1.57\sigma$; or, there is only a 0.0582 probability that the measurement will be larger than $\mu + 1.57\sigma$.

Since the probability distribution is symmetrical around the mean, in the example just quoted, there is also a probability of 0.0582 that the measurement will be less than $\mu - 1.57\sigma$. In other words, there is a 0.8836 probability that the measurement will be in the range $\mu \pm 1.57\sigma$, $1 - 2[1 - F(z)]$, and a 0.1164 probability that it will be outside that range.

3.3.3 Significance Level. In a procedure with results that are normally distributed, $1 - 2[1 - F(z)]$ represents the symmetrical probability for results in the range $\mu \pm z\sigma$, and $2[1 - F(z)]$, usually designated alpha, α, is called the significance level of the statistical test. If we hypothesize that certain results have a normal distribution with a mean μ and a standard deviation σ and we observe a result that deviates from the mean by as much as $z\sigma$, then if we say that the observed result is significantly different from the hypothesized results, there is only an α probability of being in error. $1 - 2[1 - F(z)]$ is referred to as the confidence level. It is the probability with which values are expected to be in the range $\mu \pm z\sigma$; i.e., the 95 per cent confidence level sets the range which contains 95 per cent of the distribution.

At the bottom of Table 3.12 are included the z values for even intervals of both the total probability function $F(z)$ and the significance level α. From this table it is simple to obtain the values of z corresponding to the 0.95, 0.99, 0.999 probability levels, 1 chance in 20, 1 in 100, 1 in 1,000, etc.

example 3.12. For a normally distributed variable with a mean of 87.3 and a standard deviation of 2.1, what values may be discarded with 5 per cent error? with 1 per cent error?

$\mu = 87.3$, $\sigma = 2.1$.

For 5 per cent error, $\alpha = 0.05$, $z = 1.96$.

The 95 per cent confidence range includes $\mu \pm (1.96)(2.1) = 87.3 \pm 4.1 = 83.2$ to 91.3.

For 1 per cent error, $\alpha = 0.01$, $z = 2.576$.

The 99 per cent confidence range includes $\mu \pm (2.576)(2.1) = 87.3 \pm 5.4 = 81.9$ to 92.7.

If we wish to be more certain of not being in error when we discard results, we must enlarge the range of those which are accepted.

example 3.13. With a normally distributed variable having a mean of 0.745 and a standard deviation of 0.008, what is the probability of observing a value as low as 0.735 or a value as high as 0.760?

For the low value of the variable, $z = (0.745 - 0.735)/0.008 = 1.25$, $F(z) = 0.8944$, $P(x \leq 0.735) = 1 - 0.8944 = 0.1056$.

For the high value of the variable, $z \doteq (0.760 - 0.745)/0.008 = 1.875$, $F(z) = 0.9696$, $P(x \geq 0.760) = 1 - 0.9696 = 0.0304$.

The probability that the variable would be in the range of 0.735 to 0.760 would be $1 - (0.1056 + 0.0304) = 0.8640$.

Note that the values given in Table 3.12 are the areas under the normal curve from $-\infty$ to $\mu + z\sigma$. The area from $-\infty$ to μ (or to $\mu + 0 \cdot \sigma$) is 0.50000. The curve is completely symmetrical so that the area from $-\infty$ to $\mu - z\sigma$ is the same as $1.0 - $ (area from $-\infty$ to $\mu + z\sigma$). Not all tables of the normal distribution function are given in this manner. Some tables[1] give the areas under the curve from $-\infty$ to $\mu - z\sigma$. In other words, 1.0 minus the values given in Table 3.12. Still other tables[2] give the areas from $\mu - z\sigma$ to $\mu + z\sigma$. The values in these latter tables are equal to 1.0 minus the values denoted α in Table 3.12.

The information that can be obtained from these different tabulations is the same, but it is necessary to know what integral under the normal curve is tabulated in order to use the values correctly. It is helpful to memorize one value from some table to compare the tabulated values of unfamiliar tables.

Figure 3.7 illustrates the use of the values of Table 3.12 for different types of probability problems.

3.3.4 Normal-probability Paper. There is available graph paper in which the ordinate scale is ruled proportional to the cumulative normal distribution. Data with a normal distribution when plotted on this paper will give a straight line. A simple procedure, when using this paper with small quantities of data, is to tabulate the data in order of increasing magnitude and then to locate each point at the cumulative per cent frequency corresponding to the mid-point between the cumulative frequency for that datum point and the one preceding it. This method is illustrated

---

[1] "Handbook of Chemistry and Physics," 43d ed., Chemical Rubber Publishing Co., Cleveland, Ohio, 1961.

[2] R. H. Perry, C. H. Chilton, and S. D. Kirkpatrick, eds., "Chemical Engineers' Handbook," 4th ed., McGraw-Hill Book Company, New York, 1963.

DISTRIBUTIONS

in example 3.14. With larger quantities of data, more than 50 datum points, it is usually sufficient to divide the total range of the data into 15 to 30 equal-size subgroups and then to plot the cumulative per cent frequency for each subgroup at the mid-point of that group, again locating the frequencies at the mid-points between successive groups.

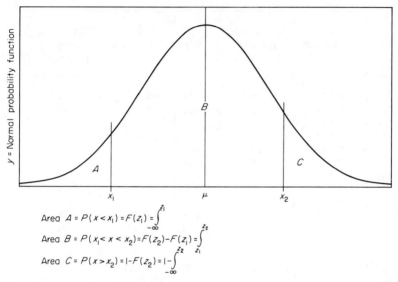

Area $A = P(x < x_1) = F(z_1) = \int_{-\infty}^{z_1}$

Area $B = P(x_1 < x < x_2) = F(z_2) - F(z_1) = \int_{z_1}^{z_2}$

Area $C = P(x > x_2) = 1 - F(z_2) = 1 - \int_{-\infty}^{z_2}$

Fig. 3.7 Normal probability areas.

The closeness of the plot to a straight line on the normal-probability paper is a measure of the closeness of the distribution of the data to a normal distribution. The intersection of the straight line with the 50 per cent point is an estimation of the mean of the data. The table of the normal curve area shows that 84.13 per cent of the area under the curve is included from $-\infty$ to 1σ beyond the mean; therefore an estimate of the standard deviation can be obtained from the difference between the 50 per cent point and the 84.13 per cent point, or the 50 per cent point and the 15.87 per cent point.

example 3.14 Table 3.13, in the first column, gives 25 numbers taken from a table of random numbers having a mean of 2.0 and a standard deviation of 1.0.[1] These might represent some analytical measurements or inspections having a similar distribution. The second column of Table 3.13 gives the same data in order of increasing magnitude. The third column gives the cumulative per cent frequency for

[1] W. J. Dixon and F. J. Massey, "Introduction to Statistical Analysis," McGraw-Hill Book Company, 1951.

each datum point, starting with the mid-point between zero and 4 per cent and increasing 4 per cent for each successive point.

These data are plotted on normal-probability paper in Fig. 3.8. A straight line through the data was estimated visually. There is fair

TABLE 3.13

Data	Ranked data	Cumulative per cent frequency	Data	Ranked data	Cumulative per cent frequency
4.142	0.425	2	1.342	2.268	54
0.736	0.736	6	4.600	2.419	58
3.610	0.832	10			
0.832	0.919	14	1.052	2.454	62
2.931	1.052	18	1.663	2.527	66
			2.704	2.704	70
2.419	1.342	22	1.625	2.912	74
1.356	1.356	26	0.425	2.931	78
1.564	1.398	30			
2.454	1.564	34	4.251	3.245	82
1.893	1.625	38	2.912	3.610	86
			1.963	4.142	90
2.268	1.663	42	0.919	4.251	94
2.527	1.893	46	1.398	4.600	98
3.245	1.963	50			

agreement between the data and a straight line, so that we may say the data approximate a normal distribution. The 50 per cent cumulative-frequency point corresponds to a datum point of about 2.19, which estimates the mean of the data. The 84.13 per cent cumulative-frequency point corresponds to a datum value of 3.35, giving an estimate of the standard deviation of $3.35 - 2.19 = 1.16$.

It is sometimes possible to obtain an approximately normal distribution by transforming data to a logarithmic basis, especially when there is a several-order magnitude range in the data. This is often the case in sieve analysis, where the material is weighed and the weight is a function of the third power of the dimensions which are separated by the sieves. A similar situation holds in analyses which determine impurities as a function of pH, where the pH is itself a logarithmic function of concentration. For cases of this nature there is available a log normal probability paper in which the abscissa is a two-cycle logarithmic scale. This type of paper was used for the Poisson distribution curves of Fig. 3.5. Except for the difference in the abscissa scale, the remarks regarding the

DISTRIBUTIONS

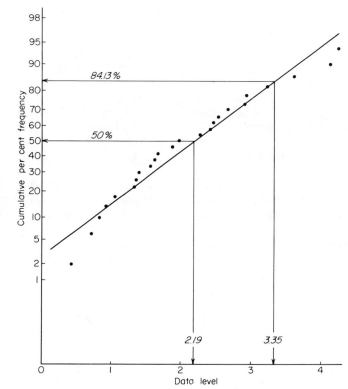

Fig. 3.8 Normal cumulative graph of data from Example 3.14.

use of the linear probability paper apply also to the logarithmic probability paper.

3.4 Summary

Distribution:	Binomial	Poisson	Normal
Mean	np	m	μ
Standard Deviation	$\sqrt{np(1-p)}$	m	σ
Probability of r, $P(r)$	$\binom{n}{r}p^r(1-p)^{n-r}$	$e^{-m}m^r/r!$	
Probability of a range of r, $P(r_1 \leqq r \leqq r_2)$	$\sum_{r_1}^{r_2}\binom{n}{r}p^r(1-p)^{n-r}$	$\sum_{r_1}^{r_2}e^{-m}m^r/r!$	$\dfrac{1}{\sigma\sqrt{2\pi}}\int_{r_1}^{r_2}e^{-(r-\mu)^2/2\sigma^2}dr$

$$F(z_2) - F(z_1), \text{ where}$$
$$z_1 = (\mu - r_1)/\sigma$$
$$z_2 = (r_2 - \mu)/\sigma$$

$$F(z) = \frac{1}{\sqrt{2\pi}}\int_{z_1}^{z_2}e^{-z^2/2}dz$$

3.5 Problems

1. What is the term containing y^5 in the expansion of $(x + y)^{15}$?

2. In a shipment of random lengths of pipe, containing 80 per cent in 20-ft lengths or longer, what is the probability of a sample of 100 lengths having as many as 30 less than 20 ft in length?   0.011.

3. In a straw poll of 500 persons, 265 favor a particular candidate. May the candidate feel 99 per cent certain of being elected? 95 per cent? What fraction of a sample of 1,000 in his favor would make him 99 per cent certain?

4. If the average red-cell count in a certain volume of blood is 20, what is the probability of a normal person having as low a count as 15? As low as 12? 0.157; 0.039.

5. Assume the probability of dying within the year for the age group 40–41 is 0.00513. In a group of 100, what is the probability that none will die within the year? That exactly 1 will die? That more than 1 will die?   0.5990; 0.3067; 0.0943.

6. For a sample of 100, with an expected percentage defective of 2 per cent, compare the probabilities of 0, 1, 2, 3, 4 defectives by the binomial and Poisson distributions.

No. of defectives	0	1	2	3	4
Binomial	0.133	0.270	0.273	0.182	0.090
Poisson	0.135	0.271	0.271	0.180	0.090

7. For a sample of 10, with an expected percentage defective of 10 per cent, compare the probabilities of 0, 1, 2, 3, 4 defectives by the binomial and Poisson distributions.

No. of defectives	0	1	2	3	4
Binomial	0.349	0.387	0.194	0.057	0.011
Poisson	0.368	0.368	0.184	0.061	0.015

8. If intelligence quotients have a mean of 96 and a standard deviation of 26, what fraction of the population can be expected to have an intelligence quotient greater than 120? Greater than 140?   18 per cent; 4.6 per cent.

9. Draw four cards at random from a deck and add the value of the four (counting the picture cards as 11, 12, and 13, and the aces as 1). Return the cards to the deck and repeat the drawing 24 times. Find the mean and the standard deviation of the 25 sums. Estimated answer: 28 and 7.5.

10. An instructor finds his class has average grades of 74 with a standard deviation of 4.7. What should the passing grade be if he does not want to fail more than 10 per cent?

chapter four
MEASURES OF VARIABILITY

It is the general practice to use one number to describe a process or technique. The octane number is 92, the carbon content is 15 parts per million, the yield is 12,000 barrels per day, the purity is 99.44 per cent, etc. When we say the nickel in the steel is 18 per cent, could it be 18.02 per cent, 18.2 per cent? Could it be as high as 19 per cent? When we say the average income is $5,300 a year, do we mean that more people have this amount of income than have any other amount, or do we mean that half the people have $9,000 and the other half have $1,600? Or do we mean something else?

Two numbers are usually necessary to give a good description of an analysis or a measurement: a number around which most of the data cluster and a number which gives a measure of the spread of the data. In this chapter we shall describe the different statistics commonly used for these two purposes and shall discuss in some detail their calculation and limitations.

4.1 Measures of Central Tendency

4.1.1 The Mean. In taking data or making measurements it is usual to find that the results cluster around some particular value so that if we plot the frequency of occurrence against the magnitude of the variable we obtain the familiar normal-distribution-type curve with a peak toward the center and a trailing off of frequency as the deviation from this central value increases. Any number which tends to represent this grouping tendency is called an average. There are several of these numbers, as we shall see, but the one that is most used in engineering is the *mean*.

The mean is the familiar arithmetic average, designated $\bar{x}$ and equal to $\Sigma x/n$. The mean is what people usually have in mind when they say average. But since the other central-tendency measures are also averages, and since the term average is sometimes used when the intent is ostensibly not to make clear what measure has been used, we shall employ the term *average* in a generic sense to indicate any central-tendency measure. The term mean will be used for the usual arithmetic average.

The mean is used for three important reasons. When we come to discuss the variation of the data, we shall do so in terms of the square of the deviations from some central value. The mean is the value from which the sum of the squares of the deviations is a minimum. The mean of the sample is the best estimate of the mean of the population from which the sample is drawn. Perhaps most important of all, means of samples of uniform size tend to be normally distributed regardless of the type of distribution of the population from which the samples were drawn; the larger the sample size, the more nearly normal the distribution of the means. This characteristic of sample means permits the use of the normal distribution in making probability statements about the population mean even when the population is not normally distributed.

4.1.2 The Mode. The mode is the number which occurs most often. It is the "fashion": a la mode. In a plant where 1 man makes $50,000 a year, 4 men make $20,000, 50 men make $4,500, and 2 men make $2,600, the mean wage is slightly more than $6,300 while the mode is $4,500. Depending on what type of impression is desired, either of these earnings can be quoted as the average.

In a frequency plot against magnitude of the variable, the high point of the curve is at the modal magnitude, i.e., the maximum frequency. When a curve has more than one peak, even though they represent different frequencies, each peak is referred to as a mode, the lesser peaks are sometimes called "local modes," and the curve is called bimodal, trimodal, multimodal, as the case may be.

The mode is most often employed in social statistics when the information to be conveyed is that which represents the largest group. In engineering work, if the mode is used as the average to represent the data, its usage should be made clear so that no misunderstanding is involved. In particle-size analyses either by microscope count or by screen analysis,

MEASURES OF VARIABILITY

bimodal and multimodal distributions are often encountered. In cases of this kind, a single number representing average particle size is usually not sufficient and either a graphical or verbal description of the distribution should be presented.

4.1.3 The Median. The median is the value of the variable that divides the frequencies into two equal portions. In the series of values in Table 4.1, which represents analyses of samples taken from a shipment

TABLE 4.1 Analyses of Samples Drawn at Random from a Shipment of Material

80.0
76.0
80.0
80.5
78.5
79.0
80.0

$\Sigma x = 554.0$

$n = 7$

$\bar{x} = 79.14$

TABLE 4.2 Data of Table 4.1 Arranged in Order of Increasing Magnitude

76.0
78.5
79.0
80.0
80.0
80.0
80.5

of material, the mean is 79.1 and the mode is 80.0, since there are three results of this value and only one of each of the others. If the data are rearranged in order of increasing magnitude, as in Table 4.2, the center value is the median—in this case the same as the mode, 80.0. The median is the middle value when the data are arranged in order of magnitude. When there is an even number of data points, the median is the mean of the two center values.

The median is an important statistic in the branch of the subject dealing with the order of measurements. It is the value from which the sum of the absolute value of the deviations is a minimum.

For a symmetrical distribution such as the normal, the mean, the mode, and the median coincide at the mid-point of the distribution. Deviation from symmetry is called skewness; a symmetrical distribution has zero skewness. If the mean, median, and mode are not equal, and the distribution is unimodal, the median will be between the mean and the mode. If the mean is larger than the median and mode, the skewness is positive; and if the mode is larger than the median and mean, the skewness is negative. The greater the difference between the averages, the greater the skewness. We shall not use the skewness in any of the quantitative statistical methods discussed in this book.

4.1.4 The Geometric Mean.
The nth root of the product of n terms is designated the geometric mean.

$$\bar{x}_G = \sqrt[n]{x_1, x_2, x_3, \ldots, x_n} \qquad (4.1)$$

The logarithm of the geometric mean is obviously the mean of the logarithms of the individual values.

$$\log \bar{x}_G = \frac{\Sigma(\log x_i)}{n} \qquad (4.2)$$

The geometric mean is used principally in the field of economic statistics where the average growth rate or the average rate of change is desired. If a business increases steadily over a number of years, the average rate of increase can be obtained from the geometric mean of the individual yearly growth rates. If the individual yearly growth rates are R_1, R_2, R_3, etc., the average growth rate—that is, the growth rate which applied each year would give the same final value that the individual yearly growths produced—is the geometric mean $\sqrt[n]{R_1 \cdot R_2 \cdot R_3 \cdots R_n}$ and not the arithmetic mean $\sum_i R_i/n$, of the individual rates. The following example illustrates this statement.

example 4.1. An enterprise expands 10 per cent the first year, 20 per cent the second year and 50 per cent the third year. What is the average yearly expansion rate?

If the starting value was V, at the end of the first year the value is $1.10V$. At the end of the second year the value is $(1.2)(1.1)V$ or $1.32V$. At the end of the third year the value is $(1.5)(1.32)V$ or $1.98V$. The average rate of expansion is the geometric mean of the three rates:

$$\sqrt[3]{(1.10)(1.20)(1.50)} = \sqrt[3]{1.98} = 1.256$$

The arithmetic mean of the three growth rates is 1.267. (Note: the geometric mean is always less than the arithmetic mean.) As a check that the geometric mean gives the correct solution, the following table gives a comparison of the two results.

Value after	Geometric mean 1.256	Arithmetic mean 1.267
1 year	1.256	1.267
2 years	1.578	1.605
3 years	1.981	2.042

Actual value after 3 years—1.980

MEASURES OF VARIABILITY

4.1.5 The Harmonic Mean. If one machine produces 4 units/hr and a second machine produces 2 units/hr, the mean output per machine per hr is 3 units. This might be interpreted as an average time per unit of $\frac{1}{3}$ hr. However, the first machine producing 4 units/hr requires only $\frac{1}{4}$ hr/unit, and the second machine requires only $\frac{1}{2}$ hr/unit. The mean of $\frac{1}{4}$ and $\frac{1}{2}$ is $\frac{3}{8}$, i.e., an average of $\frac{3}{8}$ hr/unit. The first calculation seemed to indicate an average time per unit of $\frac{1}{3}$ hr and the second gives an average of $\frac{3}{8}$ hr/unit. Which is correct?

In calculating a mean, it is necessary that the dimensional units of the separate items be the same. The mean number of defectives in a sample would not be of much value if it was calculated from samples of different sizes. The average income of a group of persons would not mean much if it were based on weekly income for some and monthly income for others. The difference between the two averages calculated in the preceding paragraph is in their units. The first was calculated from the output for 1 hr from each machine, while the second was calculated from different times but for equal output. The first is the mean output per hr, and the second gives the mean time per output. The reciprocal of a mean will not be the same as the mean of the reciprocals.

The harmonic mean is the reciprocal of the mean of reciprocals:

$$\bar{x}_H = \frac{1}{\sum_i (1/x_i)/n} \tag{4.3}$$

Although the harmonic mean can be used in specific cases where it is applicable, for engineers not familiar with its use it is probably better to verify the dimensional units of the quantities being averaged before proceeding. The following illustrations may make this point clear.

If a vehicle travels 10 miles at 30 mph, 10 miles at 40 mph and 10 miles at 50 mph, what is the average rate of travel? One way to obtain the answer is to determine the total time required to travel 30 miles. The first 10 miles at 30 mph took 20 min. The second 10 miles at 40 mph took 15 min. And the last 10 miles took 12 min. 30 miles in 47 min is 38.3 mph average. The harmonic mean gives the same answer:

$$\bar{x}_H = \frac{3}{10/30 + 10/40 + 10/50} = 38.3 \tag{4.4}$$

The reason you cannot simply take the mean of 30, 40, and 50 mph for an answer of 40 is that each of these rates applies to a different length of time. We want the miles traveled in an average hour, and it is therefore necessary to equalize the time periods for each rate.

You take samples one at a time from a batch of material until you find a bad item. The first time you try you take 200 before you find a defective item. The second time you take 300 and the third time 400 before you find a defective item. What is the average number of defectives you can expect to find in a sample? The answer is not one in 300. In order to calculate the mean, it is necessary to have samples of the same size. The first sample has 0.5 per cent defectives, the second 0.33 per cent, and the third 0.25 per cent. These are the number of defects in samples size 100, and their mean is 0.361 per cent. The harmonic mean gives the same answer:

$$\bar{x}_H = \frac{3}{1/200 + 1/300 + 1/400} = \frac{3,600 \text{ (good items)}}{13 \text{ (defectives)}}$$

The harmonic mean and the geometric mean are not involved in any of the statistical tests that are discussed in this book. However it is important to know that calculating an average offers certain pitfalls, and as in most engineering, the dimensional units should be verified.

4.2 Measures of Variability

4.2.1 The Standard Deviation. The mean is a single number which is used to represent a group of measurements. Two groups might have the same mean but have considerable difference in their variation from item to item. For example, in Table 4.3 are listed the analyses

TABLE 4.3 Different Sets of Analyses of the Same Shipment

First analyses	Second analyses
80.0	75.0
76.0	77.0
80.0	84.0
80.5	84.5
78.5	74.5
79.0	74.0
80.0	85.0
$\Sigma x = 554.0$	554.0
$n = 7$	7
$\bar{x} = 79.14$	79.14

from Table 4.1 together with another series of analyses having the same mean. It is obvious from the data that the second set of analyses has much more variation from item to item than does the first set.

The measure of variability most amenable to statistical analysis is the standard deviation. The standard deviation has already been defined in the discussion of the normal distribution (Sec. 3.3.1). It is the square

MEASURES OF VARIABILITY

root of the mean-squared deviation of the individual measurements from their mean and is designated sigma, σ.

$$\sigma(x) = \sqrt{\frac{\Sigma(x - \bar{x})^2}{n}} \tag{4.5}$$

The symbol $\sigma(x)$ designates the standard deviation of x and does not indicate a multiplication operation. It is similar in form to the symbol dx or sine x. It is merely a definitive symbol and not an operational symbol.

As the magnitude of the variation of the data increases, the deviations from the mean, and hence the standard deviation, will also increase. For the two sets of analyses in Table 4.3, the standard deviations are 1.43 and 4.73, respectively. The units of the standard deviation are the same as the units of the original data. It is customary to report the standard deviation to three significant digits.

With a normally distributed variable, the standard deviation is a measure of the fraction of the distribution that is within any specified range of the mean. Table 3.12 gives the fraction. For instance, 95 per cent of the distribution falls within the range $\bar{x} \pm 1.96\sigma(x)$. If we have two normal distributions with the same mean, say 79.1, and with different standard deviations, say 1.43 and 4.73, then in the first case we can expect 95 per cent of the data to be in the range $79.1 \pm (1.96)(1.43) = 76.3$ to 81.9. In the second case, 95 per cent of the data would be in the range $79.1 \pm (1.96)(4.73) = 69.8$ to 88.4.

In much of the statistics that follows in this book we shall make use of the sum of squares of deviations from the mean. This term will often be referred to simply as *the sum of squares* and shall be designated $\Sigma'x^2$ instead of $\Sigma(x - \bar{x})^2$.

$$\Sigma'x^2 = \Sigma(x - \bar{x})^2 \tag{4.6}$$

$$\sigma^2(x) = \frac{\Sigma'x^2}{n} = \frac{\Sigma(x - \bar{x})^2}{n} \tag{4.7}$$

The Standard Deviation Estimated from a Sample. The standard deviation as defined in Eq. (4.5) applies to the distribution of the total population of the measurements involved. In practice all that is usually available for statistical tests is a sample of the population. If the total population were available, it would not be necessary to make probability statements about it, as it would be completely known, and the only purpose of the statistics would be to define the total mass of data in some simple terms. However, when a sample only is available, in order to make statements about the population from which the sample came, it is necessary to estimate the population parameters from statistics calculated from the sample.

The mean of a sample is an unbiased estimate of the mean of the population. The standard deviation of a sample, however, is not an unbiased estimate of the standard deviation of the population. The sum of squares of deviation of the individual sample values from the sample mean results in the minimum sum of squares of deviation. The sum of squares of deviation from any value other than the mean will be larger than from the mean. The proof of this statement can be demonstrated by replacing $\bar{x}$ with a variable in the expression $\Sigma(x - \bar{x})^2$ and equating the first derivative to zero. Since the sample mean may not be identical with the population mean, the sum of squares of deviation of the individual sample values from the sample mean will be less than the sum of squares of deviation of the individual sample values from the population mean. The standard deviation of the sample, computed from the sum of squares of deviation divided by n (the number of items in the sample), will therefore be smaller than if the sum of squares has been calculated from the true population mean. To overcome this bias, the standard deviation, estimated from the sample and designated s, is obtained by dividing the sum of squares of deviation by $n - 1$ instead of by n.

$$s(x) = \sqrt{\frac{\Sigma(x - \bar{x})^2}{n - 1}} = \sqrt{\frac{\Sigma' x^2}{n - 1}} \qquad (4.8)$$

$s(x)$ is *not* the standard deviation of the sample. It is the estimate of the population standard deviation $\sigma(x)$ calculated from the sample. A discussion of the criteria for evaluating sample estimates of population parameters can be found in several books dealing with the theory of statistics[1,2] We shall limit ourselves to the statement that the best estimate of the population standard deviation is obtained by dividing the sample sum of squares of deviation from the sample mean by one less than the number of items in the sample, and then extracting the square root, as in Eq. (4.8).

Calculating the Estimated Standard Deviation. The method of calculating $s(x)$, illustrated in example 4.2, is not only somewhat cumbersome but is also subject to possible errors, since each deviation is computed separately by subtraction from the mean and the differences are squared and summed. A somewhat simpler method of calculating the sum of squares (of deviations from the mean) and a method more amenable to the use of

---

[1] L. H. C. Tippett, "The Methods of Statistics," 4th ed., John Wiley & Sons, Inc., New York, 1952.

[2] A. M. Mood, "Introduction to the Theory of Statistics," McGraw-Hill Book Company, New York, 1950.

MEASURES OF VARIABILITY

a desk calculator makes use of one of the following algebraic identities:

$$\Sigma' x^2 = \Sigma(x - \bar{x})^2 = \Sigma x^2 - \frac{(\Sigma x)^2}{n} = \Sigma x^2 - \bar{x}\Sigma x \tag{4.9}$$

example 4.2. Estimate the population standard deviation for the analytical data of Table 4.1.

TABLE 4.4

Data	Deviations from the mean $(x - \bar{x})$	Deviations squared $(x - \bar{x})^2$
80.0	0.857	0.7344
76.0	−3.143	9.8784
80.0	0.857	0.7344
80.5	1.357	1.8414
78.5	−0.643	0.4134
79.0	−0.143	0.0204
80.0	0.857	0.7344
Sum = 554.0 Mean = 79.143	−0.001	14.3568

$$s^2(x) = 14.3568/6 = 2.3928 \qquad s(x) = 1.55$$

With a desk calculator, Σx and Σx^2 can be obtained in one series of operations by accumulating the product. If negative values are included in the original data, these can be taken in stride by reversing the sign of the counter dials (Σx), while keeping the sign of the product dials (Σx^2) positive. Since the mean of the x values is usually required in all cases, the last expression for $\Sigma' x^2$, $\Sigma x^2 - \bar{x}\Sigma x$, can be obtained in one more operation on the calculator by inserting Σx^2 in the product dials of the machine and then performing a negative multiplication of $\bar{x}$ and Σx. It actually takes somewhat longer to describe the method than to carry out the operations. For example, the sum of squares for the data of example 4.2, $\Sigma' x^2 = 14.3571$, was calculated in 80 seconds on a Friden desk calculator, and in 30 seconds on a Wang desk calculator by the author. The slight difference between this answer and the answer given in example 4.2 is due to the number of figures used in the two calculations. A discussion of the number of significant figures to be used is given in Sec. 4.4. For the data of this example, the standard deviation is the same, 1.55, whether computed from the sum of squares obtained from example 4.2 or from the value of 14.3571 given above. (Note that the sum of squares of deviations from the mean cannot be negative. Since they are squared terms they must be positive. If a negative value results from the use of Eq. (4.9), it indicates either an arithmetic error or a rounding-off error which has resulted in a negative answer.)

Coding of Calculations. The standard deviation is a measure of the distribution of the population around its mean. The larger the standard deviation, the wider the spread of the data. The curves in Fig. 4.1 are intended to illustrate the relations between means, standard deviations, and population distributions.

The ordinate of the curves is a measure of the frequency of observations whose magnitude is indicated by the scale of the abscissa. The two curves at the left have a mean of x_1. The two curves at the right have a mean of x_2. Of the left pair, the high, skinny curve has a standard deviation of 5.0, with a greater frequency of values close to the mean and a smaller frequency of observations with values having a greater deviation from the mean than the short, squat curve that has a standard deviation of 10.0. The second pair of curves at the right have the same shape as the two at the left. Again, the high, thin curve has a standard

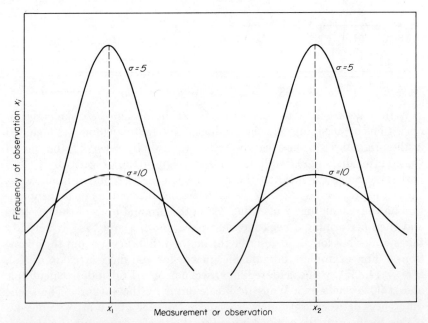

Fig. 4.1 Comparison of normal distribution of different means and different standard deviations.

deviation of 5 and the lower, wider curve has a standard deviation of 10. Only the means are different from the first pair.

The point of this is that at constant standard deviation, a change of mean only displaces the location of the distribution curve. It does not

MEASURES OF VARIABILITY

change its shape. The relative frequency of measurements with different deviations from the mean is the same. A change in standard deviation at a fixed value of the mean, on the other hand, does not change the location of the curve, but does change the shape of the curve.

The corollary of this statement is that if all the x values are shifted by a constant amount, it will have no effect on the standard deviation. This can be seen from the equation for the estimated standard deviation [Eq. (4.8)]. If we rewrite the equation in a slightly different form

$$(n - 1)s^2(x) = \Sigma(x - \bar{x})^2 = \Sigma' x^2 \tag{4.10}$$

and in place of x we substitute another variable, w, which differs from x by a constant amount, $w = x - c$, we get

$$\Sigma' x^2 = \Sigma[(w + c) - (\bar{w} + c)]^2 \tag{4.11}$$

or

$$\Sigma' x^2 = \Sigma(w - \bar{w})^2 = \Sigma' w^2 \tag{4.12}$$

$$s^2(x) = \frac{\Sigma' x^2}{n - 1} = \frac{\Sigma' w^2}{n - 1} = s^2(w) \tag{4.13}$$

Therefore, if the standard deviation is to be estimated from a series of data and it is possible to subtract some constant value from all the data points to simplify the arithmetic, this subtraction will not alter the calculated sum of squares or the estimated standard deviation. If an electric calculating machine is to be employed, it is of doubtful advantage to code the data in order to simplify the arithmetic because the coding introduces another possible source of error and the calculation by machine will not be significantly lessened. However, if the sum of squares is to be calculated longhand, coding the data will decrease the amount of work involved; or if the squares are to be taken from a table, it is sometimes essential to code the data so that the final digits that indicate the difference between datum points are not lost in rounding off.

example 4.3. The data from example 4.2 will be coded by subtracting 79 from each value and the standard deviation estimated from the coded data.

TABLE 4.5

Original data x	Coded data $w = x - 79$
80.0	1.0
76.0	−3.0
80.0	1.0
80.5	1.5
78.5	−0.5
79.0	0.0
80.0	1.0

$$\Sigma w = 1.0$$
$$\Sigma w^2 = 14.50$$
$$n = 7$$
$$\Sigma' w^2 = 14.50 - \frac{(1.0)^2}{7} = 14.3571$$
$$s^2(w) = \frac{\Sigma' w^2}{n-1} = \frac{14.3571}{6} = 2.3928$$
$$s(w) = 1.55 = s(x) \text{ of example 4.2}$$

It is sometimes useful to code the original data by multiplying or dividing by some constant, often some power of 10. If the data were fairly small, for example, 0.0076, 0.0080, etc., it might simplify the calculations to multiply by 10^3 or 10^4 so that the number of decimal places involved in the squared terms is reduced. The same is true when the data are large but the number of significant digits are only two or three. In reporting the tensile strength of steels, the results might be in thousands of pounds per square inch, in which case dividing by 10^3 would greatly simplify the subsequent arithmetic. When the original data are coded by multiplying or dividing by a constant, the calculated estimate of standard deviation, which has the same units as the original data, will also be increased or decreased by the same factor as was applied to the original data. If $w = cx$, then $s(w) = cs(x)$, and $\Sigma' w^2 = c^2 \Sigma' x^2$, so that if the standard deviation of the original data is to be estimated, the result of the coded calculation must be decoded to get back the original units.

In example 4.3, if the coded data were multiplied by 10 to eliminate the decimal so that $w' = 10w = 10(x - 79)$, then the following results would be obtained, which the student may verify for himself.

$$\Sigma w' = 10$$
$$\Sigma (w')^2 = 1{,}450$$
$$\Sigma' (w)^2 = 1{,}435.71$$
$$s^2(w') = 239.28$$
$$s(w') = 15.5 = 10 s(w) \quad \text{from example 4.3}$$

4.2.2 The Variance. The variance is the square of the standard deviation, σ^2. It is equal to the mean-squared deviation of the variable from its mean, $\Sigma (x - \bar{x})^2 / n$. From a mathematical standpoint the variance is the basic measure of the distribution. The true variances of infinite populations are additive, so that if a number of factors with different variances are affecting a measurement, the variance of the measurement is equal to the sum of the separate variances when these factors

MEASURES OF VARIABILITY

are statistically independent. This important feature of the variance is made use of in the reverse order when the total variance of a measurement is analyzed for its contributing factors. The subject of the analysis of variance is discussed in detail in Chap. 7.

The units of the variance are the square of the units of the measurements involved. It is customary, therefore, to report the distribution of measurements in terms of the square root of the variance, the standard deviation, which has the same units as the measurements.

4.2.3 The Mean Deviation. The mean of the absolute deviations of measurements from their mean is defined as the *mean deviation*.

$$M.D. = \frac{\Sigma |x - \bar{x}|}{n} \qquad (4.14)$$

The algebraic sum of deviations from the mean is of course equal to zero since

$$\bar{x} = \frac{\Sigma x}{n}$$

and

$$\frac{\Sigma(x - \bar{x})}{n} = \frac{\Sigma x}{n} - \frac{n\bar{x}}{n}$$

The mean deviation is still often used in reporting data (usually it is called the average deviation), although its usefulness as a statistical measure is limited. It has a somewhat dubious advantage in being smaller than the standard deviation, and hence gives a better first impression of the data. For a normal distribution, the standard deviation is 1.253 times the mean deviation. Actually it is $\sqrt{\pi/2}$ times as large. From this relationship it is possible to estimate the standard deviation of a measurement if the mean deviation is known from experience.

4.2.4 The Range. The difference between the largest and the smallest values is the range. It represents the absolute value of the spread of the data; all the data are included within the range. However, only two of the data points are involved in the calculation of the range, while all the data are involved in the calculation of the other measures of the spread of the data. Hence the range is not a sensitive measure of the variability of groups of data when the groups are large, $n > 10$. On the other hand, the range is extremely easy to obtain and gives a quick comparison of results.

For small samples the range is an extremely useful statistic and is often used in quality control and inspection work. Tables which give the relation between the range and the standard deviation are available in quality-control texts[1,2] and are discussed further in Chap. 11. For samples of 4, the range is about twice the standard deviation; for samples of 10 it is about three times; and for samples of 25 it is about four times. This relation provides another rapid estimate of the standard deviation.

It is also possible to report the 90 per cent range, the range that includes 90 per cent of the data; or the 95 per cent range; or the penultimate range, the range excluding the two extreme values, high and low. These modified ranges are sometimes employed in social and economic statistics where large quantities of data are available and precise mathematics are not required. In analyzing engineering data the range is usually employed only in quality-control work.

4.3 Degrees of Freedom

The denominator under the square-root sign in Eq. (4.8) for the estimated standard deviation is $n - 1$, one less than the number of items of data used to calculate $s(x)$. If only one measurement was available, it could be used as an estimate of the mean of the population from which the measurement was made, but there would be no way to estimate the standard deviation or distribution of the population. If two measurements were available, the mean would be estimated to lie midway between them and there would be only one estimate of the deviation from the mean inasmuch as the deviation of both measurements would be the same. If n measurements were involved and the mean were calculated from these measurements, there would be $n - 1$ independent estimates of the deviation from the mean, and the deviation of the last term would be fixed by having fixed the mean and the other $n - 1$ deviations; i.e., there would be $n - 1$ degrees of freedom for estimating $\sigma(x)$.

The number of independent measurements that are available for estimating a statistical parameter is called the degrees of freedom of that estimate. This term is similar to the same term in physical chemistry where the number of degrees of freedom of a system is the "number of independently variable quantities . . . which may be altered at will without producing an alteration in the number of phases. . . ."[3] In the statistical tests which are described later in this book the degrees of freedom are an important variable in establishing the significance of a calculated statistic.

[1] E. L. Grant, "Statistical Quality Control," 2d ed., McGraw-Hill Book Company, New York, 1952.
[2] ASTM Manual on Quality Control of Materials, American Society for Testing Materials, Philadelphia, Pa., 1951.
[3] J. H. Perry, C. H. Chilton, and S. D. Kirkpatrick, eds., "Chemical Engineers' Handbook," 3d ed., McGraw-Hill Book Company, New York, 1950.

4.4 Number of Significant Digits

All the statistical conclusions drawn from data will be based on numbers observed or calculated. The analyst or operator will sometimes have a "feel" for the accuracy or significance of his data. This feeling will sometimes be fairly reliable. The statistician or the engineer applying statistics will have only the data. Whether or not the statistical analysis of data should include some evaluation by the operator depends entirely on the situation under study, and no general rule can be laid down for all cases. In research and development work, the operator and the statistical analyst are often the same person, so that it is close to impossible to eliminate subjective evaluations of the data. Also, in research work it is necessary to know which data are reliable from a mechanical standpoint: which temperature effects are real and which are a result of the condenser water being shut off for half an hour; which yields are from good runs and which data were taken just after a power failure shut down the apparatus; when the standard solution was changed during the analysis; etc. In research work it is important for the statistical analyst to be in close touch with the source of the data. In production, on the other hand, it is usually better if the statistical evaluations are made independent of the production staff. Instances have been known when it was sufficient merely to tell the operating crew that the engineering department was studying an alteration in the process and a change in yield resulted without any actual alteration being made. Whether the yield was more or less depended on whether the operating crew wished to cooperate or not. Usually the purpose of a statistical study of a production process is to see how it operates under routine (normal) conditions. In this case the raw data are usually the best sources of information.

On whatever basis the data are selected, the final statistical judgment will be based on the result of some arithmetic. A number or numbers will be calculated and compared with some tabulated numbers to see whether the chance of getting the answer obtained is large or small. Not only must the arithmetic be done correctly, and this is of utmost importance, but the data must be handled in such a way that they neither give a false impression of the precision of the measurements nor are rounded off so as to give a false impression of the consistency of the results.

The following rules of mathematical behavior are offered to give a systematic procedure to the calculations. Like most rules of behavior, these should be used as a guide and not as a dictum. The discussion following each rule gives the reason for that particular rule. As experience with statistical analyses is gathered and particular situations are encountered, the engineer's judgment will dictate which rules are to be followed. The rules are meant to ensure two things: that numbers reported are significant and that no significant numbers are lost in the calculations.

One more definition before we present the rules: a *significant digit*. A significant digit has been defined as "any digit that contributes to the specification of the magnitude of the number aside from serving to locate the decimal point."[1] We will use that definition. 342 has three significant figures. 0.000342 has three significant figures. 342,000 has three, four, or five significant figures depending on whether the zeros are actual measurements or not. For example, the neolithic age of 10,000 years probably has not more than two significant figures. The smallest significant digit in a number is assumed to be accurate to plus or minus half a unit. If 10,000 years has two significant digits, the value is accurate to ± 500 years. If 342 has three significant digits, the number is accurate to ± 0.5. The true value is between 341.5 and 342.5. If 34,200 is good to three significant figures, the accuracy is ± 50 and the true value is between 34,150 and 34,250.

The accuracy of the reported number is to plus or minus half the smallest unit of measurement. If a measurement is reported as $7\frac{3}{32}$ in., and this is converted to 7.09375 in., the accuracy is still $\pm \frac{1}{64}$ in. and not ± 0.000005 in. on the assumption that the decimal figure is good to six significant digits. It is important to know the units of measurement when dealing with converted figures.

While a significant digit is one that contributes to the significance of the magnitude of the number, the significance of the number must take into account the units of measurement. In measurement data, the significance of the number is established by the smallest unit of measurement. You can not increase the number of significant digits by some conversion factor after the measurement has been made in larger units. We stress this point because it is not unusual for measurements to be made in inches or in pounds per square inch and then to be converted to millimeters of length or to millimeters of mercury pressure units with an erroneous reporting of accuracy of data. If the smallest unit of measurement is an inch, and some data are obtained to three significant digits, say 376 in., this means the accuracy is 376 ± 0.5 in. The accuracy cannot be increased by multiplying by 25.4 and reporting 9550.4 mm. If the data are so reported, the accuracy should be shown to be ± 12.7 mm ($= 25.4 \times 0.5$).

The rules that follow are based on the definition of a significant digit given before, but the assumption is made that the smallest significant number represents the smallest unit of measurement of the data reported.

Rule 1. Report all data to the smallest significant digit. Calculations and analyses made on data should be made on numbers reported to their maximum accuracy. It is usually variation in the smallest units of

[1] C. Eisenhart, M. W. Hastay, and W. A. Wallis (eds.), "Techniques of Statistical Analysis," McGraw Hill Book Company, New York, 1947.

measurement that are most important in statistical analysis of data. If a series of temperature readings of 897°, 901°, 895°, 898°, 903°, 896°, and 890° are all reported as 900°, any possible relation between a response and a variation in temperature cannot be discovered. Any rounding off, or "smoothing" in the reporting of data should not be done until the original data have been preserved with their maximum accuracy—to their least significant digits.

Rule 2. All calculations made with data—multiplication, division, squaring, extracting roots, etc.—should be carried out to at least two more significant figures than the original data. No rounding off should be done until the final results are obtained. When desk calculators are employed, the carrying of extra digits offers no burden, but when calculations are done by hand there is an inclination to drop digits to make the arithmetic easier. Do not do it. Carry two more digits than the original data until the calculations are completed.

Rule 3. In addition and subtraction, report the answer to the largest least significant unit of measurement. If you are adding data reported in tenths of an inch, centimeters, and millimeters, the sum should be reported in the largest unit—centimeters. (The absolute error of measurement is sometimes defined as half the smallest significant unit, and the foregoing rule is given as reporting sums and differences to the largest absolute error.)

Rule 4. In multiplication and division, report the result to the same number of significant digits as the measurement with the least number of significant digits that is involved in the calculation. If you measure one side of a rectangle to two significant figures and the other to four significant figures, the area is only good to two significant figures. (Remember that with data a significant digit is a function of the measurement, while with pure numbers, the number of significant digits are at the choice of the calculator. If you measure a quantity to four significant digits and then take half of it, the two in the denominator is not a factor in setting the number of significant digits of the answer. If you take π times a measurement, π can be used to as many significant figures as is desired. The accuracy of the result is a function of the accuracy of the measurement, not of π.)

Rule 5. Report the standard deviation to three significant digits. This is an arbitrary rule, but inasmuch as the statistical factor which is usually used as a multiplier of the standard deviation to obtain some result is only tabulated to three significant figures, there is no need to calculate the standard deviation to more significant digits.

Rule 6. Rounding off of reported results.

  a. If the first digit to be dropped is less than 5, the last digit retained is left unchanged.

b. If the first digit to be dropped is greater than 5 or is 5 followed by digits greater than zero, the last digit retained is increased by one.

c. If the first digit to be dropped is 5 followed by zeros, the last digit retained is left unchanged if even and is increased by one if odd.

d. In applying the foregoing rounding operations, all digits to be dropped are dropped in one operation.

$$\pi = 3.1415926535$$

By Rule 6*a*, π to three significant figures is 3.14
By Rule 6*b*, π to four significant figures is 3.142
By Rule 6*c*, $8\frac{9}{2}$ to two significant figures is 44
    $9\frac{9}{2}$ to two significant figures is 50
    π to 8 significant figures is 3.1415927 (If the last three digits were dropped one at a time the result would be 3.1415926.)

Rule 7. Report the mean to the number of significant digits according to the following schedule. This schedule is taken from the ASTM Manual on Quality Control.[1]

TABLE 4.6

Significant units	n = number of measurements		
1		2–20	21–200
2	4	4–40	41–400
5	10	10–100	101–1,000
Number of significant digits in mean	Same as individual values	One more than individual values	Two more than individual values

For example, if the measurements are made to the nearest unit, or 0.1 unit or 0.01 unit, i.e., if the last significant digit is accurate to one unit, for less than 21 measurements, report the mean to one more significant digit than the individual measurements. For more than 20 measurements, report the mean to two more significant digits than the individual measurements. (According to Rule 2, calculate the mean to two more significant digits than will be reported before rounding off.) If the measurements are made to the nearest five units, i.e., reading a high-pressure gauge graduated in 5-lb increments, so that the readings end in either the digit 5 or 0, where 0 in this case is a significant digit; for less than 10 readings, report the mean to the same number of significant digits as

[1] *Op. cit.*

MEASURES OF VARIABILITY

the original data. For 10 to 100 readings, report the mean to one more significant digit than the original measurements.

If the significant units are not known, the safest assumption is that they are single units.

4.5 Problems

1. What are the mean, mode, and median of the following series of measurements: 27.5, 27.8, 27.7, 27.5, 27.6, 27.7, 27.9, 27.7, 27.6, 27.5, 27.7, 27.8, 27.7?

2. Draw 25 five-card hands from a deck of cards (replacing each hand before drawing the next) and calculate the mean value of the five cards, counting the picture cards as 11, 12, and 13 and the aces as 1. Tabulate the 25 means and calculate the mean, the mode, and the median of the 25 means.

3. If a vehicle travels four laps on a 100-mile course at 100, 200, 300, and 400 mph, what is its average speed?

4. Calculate the standard deviation of the data of Prob. 1 both by squaring the deviations from the mean and by Eq. (4.9).

5. Estimate the standard deviation of the means of the 25 five-card hands from the data of Prob. 2.

6. Calculate the estimated standard deviation of the following measurements of the tensile strength of rubber samples:

1,210	1,210	1,196
1,180	1,200	1,188
1,189	1,203	1,219

$s(x) = 12.5$.

7. Carry out the calculations indicated in example 4.3, coding the data so that $w' = 10(x - 79)$. Calculate $s(w')$.

8. Calculate the mean deviation of the data in Prob. 1 and compare it with the estimated standard deviation. Ratio: 1.27.

chapter five

χ^2, CHI SQUARE

In general two kinds of data are encountered: data that are counted and data that represent measurements. The telephone company might record the number of poles damaged and undamaged in an area in a given time period, or the corrosion chemist might record the loss in weight or extent of penetration of test samples attached to the poles. A petroleum company might record the number of barrels damaged in shipment, or the inspection department might sample the shipment for chemical and physical analyses. The principal difference between these two types of data is that the first must occur in discrete increments of whatever the units of counting are. There is no continuity between counts. There are either 10 defective motors or 11 defective motors. There cannot be 10.42 defective motors. With measured data, on the other hand, the limitation of the reported result is the accuracy of the measurement—just how well we can read a burette or a pressure gauge or a recording potentiometer. When we report the octane number of 97.2, we mean it is somewhere between 97.15 and 97.25. If we report a weighing as 4.7273 grams, we mean this is the value to within 0.05 mg, or as close as we can read it. Data of both kinds can be tested by statistical methods for deviations from some hypothesized values. In this chapter we shall deal with counted or enumeration data. In the following three chapters we shall deal with measurement data.

χ^2 is the lower-case form of the Greek letter chi, pronounced like the "ki" in "kind." The χ^2 test was devised by Karl Pearson in 1899[1] in terms of an observed and expected ratio. The test has since been applied by statisticians to other types of enumeration data. This chapter presents methods of applying the χ^2 test to various types of engineering data. For a discussion of the derivation of the χ^2 distribution, the reader is referred to books on the theory of statistics.[2]

5.1 Definition

In the usual course of events the expected frequency does not occur every time. If you toss 10 coins, you do not expect to see 5 heads and 5 tails every time although the "expected" outcome is this even distribution. If the mean absenteeism in a plant is 1 day in 30, you do not expect every man to be out 10 days in every 300, although 1 in 30 is the "expected" frequency. It is usual to find some deviation from the expected. χ^2 is a measure of this deviation.

$$\chi^2 = \sum \frac{(\text{deviation})^2}{\text{expectation}} \tag{5.1}$$

$$\chi^2 = \sum \frac{(O - E)^2}{E} \tag{5.2}$$

where O = observed frequency
E = expected frequency

From Eq. (5.2) it can be seen that if there were no deviation, χ^2 would be zero. However, as has been pointed out, it is not reasonable to expect to observe exactly the anticipated results. The question to be answered is, how much is a reasonable deviation?

Before we answer this question, there is one more factor to be mentioned. If one sample was involved, say you were counting the number of uncut heads in a shipment of machine screws, you would expect the number to fall within some tolerable range of the allowable or "expected" number of defects. The calculated χ^2, therefore, would also be within some range. As more samples were drawn, the deviations of the observed and expected number of defects would fluctuate, but χ^2, being a function of the sum of the deviations, would steadily increase. The tolerable value of χ^2, therefore, will vary with the number of samples involved. In

---

[1] Karl Pearson, *Phil. Mag.*, (5) **50**:157 (1899).
[2] For example, A. M. Mood, "Introduction to the Theory of Statistics," McGraw-Hill Book Company, New York, 1950.

terms of statistical parameters, χ^2 varies with the degrees of freedom available for its calculation (more about these degrees of freedom below).

Table 5.1 gives the maximum values of χ^2 for different degrees of freedom that can be expected with the indicated probability. For instance, with 1 degree of freedom (the first row of values) there is only a 0.05 probability of obtaining a χ^2 equal to or larger than 3.841. With 16 degrees of freedom, there is only a 0.01 probability of obtaining a χ^2 equal to or larger than 32.000. Also with 1 degree of freedom, the probability is 0.95 that χ^2 will be 0.004 or larger, and with 16 degrees of freedom, the probability is 0.99 that χ^2 will be 5.812 or larger.

5.1.1 The Null Hypothesis. In using the χ^2 distribution to establish the significance of deviations, some basis is required for setting the "expected" results. These may be set by previous experience, or from the data themselves in tests of homogeneity as will be explained, or on an a priori basis as when a distribution is assumed to be binomial or normal. On whatever basis the expected results are established, the hypothesis is set up that the observed results conform to some expected distribution. This is the null hypothesis, symbolized

$$H_0; O = E \tag{5.3}$$

It is called a null hypothesis because it assumes there is no difference between the value calculated from the sample and the corresponding value from the population. While it is not possible to establish H_0 to be true, it is possible to calculate the probability of its being false. If a rational basis exists for setting the expected values E, then it is possible to make probability statements about deviations from E. When these deviations are calculated in the form of χ^2 as given in Eq. (5.2), the tabulated values given in Table 5.1 are the maximum values of χ^2 that can be expected at the indicated probability levels if the null hypothesis is true. If larger values are obtained than those in the table, then the null hypothesis can be rejected with the corresponding probability of being in error. For instance, if the calculated χ^2 exceeds 3.841 for an example with 1 degree of freedom, there is less than 0.05 probability (1 chance in 20) of the null hypothesis being true; hence here is only a 0.05 probability of being in error if H_0 is rejected.

The probability level at which the null hypothesis is rejected is called the significance level of the test. A value which exceeds the 0.05 significance level is usually designated with a single asterisk; at the 0.01 level a double asterisk; and at the 0.001 level a triple asterisk.

5.1.2 Degrees of Freedom. The degrees of freedom which enter into the evaluation of χ^2 are the number of independent categories of observations that are available for the calculation of χ^2. This idea is probably best illustrated by an example. For instance, if you wanted to

χ^2, CHI SQUARE

TABLE 5.1 χ^2†

Probability of a larger value of χ^2

D.F.	.99	.98	.95	.90	.80	.70	.50	.30	.20	.10	.05	.02	.01	.001
1	.0³157	.0³628	.00393	.0158	.0642	.148	.455	1.074	1.642	2.706	3.841	5.412	6.635	10.827
2	.0201	.0404	.103	.211	.466	.713	1.386	2.408	3.219	4.605	5.991	7.824	9.210	13.815
3	.115	.185	.352	.584	1.005	1.424	2.366	3.665	4.642	6.251	7.815	9.837	11.345	16.268
4	.297	.429	.711	1.064	1.649	2.195	3.357	4.878	5.989	7.779	9.488	11.668	13.277	18.465
5	.554	.752	1.145	1.610	2.343	3.000	4.351	6.064	7.289	9.236	11.070	13.388	15.086	20.517
6	.872	1.134	1.635	2.204	3.070	3.828	5.348	7.231	8.558	10.645	12.592	15.033	16.812	22.457
7	1.239	1.564	2.167	2.833	3.822	4.671	6.346	8.383	9.803	12.017	14.067	16.622	18.475	24.322
8	1.646	2.032	2.733	3.490	4.594	5.527	7.344	9.524	11.030	13.362	15.507	18.168	20.090	26.125
9	2.088	2.532	3.325	4.168	5.380	6.393	8.343	10.656	12.242	14.684	16.919	19.679	21.666	27.877
10	2.558	3.059	3.940	4.865	6.179	7.267	9.342	11.781	13.442	15.987	18.307	21.161	23.209	29.588
11	3.053	3.609	4.575	5.578	6.989	8.148	10.341	12.899	14.631	17.275	19.675	22.618	24.725	31.264
12	3.571	4.178	5.226	6.304	7.807	9.034	11.340	14.011	15.812	18.549	21.026	24.054	26.217	32.909
13	4.107	4.765	5.892	7.042	8.634	9.926	12.340	15.119	16.985	19.812	22.362	25.472	27.688	34.528
14	4.660	5.368	6.571	7.790	9.467	10.821	13.339	16.222	18.151	21.064	23.685	26.873	29.141	36.123
15	5.229	5.985	7.261	8.547	10.307	11.721	14.339	17.322	19.311	22.307	24.996	28.259	30.578	37.697
16	5.812	6.614	7.962	9.312	11.152	12.624	15.338	18.418	20.465	23.542	26.296	29.633	32.000	39.252
17	6.408	7.255	8.672	10.085	12.002	13.531	16.338	19.511	21.615	24.769	27.587	30.995	33.409	40.790
18	7.015	7.906	9.390	10.865	12.857	14.440	17.338	20.601	22.760	25.989	28.869	32.346	34.805	42.312
19	7.633	8.567	10.117	11.651	13.716	15.352	18.338	21.689	23.900	27.204	30.144	33.687	36.191	43.820
20	8.260	9.237	10.851	12.443	14.578	16.266	19.337	22.775	25.038	28.412	31.410	35.020	37.566	45.315
21	8.897	9.915	11.591	13.240	15.445	17.182	20.337	23.858	26.171	29.615	32.671	36.343	38.932	46.797
22	9.542	10.600	12.338	14.041	16.314	18.101	21.337	24.939	27.301	30.813	33.924	37.659	40.289	48.268
23	10.196	11.293	13.091	14.848	17.187	19.021	22.337	26.018	28.429	32.007	35.172	38.968	41.638	49.728
24	10.856	11.992	13.848	15.659	18.062	19.943	23.337	27.096	29.553	33.196	36.415	40.270	42.980	51.179
25	11.524	12.697	14.611	16.473	18.940	20.867	24.337	28.172	30.675	34.382	37.652	41.566	44.314	52.620
26	12.198	13.409	15.379	17.292	19.820	21.792	25.336	29.246	31.795	35.563	38.885	42.856	45.642	54.052
27	12.879	14.125	16.151	18.114	20.703	22.719	26.336	30.319	32.912	36.741	40.113	44.140	46.963	55.476
28	13.565	14.847	16.928	18.939	21.588	23.647	27.336	31.391	34.027	37.916	41.337	45.419	48.278	56.893
29	14.256	15.574	17.708	19.768	22.475	24.577	28.336	32.461	35.139	39.087	42.557	46.693	49.588	58.302
30	14.953	16.306	18.493	20.599	23.364	25.508	29.336	33.530	36.250	40.256	43.773	47.962	50.892	59.703

† Reprinted from Table IV of R. A. Fisher and Frank Yates, "Statistical Tables for Biological, Agricultural and Medical Research," Oliver & Boyd, Ltd., Edinburgh and London, 1953, by permission of the authors and publishers.

test a coin for bias by the χ^2 test and tossed it 1,000 times, the expected number of both heads and tails would be 500. If you observed 421 heads, giving a head deviation of 79, then the number of tails must be 579, since the total tosses is 1,000, and the tail deviation is also 79. Only one of these deviations can be independently established; hence there is only 1 degree of freedom for calculating χ^2 in this case.

example 5.1. A company using a large number of small motors installs 25 per cent of type A and 75 per cent of type B. After a certain period there have been 474 burnouts, 68 of type A and 406 of type B. Is there evidence of a difference between the two types?

If there were no difference between the motors, the expected number of burnouts would be in the same ratio of the number of installations: $(0.25)(474)$ of $A = 118.5$, and $(0.75)(474)$ of $B = 355.5$. χ^2 from Eq. (5.1) equals $\Sigma[(\text{deviation})^2/\text{expectation}]$:

$$\chi^2 = \frac{(68 - 118.5)^2}{118.5} + \frac{(406 - 355.5)^2}{355.5} = 21.52 + 7.17 = 28.69^{***}$$

Since the total number of burnouts is fixed at 474, and since there are only two categories to make up this total, there is only 1 degree of freedom. Referring to Table 5.1, we see that with 1 degree of freedom, there is only a 0.001 probability of χ^2 being larger than 10.827. Our χ^2 is 28.69 on the basis of the hypothesis that there is no difference between the motors. Hence there is less than 0.001 probability that these data are compatible with the hypothesis.

5.2 Correction for Continuity

χ^2 is a continuous mathematical function. When χ^2 is calculated from enumeration data which perforce must be in discrete increments, a bias is incurred which results in too high a χ^2 value or in an underestimate of the probability. This bias may be overcome when 1 degree of freedom is involved by subtracting 0.5 from each deviation before squaring; i.e.,

$$\chi^2 \text{ adjusted} = \sum \frac{(\text{deviation} - 0.5)^2}{\text{expectation}} \quad (5.4)$$

When more than 1 degree of freedom is involved, the bias, while still present, is not as pronounced and cannot be corrected in this simple manner. Also, as can be seen from the magnitude of the adjustment, when the unadjusted deviation is large, the correction will result in a relatively small change in the ultimate χ^2 value. The adjustment to χ^2 always results in a decrease in the calculated value. Therefore if the unadjusted value is not significant, i.e., if the corresponding probability is not so small as to result in a rejection of the hypothesis, then the adjusted value will certainly not be significant. This discussion brings up another point which applies to all the statistical tests for significance. If the probability is just at the 0.05 level, it is important to bear in mind that the statistic you have calculated is only an estimate. It is safer, and wiser, to try to substantiate a close statistical judgment with more data rather than adhere rigorously to a 0.05 probability level for all actions.

One other limitation of the χ^2 test when applied to discrete data is that the test should not be applied to classes of data in which the expected result is less than 5. On the basis of the binomial distribution with an event with equal probability of occurring or not occurring if the expected frequencies are 1, 2, 3, or 4, the probability of observing zero is 0.5, 0.25, 0.125, or 0.0625, respectively, so that in a single classification test, with an expected count of less than five, it is not possible to establish a significant difference. There is no possible value with a probability less than 0.05. When a classification contains less than five expected observations, it is often possible to combine several classifications so that the limit of five is exceeded and the test may be carried out.

This rule, like other statistical rules, is not an injunction but an instruction. It is to be used with judgment. When there is a large number of degrees of freedom, 10 or more, it is possible to tolerate one or two classes with expected values of less than 5. Until a degree of statistical judgment can be applied, it probably will be best to follow the rule of not using the χ^2 test for classes with expected values less than 5, and of combining those classes which have values less than 5 so that the χ^2 can be applied.

example 5.2. In a shipment of material supposed to be not more than 10 per cent defective, you find 7 defectives in a sample of 50 items. Is this grounds for concern?

On the basis of the hypothesis that the material contains 10 per cent defectives, the expected number of defectives in the sample is 5.0. The observed number of defectives is 7.0. The adjusted χ^2 therefore is calculated as in Table 5.2 on p. 88.

This χ^2 value of 0.50, with 1 degree of freedom, corresponds to a probability value between 0.3 and 0.5. In other words, there is a probability of about 0.4 of obtaining a number of defectives as large as that observed in a sample of the size when the population does in fact have only 10 per cent defectives.

Note: This problem could be solved by use of the binomial distribution with $p = 0.10$, $n = 50$; find $P(r \geq 7)$. *Ans.:* $P = 0.23$. It could also be solved by use of the Poisson distribution, $m = 5$; find $P(r \geqq 7)$. Ans.: $P = 0.24$.

TABLE 5.2

	Observed	Expected	Deviation	Deviation − 0.5	(Deviation − 0.5)²/expected
Defective........	7	5	2	1.5	0.45
Good..........	43	45	2	1.5	0.05
					Adjusted $\chi^2 = 0.50$

5.3 Effect of Sample Size

The χ^2 test may be applied to observed and expected *ratios* when the total number involved is included in the calculation. The procedure is outlined in Sec. 5.4. The χ^2 test cannot be applied to observed and expected *percentages* unless the sample size is exactly 100. The size of the sample is an important variable in the calculation of χ^2. For example, in Table 5.3 are given the values of χ^2 and adjusted χ^2 for a series of samples in each of which was observed 14 per cent defectives when the expected defectives was 10 per cent. The χ^2 values vary from 0.5 for sample size 50 to 17.3 for sample size 1,000, and in each case 1 degree of freedom is involved. At a sample size of about 200, the deviation between 10 and

TABLE 5.3

Expected defectives = 10%. Observed defectives = 14%.

Sample size	Expected	Observed	Deviation	χ^2	Adjusted χ^2
50	5	7	2	0.89	0.50
100	10	14	4	1.78	1.36
200	20	28	8	3.56	3.12
500	50	70	20	8.89	8.44**
1,000	100	140	40	17.78	17.33***

14 per cent defectives becomes significant, less than an 0.05 probability of satisfying the hypothesis. With sample size of 500 or more, there is less than 0.01 probability.

It will be noted that the unadjusted χ^2 value varies directly with the size of the sample. From this relationship it is possible to anticipate how large a sample would be required under conditions of constant observed to expected ratio to achieve a significant χ^2 value.

Note that the calculation of χ^2 involves the sum of the squares of the deviations of all categories divided by the expected values. In a type of problem like the illustration where the expected number of defectives in a sample of 200 is 20, the expected number of nondefectives is 180. If the observed number of defectives is 28, then the number of nondefectives is 172. χ^2 is calculated from the deviations of both of these values; in this case, since there is 1 degree of freedom, both deviations are adjusted by 0.5. The calculation therefore is

$$\chi^2 = \frac{(28 - 20 - 0.5)^2}{20} + \frac{(180 - 172 - 0.5)^2}{180}$$

$$= \frac{(7.5)^2}{20} + \frac{(7.5)^2}{180} = 2.8125 + 0.3125 = 3.1250$$

The other values of Table 5.3 may be checked by following a similar calculation.

5.4 χ^2 for Ratios

A two-category test may often be expressed in form of an expected ratio. When the expected defectives is 10 per cent, the ratio of defectives to nondefectives is 1/9, or the inverse ratio of nondefectives to defectives is 9. If the observed number in the category identified with the numerator of the ratio is designated a, and the observed number in the category identified with the denominator of the ratio is designated b, so that $a + b = n$, the sample size, and if the expected ratio is r, then Eq. (5.2) for χ^2 may be written

$$\chi^2 = \frac{(a - rb)^2}{rn} \qquad (5.5)$$

In the illustration of a sample of 200 containing 28 defectives and 172 nondefectives when the expected defectives was 10 per cent, $r = 1/9$, $a = 28$, $b = 172$, $n = 200$.

$$\chi^2 = \frac{[28 - (1/9)(172)]}{(1/9)(200)} = 3.56$$

Similarly, if we define r as the ratio of nondefectives to defectives, that is, $r = 9$, then $a = 172$, $b = 28$, $n = 200$.

$$\chi^2 = \frac{[172 - (9)(28)]^2}{(9)(200)} = 3.56$$

The adjustment for continuity may be applied to the formula of Eq. (5.5) by writing it

$$\chi^2 = \frac{[|a - rb| - (r + 1)/2]^2}{rn} \tag{5.6}$$

In the second illustration above

$$\chi^2 = \frac{[|172 - (9)(28)| - 10/2]^2}{(9)(200)}$$

$$= \frac{(80 - 5)^2}{1,800} = 3.1250 \quad \text{as before}$$

example 5.3. If a school usually passes two-thirds of a class in a particular subject and a new instructor fails 14 out of a class of 30, is this in reasonable conformity with the 2/3 experience?

The expected ratio of passes to failures is 2/1, that is $r = 2$.

The observed number of passes is 16, and the observed number of failures is 14, the total being 30.

Using Eq. (5.6),

$$\chi^2 = \frac{[|16 - (2)(14)| - 3/2]^2}{(2)(30)}$$

$$= \frac{110.25}{60} = 1.8375$$

Table 5.1 indicates that with 1 degree of freedom a value of χ^2 as large as 1.8375 can be expected between 10 and 20 per cent of the time when the original hypothesis is satisfied. In other words, if we reject the hypothesis and state that the instructor's grades do not conform to the school experience, we have between 10 and 20 per cent chance of being in error.

5.5 Contingency Tables

The χ^2 test has been applied to data which are compared with some a priori distribution—set by previous experience, or some quality standard, or simple wishful thinking. At any rate, the data are compared to some standards set from a source outside the data. χ^2 can also be applied to test whether a set of data is homogeneous without reference to any outside standard. The expected results are contingent solely on the arrangement of the observed data. The data are set up in a table of various categories or classes, and the distribution of the data in the table is tested for homogeneity contingent on the totals of the different classifications. Hence the title: contingency table.

Tests of this kind can conveniently be broken down into different-size tables depending on the number of categories involved.

5.5.1 The 1 × n Table. The 1 × n table indicates that there is one classification category: number of units produced, number of accidents experienced, number of sales made, etc., and there are n individuals to be tested. For example, if a new test reactor was installed and three different shifts produced 4, 4, and 10 runs each, you might question whether the last shift was significantly better than the other two. If there were no difference between the shifts, the runs per shift would be one-third of the total, six each. A 1 × n table to test this hypothesis is shown in Table 5.4. The fourth column gives the deviations from the expected values

TABLE 5.4

Shift	Observed	Expected	Deviation	(Deviation)²/expected
A	4	6	2	0.667
B	4	6	2	0.667
C	10	6	4	2.667
				$\chi^2 = 4.001$

based on a perfectly homogeneous table, and the fifth column gives the contributions to χ^2, (deviation)²/expectation, as set up in Eq. (5.1). There is one restriction on the χ^2 calculation in that the total of the expected values must equal the total of the observed values. Since there are three shifts, two of the observed values may be independently varied; i.e., there are 2 degrees of freedom in calculating χ^2. The χ^2 value of 4.001 is compared with the values in Table 5.1 at 2 degrees of freedom. It lies between the 0.20 and the 0.10 probability level. There is a probability of between 0.10 and 0.20 of getting a value as large as 4.001 if the shifts were producing equally. Put another way, if we say that the third shift is better than the other two on the basis of these data, there is between 0.10 and 0.20 probability of being in error.

Note that since there are 2 degrees of freedom, the 0.5 adjustment to the deviation is not applied in calculating χ^2.

Although the distribution of the results in Table 5.4 might appear to be significantly different from a uniform distribution, the χ^2 test indicates that we cannot be 95 per cent certain. If the same distribution had applied to 27 runs instead of 18, so that the numbers were 6, 6, and 15, then the χ^2 value, which the student can check for himself, would be 6.0. This value referred to Table 5.1 at 2 degrees of freedom indicates a probability just slightly less than 0.05. Again we see that the amount of data is important in establishing the significance of the test. The same distribution which for 18 data points could not be judged nonhomogeneous could be so judged when applied to 27 points.

5.5.2 The 2 × 2 Table. The $1 \times n$ χ^2 contingency table finds many applications in engineering work. An arrangement that finds even more general application is one in which two different categories are measured at two conditions: Does the before or after treatment show a larger number of failures or successes? Is the evidence of absence of lung cancer more prevalent with smokers or nonsmokers? When a treatment or lack of treatment is measured by presence or absence or some result, the data can usually be tabulated in one of the four possible categories and the whole table tested for homogeneity by χ^2 to see whether the treatment has had a significant effect. The application might best be illustrated by an example.

Table 5.5 shows the results of 273 runs made to test a modification of a catalyst treatment. The modified treatment has 17 per cent unsuccessful runs while the normal treatment has 27 per cent. The unsophisticated opinion might be that the modified treatment has a significantly lower number of unsuccessful runs. We shall see.

TABLE 5.5

Treatment	Successful runs	Unsuccessful runs
Modified......	115	24
Normal.......	98	36

If the table were homogeneous, i.e., if the data were independent of the classifications and only dependent on the totals, then the numbers in each group would be proportional to the totals of the corresponding rows and columns. In Table 5.5, the total runs made with the modified treatment is 139 and the total runs made with both treatments is 273. The expected number of successful runs made with the modified treatment, if the results were independent of the treatments, would be the product of the total number of successful runs and the ratio of modified-treatment runs to total runs: $(213)(139/273) = 108.5$. The number that would appear in each section of the table on the basis of homogeneity can be calculated in a similar manner. Table 5.6 repeats the data of Table 5.5, and the expected values are shown in parentheses.

Note that the totals of the expected results equal the totals of the observed results for each row and column: $108.5 + 30.5 = 139$. Since the totals of the rows and columns are fixed by the original data, only one of the numbers can be independently varied. The other three must be adjusted accordingly to keep the totals fixed. Therefore in comparing the deviations to the expected values by means of χ^2 in a 2 × 2 table,

TABLE 5.6

Treatment	Successful runs	Unsuccessful runs	Total
Modified......	115 (108.5)	24 (30.5)	139
Normal.......	98 (104.5)	36 (29.5)	134
Total.......	213	60	273

there is only 1 degree of freedom. With 1 degree of freedom, we apply the 0.5 adjustment to the deviations in calculating χ^2 and use Eq. (5.4). Then

$$\chi^2 = \frac{(115 - 108.5 - 0.5)^2}{108.5} + \frac{(104.5 - 98 - 0.5)^2}{104.5}$$

$$+ \frac{(30.5 - 24 - 0.5)^2}{30.5} + \frac{(36 - 29.5 - 0.5)^2}{29.5} = 3.075$$

Comparing this value with Table 5.1, at 1 degree of freedom, we see that there is somewhat greater than a 0.05 probability of obtaining a χ^2 value this large when the original hypothesis is true: that the data are independent of the categories; or that the modification of the treatment is not significantly effective in producing successful runs. If we reject the hypothesis and say on the basis of these data that the modification is effective, we have a greater than 5 per cent chance of being wrong. It would be better to withhold judgment until more data are available, or at least indicate the probability of being in error if we make the statement.

5.5.3 The $r \times c$ Table. The argument for the 2 × 2 table can be extended to any number of rows and columns designating categories. The data arranged in such a table are tested against the hypothesis that they are independent of the categories. The expected number in each box of the table on the basis of homogeneity would be equal to the ratio of the product of the row and column total to the grand total. The general arrangement is illustrated in Fig. 5.1. χ^2 is calculated from this table in the same manner as before, that is, $\chi^2 = \Sigma[(\text{observed} - \text{expected})^2/\text{expected}]$, where the expected values are calculated from the table totals. The number of degrees of freedom for the general table is equal to the product of one less than the number of rows by one less than the number of columns.

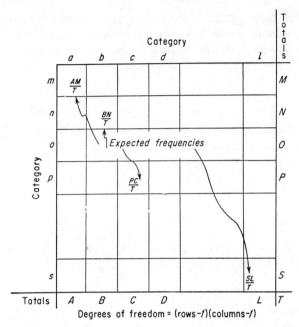

Fig. 5.1 Contingency table.

example 5.4. Four shifts over a period of time contribute to a pilot-plant program as shown in the following table. We want to know whether the results (as a group) are what might be expected from a normal fluctuation of experimental hazards or whether some shifts are particularly proficient or deficient.

To apply the χ^2 test to this 5×4 table (5 rows and 4 columns), the number that would appear under each heading if the table were com-

DATA

Category	Shift				Total
	A	B	C	D	
Runs started.............	20	18	14	12	64
Runs continued...........	16	14	20	14	64
Runs discontinued.........	1	0	2	1	4
Runs terminated successfully..	9	4	12	11	36
Runs terminated in failure....	2	16	8	6	32
Total.................	48	52	56	44	200

pletely homogeneous is calculated from the row and column totals as in Fig. 5.1. The "expected" value for shift A in the heading "runs started" is $(64)(48)/200 = 15.4$. The number for shift C for "runs discontinued" is $(56)(4)/200 = 1.1$. And so on for each observed value. Table 5.7 repeats the observed values of the data and includes in parentheses the calculated values based on homogeneity, i.e., the expected values.

TABLE 5.7

Category	Shift				Total
	A	B	C	D	
R.S.........	20 (15.4)	18 (16.6)	14 (17.9)	12 (14.1)	64 (64.0)
R.C.........	16 (15.4)	14 (16.6)	20 (17.9)	14 (14.1)	64 (64.0)
R.D.........	1 (1.0)	0 (1.0)	2 (1.1)	1 (0.9)	4 (4.0)
R.T.S........	9 (8.6)	4 (9.4)	12 (10.1)	11 (7.9)	36 (36.0)
R.T.U........	2 (7.7)	16 (8.3)	8 (9.0)	6 (7.0)	32 (32.0)
Total.......	48 (48.0)	52 (52.0)	56 (56.0)	44 (44.0)	200 (200.0)

χ^2 is calculated according to the usual formula, taking the sum of the squares of the deviations between the observed and the expected values divided by the expected values. Before making the calculation, however, note that all the expected values in the row "runs discontinued" are less than 5. In fact, they are all less than 2. Since this many low values might bias the results, this category can either be discarded or it can be combined with some other category. We shall combine its values with "runs terminated successfully" to give a modified third row to Table 5.7 as shown in Table 5.8. Since none of the other row totals is changed, there is no change in any of the other calculated values in Table 5.7.

χ^2 is now calculated from the modified table by summing the contribution of each deviation as in Table 5.9.

The modified contingency table, with the "runs discontinued" category eliminated, has four rows and four columns, hence the χ^2 is calculated with $(4-1)(4-1) = 9$ degrees of freedom. A χ^2 value of 20.80 calculated with 9 degrees of freedom can only be expected with a probability between 0.01 and 0.02 (from Table 5.1) if the hypothesis of homogeneity were valid. We can therefore with reasonable assurance reject the hypothesis.

In addition to being able to state that the different shifts did not produce similar results, an inspection of Table 5.9 permits an estimate of where the discrepancy lies. The three major contributions to χ^2 are shift A, runs terminated unsuccessfully, 4.22; shift B, runs terminated

TABLE 5.8

Category	Shift				Total
	A	B	C	D	
R.T.S.......	10 (9.6)	4 (10.4)	14 (11.2)	12 (8.8)	40 (40.0)

successfully, 3.92; and shift B, runs terminated unsuccessfully, 7.14; making a total of 15.28, or 73 per cent of the total χ^2. Although there is no way of testing these individual values for significance, they indicate the major sources of the significance of the total χ^2 value. Referring back to Table 5.7, we can see that shift A had less than the typical number of runs terminated unsuccessfully and shift B had both more

TABLE 5.9

Shift	Category	χ^2 contribution
A	R.S.	$(4.6)^2/15.4 =$ 1.37
A	R.C.	$(0.6)^2/15.4 =$ 0.02
A	R.T.S.	$(0.4)^2/\ 9.6 =$ 0.02
A	R.T.U.	$(5.7)^2/\ 7.7 =$ 4.22
B	R.S.	$(1.4)^2/16.6 =$ 0.12
B	R.C.	$(2.6)^2/16.6 =$ 0.41
B	R.T.S.	$(6.4)^2/10.4 =$ 3.92
B	R.T.U.	$(7.7)^2/\ 8.3 =$ 7.14
C	R.S.	$(3.9)^2/17.9 =$ 0.85
C	R.C.	$(2.1)^2/17.9 =$ 0.25
C	R.T.S.	$(2.8)^2/11.2 =$ 0.76
C	R.T.U.	$(1.0)^2/\ 9.0 =$ 0.11
D	R.S.	$(2.1)^2/14.1 =$ 0.31
D	R.C.	$(0.1)^2/14.1 =$ 0.00
D	R.T.S.	$(3.2)^2/\ 8.8 =$ 1.16
D	R.T.U.	$(1.0)^2/\ 7.0 =$ 0.14
		$\chi^2 = 20.80$*

runs terminated unsuccessfully and less runs terminated successfully than was typical.

It is not always possible to find a few categories in an $r \times c$ table which make the major contribution to a significant χ^2 value. The discrepancies may be distributed fairly evenly throughout the table. If the χ^2 is significant, it is worth investigating to see whether some particular categories make a particularly large contribution, and then take whatever action is indicated.

5.6 Additivity of χ^2

A valuable feature of the χ^2 distribution is that the sum of n χ^2 values is itself distributed as χ^2 with n degrees of freedom. This characteristic of χ^2 permits the evaluation of several tests in addition to the evaluation of each individual test. At first reading this feature may not sound like an advantage, but a reference to Table 5.1 will indicate how it may be used. The 0.05 probability of χ^2 for 10 degrees of freedom is 18.307. The probability level of χ^2 for 1 degree of freedom corresponding to a value of 1.8307 is between 0.10 and 0.20. We could therefore have 10 individual

TABLE 5.10

Sample	Type A	Type B	χ^2
1	73	27	1.805 †
2	74	26	2.420
3	73	27	1.805
4	72	28	1.280
5	74	26	2.420
6	73	27	1.805
7	72	28	1.280
8	74	26	2.420
9	73	27	1.805
10	74	26	2.420
Total.......	...	...	19.460*

† $[73 - 2(27)]^2/(2)(100) = 1.805$.

tests, with 1 degree of freedom each, all having χ^2 values of 1.8307, and we would consider none of the deviations significant. The sum of the 10 χ^2's would, however, be significant, at the 0.05 level, indicating that while separately none of the individual data would warrant a rejection of the test hypothesis, in the aggregate the hypothesis could be rejected.

If individual χ^2 values are to be added, the 0.5 adjustment of the deviation should not be applied as it will overcorrect the totaled χ^2 value.

Suppose that in sampling batches of mixed pellets, estimated to be made up of 2 parts of type A and 1 part of type B, the counts listed in Table 5.10 for samples of 100 pellets were observed. χ^2 for each sample is calculated from Eq. (5.4) where the expected ratio is 2/1 and the sample size is 100. The individual χ^2 values are also given in Table 5.10. None of these χ^2 values exceeds the 0.05 probability value of 3.841, so that any individual sample would be acceptable on this basis. However, the sum of the 10 χ^2's is 19.460, which exceeds the 0.05 probability value of 18.307 for 10 degrees of freedom. This value indicates that while no particular sample is suspect, the material in general is not conforming to the 2/1 ratio.

5.7 Too Good an Agreement

Table 5.1 shows the probability of getting χ^2 values as large as those listed for the number of degrees of freedom indicated. Both high and low probabilities are given. For 1 degree of freedom, for example, the probability is only 0.05 of obtaining a χ^2 value larger than 3.841, and the probability is 0.95 that the χ^2 value will be larger than 0.004; i.e., there is only 0.05 probability of obtaining a value less than 0.004. On the basis of

TABLE 5.11

Sample	Type A	Type B	χ^2
1	68	32	0.080 †
2	69	31	0.245
3	68	32	0.080
4	69	31	0.245
5	69	31	0.245
6	68	32	0.080
7	68	32	0.080
8	69	31	0.245
9	69	31	0.245
10	68	32	0.080
Total......	...	...	1.625

† $[68 - 2(32)]^2/(2)(100) = 0.080$.
 $(68 - 66.66)^2/66.66 + (33.33 - 32)^2/33.33 = 0.080$.

some hypothesis under examination, a χ^2 value is calculated from the data. When the calculated χ^2 exceeds a value that could be expected with only a low probability, we reject the hypothesis with the assurance that there is only this low probability of being in error. On the other hand, if the χ^2 is less than some value that would be expected with a high probability, then the source of data should be examined for a forced conformity to the hypothesis, since the agreement with the hypothesis is better than could be expected by chance.

If in sampling the batches of mixed pellets illustrated in the previous example the results had been those given in Table 5.11, the individual sample χ^2 values would all be acceptable but the total of the 10 χ^2's with 10 degrees of freedom is 1.625—less than the 0.99 probability value from Table 5.1. There is less than one chance in a hundred of getting a χ^2 value this low with random sampling from populations that had a 2/1 ratio of the pellets in question. The distribution from sample to sample lacks the variation that would normally be expected and suggests either that the data were made to conform to the expected ratio and did not represent actual samples, or that in some way the sampling was influenced by the previous pellets drawn so that the samples of 100 were forced to conform to the 2/1 ratio.

5.8 Goodness-of-fit

χ^2 can be used to check the agreement of a set of observations with some probability distribution by comparing the frequency of the observed data

TABLE 5.12 Expected Distribution of "Heads" in 100 Tosses of 10 Coins

Heads	Expected frequency
0	0.10†
1	0.98
2	4.39
3	11.72
4	20.51
5	24.61
6	20.51
7	11.72
8	4.39
9	0.98
10	0.10
Total	100.01

† $P(r = x) = \binom{10}{x} (0.5)^x (0.5)^{10-x}$

Expected frequency in 100 tosses = $100 P(r = x)$

with the "expected" frequencies calculated from the mathematical formula for the distribution to be tested. If 10 coins are tossed 100 times, the expected frequency of "heads" can be calculated from the binomial distribution formula as outlined in Chap. 3. The expected frequencies would be those shown in Table 5.12. Data from an actual experiment (in coin tossing, or other) can be compared with this calculated distribution on the hypothesis that the probability, $p = 0.5$, applies to the experimental data. χ^2 is then calculated from the deviations between the observed and expected frequencies, and the hypothesis accepted or rejected based on the calculated value of χ^2.

To illustrate the use of the goodness-of-fit test, the following data on 100 actual tosses were obtained for comparison with the theoretical distribution given in Table 5.12.

TABLE 5.13 Observed Frequency of "Heads" in 100 Tosses of 10 Coins

No. of heads	No. of occurrences
0	0
1	0
2	2
3	10
4	22
5	32
6	24
7	7
8	3
9	0
10	0
Total	100

To use χ^2 to check the hypothesis that the probability of "heads" in the actual experiment of coin tossing was 0.5, we determine the difference between the observed and expected frequency for each category (count of heads) and calculate its contribution to χ^2. The calculation is illustrated in Table 5.14. The expected frequencies at the ends of the distribution are all less than five so these frequencies are pooled to avoid bias in the calculation of χ^2.

The calculated χ^2 value of 8.39 is compared with the values in Table 5.1 at 6 degrees of freedom. The calculated χ^2 lies between the 0.20 and 0.30 probability level, at approximately 0.21 probability. The interpretation is that there is only about a 21 per cent chance of observing a deviation this large if the probability of "heads" for all the coins was 0.5. If we reject the hypothesis that p for all coins was 0.5, we have about a 20 per cent chance of being in error.

A word about the degrees of freedom. In this example there are seven categories of counts used in the calculation of χ^2. The total of all counts must be 100, so there are 6 degrees of freedom remaining for the calculation of χ^2. The total number of "heads" in the 1,000 tosses was 499. If the probability of "heads" had been taken from the data and set equal to

499/1,000, an additional degree of freedom would have been taken up by this calculation, and 5 degrees of freedom would have remained to apply to the χ^2 calculation. However, inasmuch as the probability of 0.5 was applied without reference to the actual data, no degrees of freedom were involved.

TABLE 5.14

No. of heads	Observed frequency	Theoretical frequency	Difference	Contribution to χ^2
0, 1, 2	2	5.47	3.47	2.20†
3	10	11.72	1.72	0.25
4	22	20.51	1.49	0.11
5	32	24.61	7.39	2.22
6	24	20.51	3.49	0.59
7	7	11.72	4.72	1.90
8, 9, 10	3	5.47	2.47	1.22
		Total χ^2		8.39

† $(5.47 - 2)^2/5.47 = 2.20$

The procedure of using χ^2 for comparing an actual distribution of data with a particular probability distribution can be applied to any of the frequency distributions discussed in this text. The following two examples illustrate the application of the goodness-of-fit test to the Poisson and to the normal distributions.

In the discussion of Poisson distribution, it was pointed out that with events such as accidents, where the probability is small but the total number of chances is large, the frequency of incidents can be expected to follow the Poisson distribution. In a problem of this type, the probability or expected frequency is calculated from the data: total number of accidents/total man-hours worked; total number of vehicle accidents/total mileage; total number of rivet failures/total number of rivets, etc. The calculation of the mean number of occurrences from the actual data utilizes one degree of freedom (puts one more restriction on the calculation). Fixing the total number of events utilizes another degree of freedom, so that the final degrees of freedom for the calculation of χ^2 are two less than the number of categories of data classification. The following example illustrates the application to some nonengineering data.

example 5.5. Table 5.15 lists the number of batters hit by individual pitchers during the 1955 baseball season and the number of pitchers hitting each number of batters. The data[1] are only for pitchers who

[1] 1956 Baseball Almanac, A. S. Barnes and Company, Inc., New York, 1956.

functioned in 154 or more innings. The total number of pitchers is 52. The total number of batters hit is 223. The "expected" number of batters to be hit by each pitcher is $223/52 = 4.29$. Using this value for m in the Poisson probability formula, $P(r) = m^r e^{-m}/r!$, it is possible to calculate the probability of a pitcher hitting 0 batters, hitting 1 batter, or hitting r batters.

Since 52 pitchers are involved, 52 times the probability calculated as described will give the number of pitchers "expected" to hit each number of batters. This is done in column 5 of Table 5.15. The difference between this "expected" number and the observed number of pitchers hitting each number of batters is used in the χ^2 calculation. This is done in the last column of Table 5.15.

In order to calculate χ^2, it is necessary to pool those categories with expected frequencies less than about 5. The first two and the last seven are pooled, and the calculation of χ^2 follows from the remaining seven

TABLE 5.15 Poisson Distribution Applied to Pitchers Hitting Batters

No. of pitchers............ 52
No. of batters hit......... 223

No. of batters hit by single pitcher r	No. of pitchers hitting exactly r batters f	Product $r \cdot f$	Poisson probability of exactly $r = P(r)$ $\dfrac{e^{-m} \cdot m^r}{r!}$	Poisson prediction of no. of pitchers $P(r) \cdot \Sigma f$	Deviation $|O - E|$ $|f - P(r) \cdot \Sigma f|$	Contribution to χ^2
0	1	0	0.0137	0.71	6.23	10.29
1	9	9	0.0589	3.06		
2	10	20	0.1262	6.56	3.44	1.80
3	10	30	0.1805	9.38	0.62	0.04
4	3	12	0.1935	10.06	7.06	4.95
5	6	30	0.1659	8.63	2.63	0.80
6	1	6	0.1186	6.17	5.17	4.33
7	2	14	0.0727	3.76		
8	0	0	0.0390	2.03		
9	5	45	0.0186	0.97	4.57	2.81
10	2	20	0.0080	0.41		
11	1	11	0.0031	0.16		
12	0	0	0.0011	0.06		
13	2	26	0.0004	0.02		
	52	223		52.00		25.02

$$m = \Sigma rf/\Sigma f = 223/52 = 4.29$$
$$\chi^2 = 25.02**$$
$$\chi^2_{5, 0.01} = 15.09$$

categories. This calculation is shown in the last two columns of the table.

The χ^2 value of 25.02 is compared with the tabulated value of 15.086 for 5 degrees of freedom (7 categories less 1 for the total and 1 for the calculation of m) at the 0.01 level. This result indicates that the number of batters hit by a pitcher does not follow the distribution expected if the chance of hitting a batter was the same for all pitchers. The largest contribution to the χ^2 is from pitchers having hit one batter, the number being larger than would be expected. This may indicate that after having hit one batter, the pitchers are more careful in subsequent games (or the batters are more wary).

The next example illustrates the application of the goodness-of-fit χ^2 test to some data to be compared to a normal distribution. With the binomial and Poisson distributions, the probability of an integer number of events is calculated directly from the distribution function. The normal function, however, gives the probability for events falling within a specified range of values. The probability is calculated as the area under a curve between two limits. The probability for any specific value, which would correspond to the area under the curve at one point, would be zero. Actually the probability for any specific value is the area under the curve within the limits of accuracy of the particular measurement. If the measurements are made to single units, the probability of a specific value, say 27, would be the area under the proper normal curve between the limits of 26.5 and 27.5. If the measurements were made to one decimal place, the probability of 27.0 would correspond to the area between 26.95 and 27.05.

The tabulation of normal curve data given in Table 3.12 is the area from $-\infty$ to different values of the standard deviate, z. z is defined as the difference between the variable value and the mean divided by the standard deviation: $z = (\mu - x)/\sigma$. If you want to find the area under the normal curve between two values, say 26.5 and 27.5, you would calculate the two corresponding z values and then calculate the difference between the two $F(z)$ values from Table 3.12. This procedure holds when both values are on the same side of the mean, either larger or smaller than the mean. When the two values are on opposite sides of the mean, the area under the normal curve between them, from the tabulation of Table 3.12, is equal to $F(z_1) = F(z_2) - 1$.

The area under the normal curve between two values times the total number of observations gives the "expected" number of observations to fall within this range when the measurements are normally distributed. Example 5.6 illustrates the application of the normal distribution approximation to a set of data.

example 5.6. One of the large oil companies publishes a company calendar and includes for each day the names of all its employees with birthdays on that day. We will test whether the number of birthdays per day is normally distributed. In Table 5.16 the data on the frequencies of birthdays occurring each day are used in a goodness-of-fit test applied to a normal distribution calculation.

The total number of employees is 12,934 (actually 12,940, but six birthdays were dropped as described in the next paragraph).

The total number of days is 365. (There were six birthdays on February 29, the 366th day, but these were not included in the calculation.)

The mean frequency of birthdays per day is $12{,}934/365 = 35.44$.

The standard deviation of the frequencies, calculated according to Eq. (4.5), is 6.15. (In calculating the standard deviation, n, rather than $n - 1$, was used in the denominator inasmuch as we are calculating the actual standard deviation of the total population rather than estimating it from a sample.)

The hypothesis we will check is that the frequencies are distributed normally with a mean of 35.44 and a standard deviation of 6.15.

The range of frequencies is from a low of 17 (on March 22) to a high of 53 (on November 4). If each frequency were taken as a category, there would be $53 - 17 + 1 = 37$ categories. For tests to determine the normality of a distribution, 10 to 25 categories are sufficient to give reliable results and to keep the arithmetic within bounds.

The calculation of the goodness-of-fit test is best described with reference to Table 5.16 in which the calculation is made.

Column 1 numbers the categories.

Column 2 gives the frequency range included in each category.

Column 3 extends the frequency range of each category to one half unit beyond the measurement (count) of the data. This has the effect of converting the variable into a continuous measurement so that the normal distribution function can be applied.

Column 4 tabulates the z value corresponding to each category that is smaller than the mean. The z value is calculated for the upper limit of the category taken as a continuous measurement, i.e., corresponding to the value midway between the actual upper limit and the lower limit of the next category. $z = $ (mean $-$ upper limit)/standard deviation.

Column 5 tabulates the z values for the categories larger than the mean, with the z values calculated for the lower limits of the categories. $z = $ (lower limit $-$ mean)/standard deviation.

TABLE 5.16 Goodness-of-fit Test of Normality for Birthday Frequency Data
Mean = 35.44 birthdays/day
Standard deviation = 6.15

Category no.	Frequency range	Range limits	Upper limit z	Lower limit z	$F(z)$	Probability for the range	No. in the category Expected	No. in the category Observed	Contribution to χ^2
1	17–22	$-\infty$–22.5	2.10[a]	....	0.9821	0.0179	6.5[d]	5	0.35[e]
2	23–24	22.5–24.5	1.78	....	0.9625	0.0196	7.2	14	6.42
3	25–26	24.5–26.5	1.45	....	0.9265	0.0360	13.1	8	1.99
4	27–28	26.5–28.5	1.13	....	0.8708	0.0557	20.3	19	0.08
5	29–30	28.5–30.5	0.80	....	0.7881	0.0827	30.2	24	1.27
6	31–32	30.5–32.5	0.48	....	0.6884	0.1037	37.9	47	2.18
7	33–34	32.5–34.5	0.15	....	0.5596	0.1248	45.6	47	0.04
8	35–36	34.5–36.5	....	0.17[b]	....	0.1271[c]	46.4	48	0.06
9	37–38	36.5–38.5	....	0.51	0.5675	0.1275	46.5	51	0.44
10	39–40	38.5–40.5	....	0.82	0.6950	0.0989	36.1	32	0.47
11	41–42	40.5–42.5	....	1.15	0.7939	0.0810	29.6	17	5.36
12	43–44	42.5–44.5	....	1.47	0.8749	0.0543	19.8	20	0.00
13	45–46	44.5–46.5	....	1.80	0.9292	0.0349	12.7	18	2.21
14	47–48	46.5–48.5	....	2.12	0.9641	0.0189	6.9	9	0.64
15	48–53	48.5–∞	....	....	0.9830	0.0170	6.2	6	0.01
						1.0000	365.0	365	21.52

[a] $(35.44 - 22.5)/6.15 = 2.10$
[b] $(36.5 - 35.44)/6.15 = 0.17$
[c] $0.5596 + 0.5675 - 1 = 0.1271$
[d] $(0.0179)(365) = 6.5$
[e] $(6.5 - 5)^2/6.5 = 0.35$

Column 6 tabulates the $F(z)$ values from Table 3.12 for each z value in Columns 4 and 5.

Column 7 gives the probabilities for each category range. These probabilities are equal to the difference between successive $F(z)$ values for each frequency range excepting that including the mean value. The probability for the category including the mean value is equal to one less than the sum of the two $F(z)$ values for the categories on either side of the mean.

Column 8 gives the "expected" number in each category for a normally distributed variable, equal to 365 times the probability for each category.

Column 9 gives the actually observed frequencies.

Column 10 gives the contribution to χ^2: (expected-observed)^2/expected.

The total χ^2 is 21.52. This value is compared to the tabulated values from Table 5.1 at 12 degrees of freedom. The 0.05 value at 12 degrees of freedom is 21.026 indicating that there is less than an 0.05 probability of a deviation as large as that observed in the example if the data were really normally distributed. The 12 degrees of freedom were obtained from the fact that there were 15 categories with one restriction for the total, one for the mean, and one for the standard deviation, all calculated from the data.

The conclusion, with less than 0.05 chance of error, is that the birthdays are not normally distributed.

5.9 ′ χ^2 for Degrees of Freedom Greater Than 30

The χ^2 values given in Table 5.1 extend up to 30 degrees of freedom. Tables exist[1] for degrees of freedom up to 100. However, for degrees of freedom greater than 30 it is possible to obtain a close approximation of χ^2 from the fact that $\sqrt{2\chi^2}$ approaches a normal distribution with a mean of $\sqrt{2\nu - 1}$ and unit standard deviation, where ν is the number of degrees of freedom. The larger the value of $\sqrt{2\nu - 1}$, hence the larger the value of ν, the more nearly $\sqrt{2\chi^2}$ approaches a normal distribution.

We have seen, in the discussion of the normal distribution, that the probability of observing a value which differed from the mean by as much as some specific amount could be determined for a normally distributed variable by the ratio of this difference to the standard deviation. This ratio was called the standard deviate, and the probability values were given in Table 3.12. Applying this relation to $\sqrt{2\chi^2}$, which is normally

[1] G. W. Snedecor, "Statistical Methods," 5th ed., Iowa State College Press, Ames, Iowa, 1956.

distributed and has a mean of $\sqrt{2\nu - 1}$ and a standard deviation of 1, we obtain

$$\sqrt{2\chi^2} = \sqrt{2\nu - 1} \pm (z)(1) \tag{5.7}$$

The values of z are given in Table 3.12 for various probabilities, so that Eq. (5.7) can be solved for χ^2 at any desired probability level.

$$\chi^2 = \frac{(\sqrt{2\nu - 1} \pm z)^2}{2} \tag{5.8}$$

For example, to determine the 0.05 probability value of χ^2 for 50 degrees of freedom, we need the value of z corresponding to the 0.95 probability level, i.e., the value beyond which lies only 5 per cent of the area under the probability curve. This value from Table 3.12 is 1.645. ν, the degrees of freedom, is 50. χ^2 therefore is calculated as follows:

$$\chi^2_{0.05} = \frac{(\sqrt{99} + 1.645)^2}{2}$$

$$= \frac{(11.59487)^2}{2}$$

$$= 67.22$$

To determine the χ^2 value corresponding to the 0.95 probability level, the negative signs of Eq. (5.7) and (5.8) are used; i.e., we want the value beyond which lies 95 per cent of the area under the probability curve.

$$\chi^2_{0.95} = \frac{(\sqrt{99} - 1.645)^2}{2}$$

$$= \frac{(8.30587)^2}{2}$$

$$= 34.49$$

5.10 Problems

1. Carbon-residue analyses are run in a heated metal block in which four test cells are placed simultaneously and assumed to come to the same temperature. In a series of several hundred tests, the following obviously bad results were obtained in each cell location. Is there evidence of inconsistency in these results?

Cell location number	1	2	3	4
Number of bad results	46	33	37	49

$\chi^2 = 4.091$, $\nu = 3$.

2. You match coins with a friend, and he wins 85 times out of 140. Is there less than 0.05 probability that he's your friend? $\chi^2 = 6.006$, $\nu = 1$.

3. In a factory poll among 780 workmen as to whether seniority or production rating should hold preference, the following results were obtained from workmen with more than 5 years seniority and with less than 5 years seniority. Is there evidence that the workmen were influenced by their own seniority?

	Favor rating	Favor seniority
Senior men........	100	110
Nonsenior men.....	305	265

$\chi^2 = 1.886$, $\nu = 1$.

4. Snedecor[1] gives the following values of χ^2 for 100 degrees of freedom: 124.34 at 0.05 probability, 135.81 at 0.01 probability. Check these values by the approximate method of Sec. 5.10.

5. The following data give the number of analyses run by three different operators during three different shifts. Is there evidence that some of the operators perform more proficiently during some shifts than others?

Shift	Operator		
	1	2	3
Day..........	17	17	12
Afternoon......	11	9	13
Night..........	11	8	19

$\chi^2 = 5.430$, $\nu = 4$.

6. A company operating a fleet of automobiles has a record of flat tires occurring in one 60-day period. Of 210 cars, 84 had one flat tire, 6 had two flat tires, and 3 cars had three flat tires. The balance had no flat tires. Is there evidence that the probability is the same for all cars? $\chi^2 = 13.55$, $\nu = 2$.

[1] Snedecor, *op. cit.*

chapter six

THE *t* TEST

If the three phases of statistical analysis that were of most value to the engineer studying and analyzing data had to be named, the author's choice would be the *t* test, the analysis of variance, and a study of correlation and regression. Others with different experience might select other branches: design of experiments, propagation of error, a study of sampling. The wisest course might be to avoid selecting any particular phases as being most important. From some experience both with student engineers learning statistics and with the application of statistics to engineering data, we are inclined to support the case for these three branches of statistics.

The *t* test, which is the subject of this chapter, deals with the estimation of a true value from a sample and the establishing of confidence ranges within which the true value can be said to lie. The analysis of variance, covered in Chap. 7, deals with the study of the scatter of data and the contributions to this scatter. Regression and correlation, which deal with the variation of one variable with another, or with more than one other, are dealt with in Chap. 9.

6.1 Introduction

6.1.1 Definition. The *t* distribution can be explained in terms of what we hope is now the familiar standard deviate of the normal distribution. The standard deviate z has been described in Chap. 3 as a measure of the deviation of a variable from its mean in units of the standard deviation. The equation for z from Chap. 3 is

$$z = \frac{|x - \mu|}{\sigma} \tag{3.17}$$

With a normally distributed variable, stating the value of z is equivalent to giving the relative frequency (or the probability) of the occurrence of a value which deviates from the mean by as much as $|x - \mu|$. These are the probabilities given in Table 3.12.

t can be defined in a similar manner, being the difference between the mean of a sample and the true mean of the population from which the sample was drawn, divided by the estimated standard deviation of the mean. Thus, if μ designates the true mean, t may be written

$$t = \frac{|\bar{x} - \mu|}{s(\bar{x})} \tag{6.1}$$

The similarity of t to the normal deviate z is mentioned merely for descriptive purposes. The t distribution is not a normal distribution, although it approaches the normal under certain conditions. It is an independent mathematical function and depends on the number of measurements involved in the calculation of $s(\bar{x})$. If n is the number in the sample, ranging from 2 to ∞, t has a different distribution for each value of n, becoming equivalent to the normal distribution when n becomes infinite.

n enters the formula for t in the form of the *degrees of freedom* available for the calculation of the standard deviation. The degrees of freedom vary for different usages of t, as will be explained when specific problems are discussed. The values of t for different degrees of freedom are given in Table 6.1, with the probability levels for observing larger absolute values than those tabulated. For example, with 5 degrees of freedom, there is only a 0.05 probability of observing a t larger than 2.57 or smaller than -2.57. With 17 degrees of freedom, there is only a 0.02 probability of observing an absolute t value larger than 2.57. The t values in the last line of the table, for ∞ degrees of freedom, correspond to the values for the normal distribution. Detailed discussion of the use of this table will follow in the next few paragraphs.

6.1.2 Calculation of $s(x)$ and $s(\bar{x})$. $s(x)$ designates the estimate of the population standard deviation from the sample $x_1, x_2, \ldots, x_n$. The estimated standard deviation can be calculated or expressed in any of the

TABLE 6.1 t†

α, probability of a larger absolute value of t

D.F.	.9	.8	.7	.6	.5	.4	.3	.2	.1	.05	.02	.01	.001
1	.158	.325	.510	.727	1.000	1.376	1.963	3.078	6.314	12.706	31.821	63.657	636.619
2	.142	.289	.445	.617	.816	1.061	1.386	1.886	2.920	4.303	6.965	9.925	31.598
3	.137	.277	.424	.584	.765	.978	1.250	1.638	2.353	3.182	4.541	5.841	12.941
4	.134	.271	.414	.569	.741	.941	1.190	1.533	2.132	2.776	3.747	4.604	8.610
5	.132	.267	.408	.559	.727	.920	1.156	1.476	2.015	2.571	3.365	4.032	6.859
6	.131	.265	.404	.553	.718	.906	1.134	1.440	1.943	2.447	3.143	3.707	5.959
7	.130	.263	.402	.549	.711	.896	1.119	1.415	1.895	2.365	2.998	3.499	5.405
8	.130	.262	.399	.546	.706	.889	1.108	1.397	1.860	2.306	2.896	3.355	5.041
9	.129	.261	.398	.543	.703	.883	1.100	1.383	1.833	2.262	2.821	3.250	4.781
10	.129	.260	.397	.542	.700	.879	1.093	1.372	1.812	2.228	2.764	3.169	4.587
11	.129	.260	.396	.540	.697	.876	1.088	1.363	1.796	2.201	2.718	3.106	4.437
12	.128	.259	.395	.539	.695	.873	1.083	1.356	1.782	2.179	2.681	3.055	4.318
13	.128	.259	.394	.538	.694	.870	1.079	1.350	1.771	2.160	2.650	3.012	4.221
14	.128	.258	.393	.537	.692	.868	1.076	1.345	1.761	2.145	2.624	2.977	4.140
15	.128	.258	.393	.536	.691	.866	1.074	1.341	1.753	2.131	2.602	2.947	4.073
16	.128	.258	.392	.535	.690	.865	1.071	1.337	1.746	2.120	2.583	2.921	4.015
17	.128	.257	.392	.534	.689	.863	1.069	1.333	1.740	2.110	2.567	2.898	3.965
18	.127	.257	.392	.534	.688	.862	1.067	1.330	1.734	2.101	2.552	2.878	3.922
19	.127	.257	.391	.533	.688	.861	1.066	1.328	1.729	2.093	2.539	2.861	3.883
20	.127	.257	.391	.533	.687	.860	1.064	1.325	1.725	2.086	2.528	2.845	3.850
21	.127	.257	.257	.532	.686	.859	1.063	1.323	1.721	2.080	2.518	2.831	3.819
22	.127	.256	.390	.532	.686	.858	1.061	1.321	1.717	2.074	2.508	2.819	3.792
23	.127	.256	.390	.532	.685	.858	1.060	1.319	1.714	2.069	2.500	2.807	3.767
24	.127	.256	.390	.531	.685	.857	1.059	1.318	1.711	2.064	2.492	2.797	3.745
25	.127	.256	.390	.531	.684	.856	1.058	1.316	1.708	2.060	2.485	2.787	3.725
26	.127	.256	.390	.531	.684	.856	1.058	1.315	1.706	2.056	2.479	2.779	3.707
27	.127	.256	.389	.531	.684	.855	1.057	1.314	1.703	2.052	2.473	2.771	3.690
28	.127	.256	.389	.530	.683	.855	1.056	1.313	1.701	2.048	2.467	2.763	3.674
29	.127	.256	.389	.530	.683	.854	1.055	1.311	1.699	2.045	2.462	2.756	3.659
30	.127	.256	.389	.530	.683	.854	1.055	1.310	1.697	2.042	2.457	2.750	3.646
40	.126	.255	.388	.529	.681	.851	1.050	1.303	1.684	2.021	2.423	2.704	3.551
60	.126	.254	.387	.527	.679	.848	1.046	1.296	1.671	2.000	2.390	2.660	3.460
120	.126	.254	.386	.526	.677	.845	1.041	1.289	1.658	1.980	2.358	2.617	3.373
∞	.126	.253	.385	.524	.674	.842	1.036	1.282	1.645	1.960	2.326	2.576	3.291

† Reprinted from Table III of R. A. Fisher and Frank Yates, "Statistical Tables for Biological, Agricultural and Medical Research," Oliver & Boyd, Ltd., Edinburgh and London, 1953, by permission of the authors and publishers.

following identical forms:

$$s(x) = \sqrt{\frac{\Sigma(x-\bar{x})^2}{n-1}} \quad (6.2)$$

$$= \sqrt{\frac{\Sigma' x^2}{n-1}} \quad (6.3)$$

$$= \sqrt{\frac{\Sigma x^2 - (\Sigma x)^2/n}{n-1}} \quad (6.4)$$

$$= \sqrt{\frac{\Sigma x^2 - \bar{x}\Sigma x}{n-1}} \quad (6.5)$$

$$= \sqrt{\frac{\Sigma x^2 - n\bar{x}^2}{n-1}} \quad (6.6)$$

The use of $n-1$ in the denominator in each case, as explained in Chap. 4, is to offset the bias due to calculating the deviations of x from $\bar{x}$, the sample mean, rather than from the unknown true population mean μ.

$s(\bar{x})$ designates the estimated standard deviation of the means of samples of size n drawn from the population which is estimated to have a standard deviation of $s(x)$. These two quantities are related in the following way:

$$s(\bar{x}) = \frac{s(x)}{\sqrt{n}} \quad (6.7)$$

The relation of Eq. (6.7) is based on an important theorem of statistics. This theorem states that for any population, regardless of its distribution, if it has a finite variance σ^2, the variance of means of samples of size n drawn from this population will be equal to the population variance divided by the sample size:

$$\sigma^2(\bar{x}) = \frac{\sigma^2(x)}{n} \quad (6.8)$$

The proof of this theorem can be found in texts on the theory of statistics.[1] An intuitive approach to the relation can be had by imagining (or actually) sampling card hands from a deck of cards. A deck has a rectangular distribution: four cards of each denomination from 1 to 13. Individual cards will deviate from the mean of 7 over a range from −6 to +6. If the means of groups of cards are observed, these can be expected to deviate from the population mean of 7 much less than the deviation of the individ-

[1] A. M. Mood, "Introduction to the Theory of Statistics," McGraw-Hill Book Company, New York, 1950.

ual cards. If the deviation is measured as the variance, $\Sigma(\bar{x} - 7)^2/$(no. of groups) $= \Sigma[(\Sigma x/n) - 7]^2/$(no. of groups), it can be seen that the variance will decrease as n increases. The actual relation of Eq. (6.8) can be estimated by obtaining data from 10 or 12 card hands.

A second important theorem, which may not be so readily accepted intuitively, is the central-limit theorem. This theorem states that if a population has a finite variance σ^2 and a mean μ, then the distribution of means of samples of size n drawn from the population approaches the normal distribution with a mean μ and a variance σ^2/n as the sample size n increases. Although the proof of this theorem is somewhat beyond elementary statistics theory, the approach to normality of the distribution of means can be demonstrated by observing a fairly large number, 50 or more, card hands. A perhaps somewhat easier example is to take the means of successive five-digit groups from a table of random numbers. The agreement of the observed distribution with the normal may be checked by the χ^2 method of the last chapter.

The value of these two theorems is that they permit the calculation of the estimated standard deviation of the mean from the estimated standard deviation of the sample and they permit the application of the t distribution to the sample mean regardless of the distribution of the population from which the sample was drawn.

6.1.3 Degrees of Freedom. The degrees of freedom for the t test have the same meaning as was employed in the χ^2 test: the number of independent measurements that are available for the calculation. In the t test the degrees of freedom apply to the calculation of the estimated standard deviation. If the standard deviation is known, or assumed to be known, from large amounts of previous data, then there is no restriction on the estimation of the standard deviation and the t values for ∞ degrees of freedom are employed. These, as mentioned, correspond to the normal-distribution values.

If the standard deviation is estimated from data other than those under test, then the degrees of freedom corresponding to those involved in the estimate are the proper ones to use in determining the t value.

For the ordinary example, when the standard deviation is estimated from n values of the measurement, it is necessary to calculate the deviations from the sample mean. Since the determination of the mean involves fixing the sum of the n measurements, only $n - 1$ of them may be independently varied without changing the sum. On this basis there are $n - 1$ degrees of freedom available for estimating the variance or standard deviation. Examples employing different numbers of degrees of freedom are given throughout this chapter.

6.1.4 The Null Hypothesis. The execution of the t test, like that of the χ^2 test, involves setting up a hypothesis and calculating t on the basis that the hypothesis is true. Several different forms of equations for t will be given, each with its own form of the null hypothesis. The distribution of t is the same for them all, and the values in Table 6.1 are the maximum that can be expected at the indicated probability levels if the null hypothesis is true. For example, a t calculated with 5 degrees of freedom has only a 0.05 probability of being larger than 2.57, a 0.02 probability of being larger than 3.36, and a 0.01 probability of being larger than 4.03, *if the null hypothesis is true.* If, then, a t value of 2.80 is observed for an example of this type, we can say with less than 0.05 chance of being wrong that the null hypothesis is false, since the occurrence of a t of 2.80 is an event which has less than 0.05 chance of happening if the hypothesis were true.

When we reject the null hypothesis at the 0.05 level on the basis of the t test (or any other statistical test), we accept the risk of a 5 per cent chance of being in error. This error, of falsely rejecting the null hypothesis, is called an error of type I. It is also possible to make an error of another kind. If the data do not satisfy the null hypothesis but give a t value less than the critical value for the test, then the hypothesis will be incorrectly accepted as being true. This false acceptance of the null hypothesis is an error of type II. The probability level for the rejection of the null hypothesis is called the significance level of the test. We say there is a significant difference from the null hypothesis with the indicated probability of being wrong. If this significance level is set so low that there is small chance of an error of type I, which means using a larger critical value of the statistic in question, then there is more chance of an error of type II. It is not possible to reduce both errors simultaneously in a single test. If the significance level, usually designated α, is set, then the probability of type II errors, designated β, is also set. A decrease in α causes an increase in β, and vice versa. With a fixed value of α, β can only be decreased by increasing the sample size. Some tables showing the relation between the probabilities of the two types of error and the sample size are given in the last section of this chapter.

It would appear that the best general test would be that which makes the probabilities of both types of errors a minimum. For practical purposes, with the ordinary size of samples encountered, the 0.05 to 0.01 probabilities for errors of type I usually give the optimum, i.e., the minimum, probability of both types of errors. There are occasions when one type of error is more important to avoid than the other. In pharmaceutical work it is more important not to accept defective material than not to reject good material. In cases where costs are relatively high compared with the inconvenience of using defective material, it is more essential not to reject good material than not to accept defective material. The rejection of good material and the acceptance of defective material correspond to type I and type II errors, respectively. If more precise evaluation of the two types of errors is required than that given in Sec. 6.5, reference should be made to texts dealing with quality control,[1] or to more advanced statistics texts.[2]

6.2 Estimation of the True Mean

6.2.1 The Hypothesis: $\mu = m$. This is the first of three null hypotheses which will be used with the t test. Several measurements are made, $x_1, x_2, \ldots, x_n$, and their mean $\bar{x}$ is calculated. The mean $\bar{x}$ is an estimate of the true population mean μ. The hypothesis is set up that these measurements could have come from a population with some specific mean value m. A vendor claims his product is 99.44 per cent pure, and several samples from a shipment have a mean purity of 99.36 per cent. Is this significantly different from the claimed purity? Specifications for a certain product call for 18.0 per cent nickel. Several analyses of a batch of material give a mean nickel content of 18.14 per cent. Does the material meet specifications?

The null hypothesis for a test of this nature is that the true mean μ is equal to the specified value m.

$$H_0; \mu = m \tag{6.9}$$

In order to test this hypothesis, we calculate a mean, $\bar{x}$, and estimate $\sigma/\sqrt{n}$ by calculating $s(\bar{x})$.

6.2.2 The t Test. To test this particular hypothesis, the t equation is written as in Eq. (6.1), substituting the specified value m for the true mean μ, or in the following form, with $s(x)/\sqrt{n}$ substituted for $s(\bar{x})$ from Eq. (6.7).

$$t = \frac{|\bar{x} - m|}{s(x)\sqrt{1/n}} \tag{6.10}$$

The estimated standard deviation $s(x)$ may be calculated from any of formulas (6.2) to (6.6). If a calculating machine is used, Eq. (6.5) is probably the easiest since Σx^2 and Σx may be obtained in one series of operations by accumulating the product of the successive squaring operations. If no machine is available, Eq. (6.2) is probably the simplest to use, obtaining the squares of the differences from the mean from a table of squares. This is especially true when the mean is a whole number and the

---

[1] E. L. Grant, "Statistical Quality Control," 2d ed., McGraw-Hill Book Company, New York, 1952.

[2] C. Eisenhart, M. W. Hastay, W. A. Wallis (eds.), "Techniques of Statistical Analysis," McGraw-Hill Book Company, New York, 1947.

differences contain only two or three digits. A more detailed discussion of the calculation of the estimated standard deviation is given in Chap. 4.

The t value calculated from Eq. (6.10) [or Eq. (6.1)] is compared with the tabulated values of t from Table 6.1 at the number of degrees of freedom corresponding to $n - 1$, $n - 1$ being the number of independent measurements that are available for the calculation of $s(x)$. Before the calculation is made, however, some significance level representing the acceptable risk of making an error should be decided upon for rejection of the null hypothesis. If the tabulated value of t for this level is exceeded, then the hypothesis can be rejected—the data giving a mean of $\bar{x}$ can be said not to have come from a population with a mean of m—with the accepted risk of being wrong.

The reason for setting the significance level before the calculations are made is to avoid letting personal desires, conscious or otherwise, influence the statistical judgment. It is often possible to collect more data with which to substantiate or reverse a close statistical judgment. But when more data are not available, it is the better part of valor to stick to the preset significance levels, remembering of course that at the 0.05 level you will reject the hypothesis when it is true 1 time in 20.

example 6.1. A gasoline product in a pilot-plant process is supposed to have an octane number, clear, of 87.5. The data in Table 6.2 are 11 daily analyses with a mean of 87.1. We wish to test whether these data might have come from a population with a mean of 87.5.

The null hypothesis therefore is H_0; $\mu = 87.5$.

We shall set a significance level of 0.05; that is, if there is 0.05 or less probability of being in error, we shall reject H_0; otherwise we shall accept it.

To calculate t, from Eq. (6.1), we calculate $\bar{x}$ and the estimated standard deviation of the mean, $s(\bar{x})$. The calculation using Eq. (6.3) is shown in Table 6.2.

Since there are 11 pieces of data, there are 10 degrees of freedom for this test. The corresponding t, at the 0.05 level from Table 6.1, is

$$t_{0.05,10} = 2.228$$

The calculated value of t does not exceed the tabulated value at the significance level at which we had decided to reject the hypothesis; therefore, pending further data, we accept the statement that the pilot plant is producing gasoline with an octane number of 87.5, even though the mean of the analyses is 87.1. Actually, what we are saying is that with as much scatter as there is in these data, we cannot really tell the difference between 87.1 and 87.5 from 11 analyses.

TABLE 6.2

Observed octane numbers	Calculation of $s(\bar{x})$
88.6	$n = 11$
86.4	$\Sigma x = 958.1$
87.2	$\Sigma x^2 = 83{,}456.83$
88.4	$\bar{x} = \dfrac{\Sigma x}{n} = 87.10$
87.2	
86.8	$\Sigma' x^2 = \Sigma x^2 - \bar{x}\Sigma x$
86.1	$\quad = 83{,}456.83 - (87.10)(958.1)$
87.3	$\quad = 6.32$
86.4	$s^2(x) = \dfrac{6.32}{10} = 0.632$
86.6	
87.1	$s^2(\bar{x}) = \dfrac{0.632}{11} = 0.05745$
	$s(\bar{x}) = 0.2397$

From Eq. (6.1), assuming H_0 to be true, i.e., that $\mu = m$,

$$t = \frac{|\bar{x} - m|}{s(\bar{x})} = \frac{|87.1 - 87.5|}{0.2397} = 1.669$$

We shall demonstrate the solution of the problem from these data in two other ways to illustrate different methods of calculating the sum of squares. Beyond the calculation of $\Sigma' x^2$, the procedures are identical and are omitted in the following illustrations. The solution by means of Σx^2 as in Table 6.2 would probably only be used if a calculating machine was available. Note that the sum of squares, the difference between Σx^2 and $\bar{x}\Sigma x$, is 6.32, while each of these numbers has seven significant digits. If the squares to the individual x values were rounded off (or read on a slide rule) to less than two decimal places before the total was obtained, the value of $\Sigma' x^2$, and hence the value of $s^2(x)$, could be order of magnitude in error.

If a calculating machine is not available, Eq. (6.2) might be easier to use for the calculation, determining the deviation of each x value from the mean before squaring and adding. This method is illustrated in Table 6.3. You will note that $\Sigma(x - \bar{x})^2$ is identical with $\Sigma' x^2$ from Table 6.2.

If $\bar{x}$ had not ended in zero in the second decimal place, it would have been necessary to calculate $x_i - \bar{x}$ to at least two decimals, and the square, $(x_i - \bar{x})^2$ to at least four decimals to ensure the accuracy of $\Sigma' x^2$.

Another method of solution is to code the original data by subtracting a constant from each term before calculating the sum of squares. This in effect is what is done when the mean is subtracted from each term, the mean being a special constant which further simplifies the calculation. It is often easier to code the data before calculating the mean, especially when a relatively large constant can be subtracted from each value. Assuming that we do not know the mean of the data for this example, and since 87.5 is the target mean, we shall code the data by this amount. Table 6.4 illustrates the calculation using the coded data.

TABLE 6.3   Calculation of $\Sigma' x^2$ Using Eq. (6.2)

x	$(x - \bar{x})$	$(x - \bar{x})^2$	
88.6	1.5	2.25	$n = 11$
86.4	−0.7	0.49	$\Sigma x = 958.1$
87.2	0.1	0.01	$\bar{x} = 87.10$
88.4	1.3	1.69	$\Sigma(x - \bar{x})^2 = \Sigma' x^2 = 6.32$
87.2	0.1	0.01	
86.8	−0.3	0.09	
86.1	−1.0	1.00	
87.3	0.2	0.04	
86.4	−0.7	0.49	
86.6	−0.5	0.25	
87.1	0.0	0.00	
958.1		6.32	

6.2.3 The One-sided t Test. Up until now we have said nothing about the sign of the t function. The t distribution, like the normal distribution, is symmetrical. The t values, like those of the standard deviate, measure either positive or negative deviations from the mean. Equation (6.1) is written with the bar symbols to indicate that it is the absolute deviation of $\bar{x}$ from μ that is being measured.

Only positive values of t are given in Table 6.1, but the probabilities apply equally to either positive or negative values. A t value of 2.131 for 15 degrees of freedom at the 0.05 probability level means that there is a 0.025 probability of observing a t value greater than 2.131 and a 0.025 probability of observing a t value less than -2.131, when $\bar{x}$ is drawn from a population with a mean of μ and t is calculated in the form of Eq. (6.10) or (6.1). A t at the 0.10 probability level means there is half this probability, 0.05, of observing a value greater than t and half this probability of observing a value less than $-t$. When we set up a t test to establish whether the true mean can be considered equal to some theoretical mean, we agree (tacitly) to say there is a significant difference between the two if the measured mean is either greater than or less than the theoretical value as measured by the t equation, and we are willing to accept the corresponding probability of being wrong.

There are many occasions when we are only interested in establishing whether the true mean is either less than or greater than some theoretical value, but not both. If the purity of a substance is specified to be, say, 99.44 per cent, we might be interested in determining whether the mean value of a sample is significantly less than this purity and not particularly care whether the purity is higher (if we are the purchaser). Or con-

TABLE 6.4 Calculation of $\Sigma'x^2$ Using Coded Data

x	$y = (x - 87.5)$	y^2	
88.6	1.1	1.21	$n = 11$
86.4	−1.1	1.21	$\Sigma y = -4.4$
87.2	−0.3	0.09	$\bar{y} = -0.40$
88.4	0.9	0.81	$\bar{y} = (\bar{x} - 87.5)$
87.2	−0.3	0.09	$\bar{x} = \bar{y} + 87.5$
86.8	−0.7	0.49	$\phantom{\bar{x}} = (-0.4 + 87.5) = 87.1$
86.1	−1.4	1.96	$\Sigma y^2 = 8.08$
87.3	−0.2	0.04	$\Sigma' y^2 = y^2 - \bar{y}\Sigma y$
86.4	−1.1	1.21	$ = [8.08 - (-0.4)(-4.4)]$
86.6	−0.9	0.81	$ = 6.32$
87.1	−0.4	0.16	$ = \Sigma' x^2$
Total...	−4.4	8.08	

versely, if the impurity is specified as 0.56 per cent, we might be interested in testing whether the mean of a sample shows an impurity significantly greater than this and not be interested in testing whether it could be less. Tests of this type are called one-sided, one-tailed, or unsymmetrical t tests and involve only the probability of falling in one-half of the t distribution, which is half the probability given in Table 6.1.

If instead of the hypothesis of Eq. (6.9), stating that the true mean is equal to some specified value, we set up the hypothesis that the true mean is not less than a specified value, we can express this

$$H_0;\ \mu \geq m \tag{6.11}$$

and the corresponding t equation is

$$t = \frac{m - \bar{x}}{s(\bar{x})} \tag{6.12}$$

If the alternative hypothesis, that the true mean is not greater than a specified value, is to be tested, the null hypothesis may be written

$$H_0;\ \mu \leq m \tag{6.13}$$

and the corresponding t equation is

$$t = \frac{\bar{x} - m}{s(\bar{x})} \tag{6.14}$$

If either of these specific one-sided t tests is to be employed, it is essential that the test and the significance level be decided upon before the data have been studied. The critical values of t for these one-sided tests are those at the probability levels corresponding to twice the significance levels decided upon. If a 0.05 significance level is used, the hypothesis is to be rejected if the calculated t value exceeds the 0.10 level of t at the proper degrees of freedom, since the tabulated 0.10 level corresponds to a 0.05 probability for deviations of equal magnitude on either side of the mean.

example 6.2. The addition of a new accelerator is claimed to decrease the curing time of a rubber compound by more than 4 per cent. If it does not decrease the curing time by more than 4 per cent, it will not be used because of its extra cost and difficulty of handling. A series of tests is made, and the difference in curing time is measured. The hypothesis of Eq. (6.13) is formulated, i.e., that the true-mean decrease in curing time is equal to or less than 4 per cent. If we are willing to reject the hypothesis (and accept the new accelerator) with a 5 per cent chance of being wrong, we select the 0.10 probability level of t for our critical value. Table 6.5 gives the results for seven runs together with the calculation of t.

On the basis of these data, since the calculated t does not exceed the critical value of 1.943, we accept the hypothesis that true-mean decrease in curing time is not more than 4 per cent.

Note that if the mean was less than 4.0 there would be no chance of rejecting the hypothesis. The magnitude of the increase of the mean beyond 4.0 necessary to cause a rejection of the hypothesis and an acceptance of the new accelerator is dependent on $s(\bar{x})$. We cannot be sure, even though the sample mean was 4.5, because of the scatter of the data as measured by $s(\bar{x})$, that the population mean is larger than 4.0.

6.2.4 The Confidence Range. $\bar{x}$, as has been pointed out, is an estimate of the true mean μ. The t function, expressed as in Eq. (6.1), gives the distribution of deviations of $\bar{x}$ from μ in terms of relative frequencies or probabilities. If we transpose the equation and write the absolute value in terms of plus or minus, we can express the true mean in terms of the measured mean and t, thus:

$$\mu = \bar{x} \pm ts(\bar{x}) \tag{6.15}$$

THE t TEST

Verbally, the true mean can be said, with the tabulated probability of error, to be within the range of the calculated mean included in the limits of plus and minus t times the estimated standard deviation of the mean. If, as in Example 6.2, the calculated mean $\bar{x}$ is 4.5 and the estimated standard deviation of the mean $s(\bar{x})$ is 0.533, and if $s(x)$ is calculated with

TABLE 6.5

Run no.	Per cent decrease in curing time	Calculation of t
1	5.2	$n = 7$
2	6.3	$\Sigma x = 31.5$
3	3.7	$\bar{x} = 4.50$
4	6.2	$\Sigma x^2 = 153.69$
5	4.1	$\Sigma' x^2 = 11.94$
6	2.9	$s^2(x) = 1.99$
7	3.1	$s^2(\bar{x}) = 0.284$
		$s(\bar{x}) = 0.533$
		$t = \dfrac{4.50 - 4.00}{s(\bar{x})}$
		$t = \dfrac{0.50}{0.533} = 0.938$
		$t_{0.10,6} = 1.943$

6 degrees of freedom so that the corresponding t value from Table 6.1 is 2.447 at the 0.05 significance level, then from Eq. (6.15) the true mean μ can be said with 0.05 probability of error to be within the range $4.5 \pm (2.447)(0.533)$, or from 3.2 to 5.8. This range, within which the mean can be said to lie, is the confidence range of the mean from these particular data. The confidence range at the 0.05 significance level of t is usually designated the 95 per cent confidence range of the mean.

The term $ts(\bar{x})$ can be looked on as an estimation of the precision of the measurement of $\bar{x}$. Ninety-five per cent of similar measurements of $\bar{x}$ would fall within the range of $\bar{x} \pm ts(\bar{x})$, where t is taken at the 0.05 significance level. If we designate the precision limits, usually called simply the precision, by the symbol l,

$$l_{0.95,\bar{x}} = \pm t_{0.05,\nu} s(\bar{x}) \qquad (6.16)$$

where ν is the degrees of freedom associated with $s(x)$.

We shall have occasion to refer a little later to the precision of a single measurement in similar terms.

The purpose of most measurements is to obtain a number which is the best possible estimate of the true value. In using the t distribution, in the form of Eq. (6.15), to make statements about the range within which the true mean lies, we refer to the values represented by the measurements involved. No account is taken in these statistical tests for a fixed error or bias in the measurements. If titrations are made with a solution the normality of which is incorrectly known, or if pressure measurements are made with an instrument that does not zero out, or if one of the weights on an analytical balance has been worn, then there is apt to be a constant error in all measurements which will not be evident in the statistical tests unless some special provision is made in the design of the experiment to detect this bias.

On the assumption that no bias exists in the measurements, Eq. (6.15) gives an expression for locating the true mean with any desired error risk. At the 0.05 level there is 1 chance in 20 that the true mean lies outside the specified range. At the 0.01 level, there is 1 chance in 100. At the 0.001 level, there is 1 chance in 1,000. To set the range for the true mean with less and less chance of error, it is necessary to use correspondingly larger values of t, and hence wider ranges of prediction, the value of t depending on the number of degrees of freedom available for the calculation of $s(x)$ and the probability level chosen. Even when σ is assumed to be precisely known, i.e., calculated with infinite degrees of freedom, a range for the true mean with only 1 chance in 100 of an error is more than five times the standard deviation of the mean: $\pm 2.576\sigma$. The factor that can be reduced, so that under a specific test the range of the true mean can be narrowed, is $s(\bar{x})$. $s(\bar{x})$, from Eq. (6.7), is equal to $s(x)/\sqrt{n}$. The estimated standard deviation of the mean can therefore be reduced either by increasing the number of terms n included in the mean or by decreasing the estimated standard deviation of the measurements, i.e., increasing the precision—reducing the variation from measurement to measurement. This is an experimental problem, not a statistical one. However, it is important to realize how much the scatter of the measurements contributes to widening the confidence range on the mean. Much confusion would be avoided and a better understanding of processes would evolve if statements about measurements included confidence limits. If instead of reporting the mean as m it is reported as $m \pm l_{0.95}$, that is, the range of the mean is from $m - l$ to $m + l$, with a 5 per cent chance of error, a much clearer knowledge of the facts is conveyed.

6.3 Difference between the Mean and Zero

6.3.1 The Hypothesis: $\mu = 0$.
Measurements are often made in pairs to compare two processes or two different materials. The purpose of the test is to determine whether there is a significant difference between the two items under test in terms of the measurement involved or whether

the mean difference is significantly different from zero. The null hypothesis, therefore, is

$$H_0; \mu = 0 \tag{6.17}$$

This is a special case of testing whether a mean is significantly different from some specified value. There are, however, some features about this test which warrant particular mention.

In comparing paired data, the pairs do not have to be measures of the same thing, although the individual measurements in a pair will be made at the same conditions. It is the difference within pairs and not the difference between the pairs that is to be tested. If two different makes of automobile tires are to be tested for wear under actual road conditions, the wear as measured by weight loss may vary considerably during the test as different road conditions are encountered. This variation need not be reflected in a variation of difference in wear between the two tires under similar conditions. In fact, the purpose of the test would be to see whether the difference in wear between the two tires was significant when judged by the amount of variation in this difference. If two commercial catalysts are to be compared on the basis of the approach to equilibrium of some reaction, they could be compared over a range of space velocities resulting in a wide variation in the approach to equilibrium. However, the differences between the two catalysts at each set of conditions might be fairly uniform and not reflect the variation in the test conditions.

6.3.2 The t Test. The t test for this hypothesis is similar to that described in the last section. The mean μ is estimated by the mean difference within pairs of measurements. t is then calculated from the quotient of the difference between the calculated mean and zero and the estimated standard deviation of the mean difference. If we designate the difference within pairs of measurements as y, t is then calculated as follows:

$$t = \frac{|\bar{y} - 0|}{s(\bar{y})} \tag{6.18}$$

The number of degrees of freedom associated with the t value is one less than the number of differences used to calculate $s(y)$.

The t test is judged in the same manner as discussed in the previous section. If the calculated value of t from Eq. (6.18) exceeds the tabulated value at the proper degrees of freedom and at the significance level decided upon, then the null hypothesis that the mean was equal to zero can be rejected with the accepted risk of error. We say, in effect, there is a significant difference between $\bar{y}$ and zero.

example 6.3. The following data[1] show a comparison of iron determinations by two different methods on several copper electrodes containing different amounts of iron. We shall use the t test to see whether there is a significant difference between the two analytical methods. If the t value exceeds the 0.05 value, we shall consider the difference significant. The data and calculation of t are given in Table 6.6.

TABLE 6.6   Per Cent Iron

Sample	Analysis 1 (x_1)	Analysis 2 (x_2)	y' ($x_1 - x_2$)	y ($1{,}000 y'$)
1	0.063	0.063	0.000	0
2	0.051	0.050	0.001	1
3	0.067	0.062	0.005	5
4	0.056	0.060	−0.004	−4
5	0.074	0.070	0.004	4
6	0.076	0.070	0.006	6
7	0.036	0.033	0.003	3

$n = 7$ $\quad\quad \Sigma y^2 = 103 \quad\quad s^2(\bar{y}) = 1.6874$
$\Sigma y = 15 \quad\quad \Sigma' y^2 = 70.870 \quad\quad s(\bar{y}) = 1.299$
$\bar{y} = 2.142 \quad\quad s^2(y) = 11.8116$

$$t = \frac{|2.142 - 0|}{1.299} = 1.649$$

$$t_{0.05,6} = 2.447$$

The calculated t value is less than the critical value for 6 degrees of freedom (7 pairs of data), so the hypothesis, that there is no difference between the analytical methods, is accepted.

Note that the data have all been multiplied by 1,000 to simplify the calculation of $s(\bar{y})$. The reason that this adjustment of the data does not have to be recorrected before determining t is that both the numerator $\bar{y}$ and the denominator $s(\bar{y})$ would be corrected by multiplying by 10^{-3}, and this correction cancels out. (The doubting student can check this statement by calculating t from the unadjusted data.)

6.3.3   Data in Duplicate. Analyses are often run in duplicate (perhaps less often than they ought to be). The mean difference between duplicates provides an experienced analyst with an estimate of the standard deviation of a single measurement, σ. Although we shall not

[1] Charles A. Bicking, *ASTM Bull.*, September, 1954.

endeavor to prove this statement, it can be shown that the mean difference between duplicates is equal to 1.128σ, actually $(2/\sqrt{\pi})\sigma$. When an operator knows from experience that the mean difference within pairs of measurements is some value, say $\bar{d}$, he has at hand valuable information for estimating the precision of single measurements and for setting confidence ranges to the means of other duplicate measurements.

We can transpose the relation given for the mean difference between duplicates and write an equation for the standard deviation of a single measurement in terms of the mean difference.

$$\sigma = \frac{\sqrt{\pi}\,\bar{d}}{2} = 0.8862\bar{d} \tag{6.19}$$

where $\bar{d} = \Sigma|d|/n$. If we assume that a single measurement is drawn from a distribution of similar measurements affected by a large number of small random factors resulting in normal distribution of the measurement, then the table of the standard deviate (Table 3.12) gives the probability of drawing a value which deviates from the mean of the measurements as great as $|x - \mu|/\sigma$. Transposing this relation, we can set a confidence range to the mean (and hence the "true" value) from a single measurement,

$$\mu = x \pm z\sigma \tag{6.20}$$

where z is the standard deviate and the corresponding probability is $2[1 - F(z)]$ of Chap. 3. z is equal to the last row of the t table (Table 6.1), and the probability is equal to α, the significance level. This formulation is similar to setting a confidence limit to the mean in terms of the estimated standard deviation of the mean and the t function but requires the assumption of normality for the distribution in question and a knowledge of σ. These assumptions are not essential to the application of the t limits to the mean because of the fact that means tend to be normally distributed regardless of the distribution of the measurements from which they are calculated, and the increase of the t value over that of the standard deviate takes into account the estimation of the standard deviation from the amount of data at hand.

The term $z\sigma$ of Eq. (6.20) can be taken as the precision of the measurement x. When x is normally distributed with standard deviation σ, as is very frequently the case when the factors causing variations in x are small random influences, 95 per cent of similar measurements will fall within the range $\pm z\sigma$ of the true value, if z is taken at the 0.05 probability level. If, as in Eq. (6.16), the precision limits are designated by the symbol l,

$$l_{0.95,x} = \pm z_{0.05}\sigma \tag{6.21}$$

The relation of the standard deviation and the mean difference between duplicates of Eq. (6.19) permits the calculation of two handy constants (which allow statisticians to make apparently instantaneous judgments about the precision of measurements they had never seen before). If an analyst can say from past experience that the mean difference within pairs of measurements is $\bar{d}$, the 95 per cent confidence range of a single measurement x is $x \pm 1.74\bar{d}$ and the 95 per cent confidence range on the mean $\bar{x}$ of a pair of measurements is $\bar{x} \pm 1.23\bar{d}$. These two constants are arrived at from what we have already learned and Eq. (6.19) as follows:

$$\mu_{0.95} = x \pm 1.96\sigma$$
$$= x \pm (1.96)(0.8862)\bar{d}$$
$$= x \pm 1.737\bar{d} \qquad (6.22)$$

$$\mu_{0.95} = \bar{x} \pm \frac{1.96\sigma}{\sqrt{n}}$$
$$= \bar{x} \pm \frac{1.69\sigma}{\sqrt{2}}$$
$$= \bar{x} \pm \frac{(1.96)(0.8862\bar{d})}{\sqrt{2}}$$
$$= \bar{x} \pm 1.228\bar{d} \qquad (6.23)$$

Similar constants can be calculated for any desired confidence range.

example 6.4. The mean difference in the end point between duplicate ASTM distillations of gasoline in a particular laboratory is 2°F. If a single distillation of a sample results in an end point of 398°F, what are the 95 and the 99 per cent confidence ranges of the end point?

For the 95 per cent range, we can use Eq. (6.22), writing $E.P.$ for end point:

$$E.P._{0.95} = 398° \pm (1.737)(2°)$$
$$= 398° \pm 3.5°F$$

With a 5 per cent chance of error we say the end point is in the range 394.5 to 401.5°F.

For the 99 per cent range, we use the t value from Table 6.1, corresponding to the 0.01 level, and ∞ degrees of freedom, 2.576.

$$E.P._{0.99} = 398° \pm (2.576)(0.8862)(2°)$$
$$= 398° \pm 4.6°$$

With a 1 per cent chance of error we say the end point is in the range 393.4 to 402.6°F.

6.4 Difference between Two Means

6.4.1 The Hypothesis: $\mu_1 = \mu_2$.

This is the last form of the t test that will be described: to test whether the means of two different samples could have come from the same population or from populations with the same means. The efficiency of a particular bubble-cap design has a mean value of 82.7 per cent over a series of tests. Another tray design has a mean efficiency of 81.9 per cent during a similar test. Is the first design significantly better than the second? With the application of a particular corrosion inhibitor the mean corrosion rate for several samples is 0.017 in./year. A second inhibitor under a similar test has a mean corrosion rate of 0.019 in./year. Is there significant difference between these inhibitors?

This last form of the t test is designed to furnish answers to questions of this type. In order to apply the t test to the difference between two means, it is necessary to have an estimate of the standard deviation of the measurements.

6.4.2 The t Test.

The t expression to test the difference between two means is as follows:

$$t = \frac{|\bar{x}_1 - \bar{x}_2|}{\bar{s}(x)\sqrt{1/n_1 + 1/n_2}} \tag{6.24}$$

where $\bar{x}_1$ and $\bar{x}_2$ = two separate means
$\quad n_1$ and n_2 = number of samples of which $\bar{x}_1$ and $\bar{x}_2$ are, respectively, the means
$\quad \bar{s}(x)$ = pooled estimate of standard deviation from both sets of data—a new term

There are available two estimates of the standard deviation, $s(x_1) = \sqrt{\Sigma' x_1^2/(n_1 - 1)}$ and $s(x_2) = \sqrt{\Sigma' x_2^2/(n_2 - 1)}$. One of the bases of this t test is that both of these estimates are estimates of the same thing and therefore may be pooled to give a better estimate. A test, the F test, to check this assumption is discussed in the next chapter, dealing with variances. For the time being let it be sufficient to say that for two sets of analyses run by the same procedure, or for two sets of test measurements made under the same conditions, it is usual to find that the estimated standard deviations are not significantly different from set to set. The F test gives the ratio of the variances (the square of the standard deviations) that can be expected for different values of n_1 and n_2 at different probability levels. More about this later.

The pooled estimate of the standard deviation is calculated from the following equation:

$$\bar{s}(x) = \sqrt{\frac{\Sigma' x_1^2 + \Sigma' x_2^2}{n_1 + n_2 - 2}} \tag{6.25}$$

This relation is equivalent to weighting each estimated variance by the number of degrees of freedom available for its calculation:

$$\bar{s}^2(x) = \frac{(n_1 - 1)s^2(x_1) + (n_2 - 1)s^2(x_2)}{(n_1 - 1) + (n_2 - 1)} \tag{6.26}$$

Equation (6.25) follows from Eq. (6.26) since $(n_1 - 1)s^2(x_1)$ is equal to $\Sigma' x_1^2$, etc.

$\bar{s}(x)$ from Eq. (6.25) or (6.26) is the pooled estimate of the standard deviation of a single measurement. In order to employ the t test we need the standard deviation of the difference between the means. We shall see in the next chapter that the variance of a sum or difference of two measurements is the sum of the variances of the individual measurements. Hence the variance of the difference between two means is the sum of the variances of the means. Since the pooled variance estimate applies to both sets of measurements, the estimated variance of the first mean, $s^2(\bar{x}_1)$, is $\bar{s}^2(x)/n_1$. And the estimated variance of the second mean, $s^2(\bar{x}_2)$, is $\bar{s}^2(x)/n_2$. The variance of the difference between the means is the sum of these two variances, $\bar{s}^2(x)/n_1 + \bar{s}^2(x)/n_2$ or $\bar{s}^2(x)(1/n_1 + 1/n_2)$. The standard deviation of the difference between the means is the square root of this quantity, or the denominator shown in Eq. (6.24).

The application of the t test is similar to that discussed for the other forms of the t equation. The null hypothesis is set up that the two means are equal.

$$H_0; \mu_1 = \mu_2 \tag{6.27}$$

t is calculated from Eq. (6.24) and is compared with the tabulated values at the degrees of freedom equal to $n_1 + n_2 - 2$, or the sum of the total degrees of freedom available for calculating the pooled estimate of the standard deviation. If the calculated t is larger than the tabulated t at the preselected probability or significance level, then we conclude that the null hypothesis is false, the population mean estimated by $\bar{x}_1$ is significantly different from the population mean estimated by $\bar{x}_2$, with the α chance of being wrong.

example 6.5. The following data[1] represent the results of two different powdered-metal flow-rate tests on the same iron powder by the

[1] Private communication, Metals Powder Association, New York.

Metal Powder Association standard procedure. The means for the two runs are 30.0 and 30.8 sec. We wish to establish whether this difference is significant at the 0.05 level, i.e., whether this difference is larger than would ordinarily be expected 95 per cent of the time if the two sets of samples were drawn from populations with the same mean.

Table 6.7 gives the data and the calculation of t from Eq. (6.24).

Discussion. The calculated t value not only exceeds the tabulated 0.05 level but exceeds the 0.01 level. The difference between the means is greater than could be expected by chance 1 time in 100 on the basis of the scatter of the measurements within each test. It is important to remember that while it is the difference between means that is being tested, the scatter of the data as measured by the pooled estimate of the standard deviation plays an important role. When the calculated t

TABLE 6.7

	Time of run, sec.		$(y_1 - 29.0)$ x_1	$(y_2 - 29.0)$ x_2
	Test 1 y_1	Test 2 y_2		
Run 1.........	29.6	31.0	0.6	2.0
Run 2.........	30.6	30.8	1.6	1.8
Run 3.........	29.6	30.8	0.6	1.8
Run 4.........	30.0	31.0	1.0	2.0
Run 5.........	30.0	30.6	1.0	1.6
			$n = 5$	5
			$\Sigma x = 4.8$	9.2
			$\bar{x} = 0.96$	1.84
			$\Sigma x^2 = 5.28$	17.04
			$\Sigma' x^2 = 0.672$	0.112
			$s^2(x) = 0.168$	0.028

$$\bar{s}^2(x) = \frac{0.672 + 0.112}{8} = 0.098$$

$$\bar{s}(x) = 0.313$$

$$t = \frac{|1.84 - 0.96|}{0.313\sqrt{(1/5 + 1/5)}} = 4.442**$$

$$t_{0.01,8} = 3.250$$

Note: The two asterisks indicate significance at the 0.01 level; one indicates significance at the 0.05 level, and three at the 0.001 level.

value does not exceed the critical tabulated t value, we say there is no significant difference between the means. What we are really saying is that with the observed variation of the measurements involved we cannot tell whether the observed difference between the means is due to some real cause or is due to the fluctuation in the measurements. For example, if the data of Table 6.7 were changed slightly by adding 1 sec to one of the times and subtracting 1 sec from another, so that the means were not changed but the estimated standard deviation was increased, the t value could become nonsignificant and the same difference between means previously judged to be real would now have to be assumed to be attributable to the fluctuation of the data. The calculation of t from these revised data is shown in Table 6.8.

In comparing two means, it is not necessary that they be made up of the same number of terms. A mean of 14 measurements made by one method can be compared with the mean of 6 measurements made by a second method. As long as the number of measurements, represented by n_1 and n_2, are correctly applied in the equations and the degrees of freedom corresponding to $n_1 + n_2 - 2$ are used, the resulting t test will be applicable.

TABLE 6.8

	Test 1	Test 2	
Run 1	28.6	32.0	$\hat{s}^2 = 0.898$
Run 2	31.6	29.8	$\hat{s} = 0.9477$
Run 3	29.6	30.8	$t = 0.88/(0.9477)(0.633)$
Run 4	30.0	31.0	$\quad = 1.47$
Run 5	30.0	30.6	
Mean	29.96	30.84	

The t test for the difference between two means is a different test from that for the mean difference within pairs of measurements. Although, when the same number of measurements are involved for both means, the difference between means will be equal to the mean difference, since, if $n_1 = n_2 = n$, $\Sigma x_1/n_1 - \Sigma x_2/n_2 = \Sigma(x_1 - x_2)/n$, the standard deviation within the measurements is different from the standard deviation of the differences within pairs of measurements. We shall deal more fully with the standard deviation within and between measurements in the next chapter. Suffice it here to point out that pairs of measurements may be made over a wide range of conditions and the variation within measurements may be fairly large so that a small difference between means might be deemed insignificant because of a large denominator in the t test of Eq. (6.24). However, the differences within pairs may be very consistent and a t test in the form of Eq. (6.18) might show the mean difference to be highly significant. In order to apply the test to the mean difference, there must be some logical reason for pairing the data. Simply because there are the same number of measurements in two sets is not sufficient reason to pair the data. The calculation of the mean difference may present a simple method of calculating the difference between the means, but the calculation of the standard deviation of the differences of data that have no reason to be paired does not give an estimate of any useful function.

The t test for the difference between means described so far is used to establish whether either mean is larger or smaller than the other. The test can also be applied in a nonsymmetrical manner to determine whether one mean is significantly larger than the other or whether one mean is significantly smaller than the other. Often it is of particular interest to test whether a new process or material is significantly better than the old or whether a modified procedure gives a significantly larger yield—not just whether there is a significant difference between the two. Or what might be of interest is whether a new analytical procedure gives significantly less deviation from the known composition or whether a modified drying process gives significantly less moisture in the product—not simply whether the methods give different results.

The explanation of Sec. 6.2.3 dealing with one-sided t tests for the significance of a single mean applies when dealing with the difference between two means. The hypothesis to be tested is set up to establish a difference between means of one sign only, and only one side of the symmetrical t distribution is employed. The significant values are those corresponding to the $\alpha/2$ levels in the t table; i.e., the 0.05 level for the symmetrical test is the 0.025 level for the unsymmetrical test. Table 6.9 is presented to summarize the comparison between the two-sided and one-sided t tests for both the significance of a single mean and the comparison of two means.

example 6.6. A manufacturer offers a new cutting tool which is advertised to give a greater output. The installation of this tool will require a certain amount of inconvenience in adjusting conditions in the shop, but if it is really better, the supervisor does not want to turn it down. Several runs are made with the new tool, and the mean output is compared with the output of the standard tool. The hypothesis

TABLE 6.9 Comparison of Two-sided and One-sided t Tests

	Significance of single mean		
	Two-sided	One-sided	
To test whether......	$\mu \neq m$	$\mu > m$	$\mu < m$
Hypothesis, H_0......	$\mu = m$	$\mu \leq m$	$\mu \geq m$
t test...............	$|\bar{x} - m|/s(\bar{x})$	$(\bar{x} - m)/s(\bar{x})$	$(m - \bar{x})/s(\bar{x})$
Degrees of freedom...	$n - 1$	$n - 1$	$n - 1$
Error of type I......	False rejection of equal μ	False rejection of lesser or equal μ	False rejection of greater or equal μ
Chance of error......	α	$\alpha/2$	$\alpha/2$
	Comparison of two means		
	Two-sided	One-sided	
To test whether......	$\mu_1 \neq \mu_2$	$\mu_1 > \mu_2$	$\mu_1 < \mu_2$
Hypothesis, H_0......	$\mu_1 = \mu_2$	$\mu_1 \leq \mu_2$	$\mu_1 \geq \mu_2$
t test...............	$\|\bar{x}_1 - \bar{x}_2\|/s(\Delta)$	$(\bar{x}_1 - \bar{x}_2)/s(\Delta)$	$(\bar{x}_2 - \bar{x}_1)/s(\Delta)$
Degrees of freedom...	$n_1 + n_2 - 2$	$n_1 + n_2 - 2$	$n_1 + n_2 - 2$
Error of type I......	False rejection of equal means	False rejection of μ, equal or less	False rejection of μ, equal or greater
Chance of error......	α	$\alpha/2$	$\alpha/2$

$$s(\Delta) = \bar{s}(x) \sqrt{\frac{1}{n_1} + \frac{1}{n_2}}$$

is set up that the new tool is equal or inferior to the standard, and because of the inconvenience in retooling, we are willing to risk only a 0.01 chance of a false rejection of this hypothesis. In other words, there will be less than 1 per cent chance of accepting the new tool when it is only equal to or less than the old tool.

The mean output for the new tool is 88.0 units for 5 trials, and the mean output for the standard tool is 85.6 units for 5 trials. The pooled estimate of the standard deviation from the 10 trials is 3.52 units.

The calculation of t is $(88.0 - 85.6)/(3.52)(\sqrt{1/5 + 1/5}) = 1.08$. The critical t value for 8 degrees of freedom at the 0.02 level is 2.896; therefore we accept the hypothesis and reject the new tool. The t

value of 1.08 for 8 degrees of freedom corresponds to a significance level of between 0.3 and 0.4, and for a one-sided test this is equivalent to a level of between 0.15 and 0.2. Therefore, even though the mean output of the new tool exceeded the output of the standard tool by 2.8 per cent in the test, there is between 15 per cent and 20 per cent chance that the outputs for a large number of runs would in fact be equal or even less than for the standard tool.

It is of particular importance in applying one-sided t tests to set the hypothesis that is to be tested before the data are studied. Reference to Table 6.9 shows that if you wish to test whether the true mean is larger than some value, the hypothesis is set that the true mean is equal or smaller than that value. The t test permits only a quantitative rejection of an hypothesis. If the hypothesis were true, a particular frequency of deviations would occur. If an event occurs in an area of very low frequency, we say that the initial hypothesis was probably not true. The t test gives us a quantitative rule for rejecting the hypothesis. If we want to test whether a mean is greater than some given value, since the t test only gives a quantitative criteria for rejecting a hypothesis, we set the hypothesis that the true mean is less than or equal to some given value and the rejecting of this hypothesis satisfies our test. The alternative occurs when we wish to test whether a mean is less than some given value.

If we wish to test whether the true mean purity of a product is greater than 99.44 per cent, we set the hypothesis: $H_0: \mu \leq 99.44$ and calculate t from the equation $t = (\bar{x} - 99.44)/s(\bar{x})$, where $\bar{x}$ and $s(\bar{x})$ are calculated from the observed data. Only values of $\bar{x}$ greater than 99.44 can give significant values (i.e., values that will permit the rejection of the hypothesis) of t. Values of $\bar{x}$ equal to or less than 99.44 will give t values of zero or negative values, which can only result in acceptance of the hypothesis. The symmetrically alternative condition exists if the hypothesis is that the true mean is equal to or larger than 99.44. In this latter case only $\bar{x}$ values less than 99.44 can give significant t values and result in rejection of the second hypothesis.

If the value of the sample mean is known before the one-sided t test hypothesis is established, the test is no longer unbiased. One whole area of possible rejection of the hypothesis is eliminated. If a sample mean of 97.0 is measured and known, it is not possible to make a one-sided t test to see if the true mean is, say 98.0, or any other value larger than 97.0, because this would result in a negative value of t, as explained above. If a one-sided t test is to be made, the test and the hypothesis must be decided upon before the data are available. If the data are known before the test is established, use a two-sided t test, or simply report the confidence range of the true value.

6.5 Limiting Type II Errors

The application of the t test might be summed up as follows:
1. A sample or samples are obtained.
2. A hypothesis is set up regarding the value of the mean or means of the populations from which the samples were drawn.
3. The standard deviation of the population is estimated from the samples or is obtained from some other source.
4. t is calculated on the basis of the estimated standard deviation and the particular hypothesis to be tested in the form of one of the equations discussed.
5. The calculated t is compared with the tabulated values of t at the degrees of freedom involved in the estimate of the standard deviation. If the calculated t exceeds the tabulated t at some probability level, α, then there is only an α chance of being in error if the hypothesis is rejected.

As was pointed out in Sec. 6.1.4, when the calculated t is less than the tabulated value and the hypothesis is accepted, there is always the chance that the hypothesis was in fact false and has been mistakenly accepted. This kind of error was called type II, and the probability of its happening was designated β. In any particular test, the level of α can be arbitrarily set and the chance of a type I error, a false rejection of the hypothesis, can be kept as small as desired. However, under a fixed set of experimental conditions as the chance of the false rejection of the hypothesis is decreased, the chance of its false acceptance, the type II error, is increased.

Consider the following parallel situation. A large engineering concern gives its applicants for jobs a qualifying examination. With a set examination, two types of incorrect judgment can be made. A well qualified engineer might fail the examination because of nervousness, or because the examination is poorly designed for his type of experience, or because of some other reason which was not foreseen when the test was drawn up. The other misjudgment would occur if an unqualified man passed the examination by guessing the answers, or if the test was not severe enough for the job to be filled. In order to decrease the errors of the first kind, and not eliminate qualified applicants, the examination could be made easier, but this would result in more errors of the second kind—accepting men who were not qualified. If the alternative step was taken and the examination made more difficult to cut down errors of the second kind, then more errors of the first kind would be made. One way out of this problem would be to give several tests of a certain standard which would set the chance of eliminating the qualified applicant at some fixed level. The increase in the number of tests would decrease the chance of an unqualified applicant getting by.

A somewhat similar procedure is used in testing quantitative data at some fixed significance level of the t test. For a specific t test the chance of an error of type II can be decreased at a fixed probability of an error of type I by increasing the sample size. Before discussing the application of the t test to type II errors, a word will be said about the power of a test.

6.5.1 Power of a Test. The power of a statistical test is defined as the probability of rejecting the hypothesis when it is false, $1 - \beta$ (β being the probability of a false acceptance). The power, then, is one minus the probability of a type II error, and the factors causing a decrease in the value of β will increase the power of the test correspondingly. This relation might be best illustrated graphically.

Refer to Fig. 6.1. (a) illustrates a simple null hypothesis that μ, the true mean of the population, is equal to some value m_0. An α risk of a false rejection of the hypothesis sets upper and lower limits to the acceptance; i.e., if the observed $\bar{x}$ is less than m_1 or greater than m_2 the hypothesis that $\mu = m_0$ will be rejected. The illustration shows that with a normal distribution there is only an α probability of obtaining a value outside the limits set by m_1 and m_2 if the mean is actually m_0. The values of m_1 and m_2 are established from the normal deviate corresponding to α, and the standard deviation of $\bar{x}$, $\sigma/\sqrt{n}$.

Figure 6.1(b) illustrates the situation if the mean is actually at some value m, larger than m_0. (A parallel situation would apply if m was smaller than m_0 at the other end of the distribution.) If the population mean is actually m, there is a probability, indicated by β, of obtaining a value less than m_2, which would result in the acceptance of the hypothesis that the mean was m_0. This is the probability of a type II error. (If m was close enough to m_0 so that its distribution extended beyond m_1, the value of β would only correspond to the area under the curve between m_2 and m_1.)

Figure 6.1(c) shows the two distributions superimposed. The value of β is seen to be affected by several quantities: The difference between m_0 and m, the value of α, and the value of $\sigma/\sqrt{n}$. The difference between m_0 and m, here labelled d, is the difference that it is necessary to establish by the particular test involved. If the carbon in the steel is supposed to be 0.10, it is probably acceptable if it is 0.11. It might even be acceptable at 0.15. But you would probably want to be "certain," $1 - \beta$ certain, that it wasn't 0.20 when you said it was 0.10. β is the risk of accepting the hypothesis that the carbon is 0.10 when it is really 0.20, and d in this case is $0.20 - 0.10 = 0.10$. It is obvious from Fig. 6.1(c), that other things being the same, the larger the acceptable difference d, the smaller is the risk β, and hence the greater the power of the test, $1 - \beta$.

β is also dependent on the value of α. If the other quantities are fixed, the smaller the risk of rejecting material meeting the hypothesis, the larger the risk of accepting material differing as much as d from the standard value m_0. And the larger the value of α, the smaller the value of β.

Finally, β is dependent on the standard deviation of the measured mean,

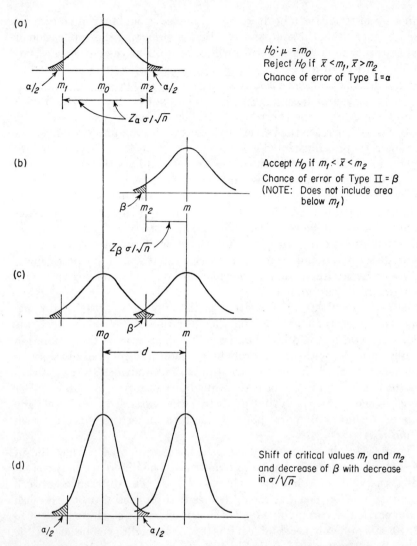

Fig. 6.1 Relation of α and β probabilities of error.

$\sigma(\bar{x}) = \sigma(x)/\sqrt{n}$. The critical values m_1 and m_2 are set by α and the standard deviation of the mean $\sigma/\sqrt{n}$, being respectively $\pm z_\alpha \, \sigma/\sqrt{n}$ units from m_0, where z_α is the standard deviate corresponding to the significance level α, and read either from Table 3.12, the table of standard deviates, or from the t table, Table 6.1, using the values corresponding to infinite degrees of freedom. For example, when α is 0.05, the critical values are ± 1.96 standard deviations from m_0; when α is 0.01, the critical values are ± 2.58 standard deviations from m_0 for the general two-sided

THE t TEST

test. β on the other hand represents probability of a one-sided deviation of $-z_\beta\sigma/\sqrt{n}$ (if m is larger than m_0 or $+z_\beta\sigma/\sqrt{n}$ if m is smaller than m_0) measured in standard deviation units, or z_β standard deviations from the maximum acceptable value of the mean.

The following equations will show the relations between the different factors for the case illustrated in Fig. 6.1.

$$z_\alpha\sigma/\sqrt{n} = m_2 - m_0 \tag{6.28}$$

$$z_\beta\sigma/\sqrt{n} = m - m_2 \tag{6.29}$$

$$d = m - m_0 \tag{6.30}$$

$$z_\beta = \frac{d}{\sigma/\sqrt{n}} - z_\alpha \tag{6.31}$$

$$\sqrt{n} = (z_\alpha + z_\beta)\sigma/d \tag{6.32}$$

Equation (6.31) shows that if d, and σ, and the sample size n, and α (hence z_α) are fixed, then z_β, (and hence β) is fixed; remembering that β represents a one-tailed probability, and values corresponding to half the significance values should be read from Tables 3.12, or 6.1. For example, if z_β equals 1.96, β equals 0.025, and if z_β equals 2.58, β equals 0.005.

Equation (6.31) also shows that as the standard deviation of the mean becomes smaller; i.e., as n increases, z_β also increases, and β therefore decreases. This is illustrated in Fig. 6.1(d) where the normal distributions are drawn with smaller variances but at the same means as the curves for the upper parts of the figure.

Equation (6.32) shows the relation between the sample size and other factors. When the α and β risks are set and when the difference d that it is necessary to detect is fixed, with a process of known variance, the sample size n required to satisfy the risks established, is set and may be calculated from Eq. (6.32).

Note that these analytic expressions are only true in this form when the α risk applies to a two-sided test, and when the β risk does not include area beyond the far limit of the α test. Suitable adjustments can be readily made in these expressions for one-tailed α risks, and for cases where the β risk does include areas in both tails of the α test.

example 6.7. A sample of four analyses is taken on an acetylene process to determine whether a yield of 8.0 per cent is obtained. The standard deviation is 0.26, and a significance level of 0.01 is selected for the test; i.e., a one in a hundred risk is to be taken of rejecting the hypothesis when it is true. What is the power of this test to distinguish a process which had a yield as high as 8.5 per cent? What sample size is necessary if the power of the test is to be 99 per cent?

In the first part we are asking for the value of $1 - \beta$ for the sample of four; and in the second part we are asking for the sample size n which will give a value of $1 - \beta$ equal to 0.99, or a β equal to 0.01.

Solution (a):

$$m_0 = 8.0$$
$$\sigma = 0.26$$
$$n = 4$$
$$m = 8.5$$
$$m - m_0 = d = 0.5$$
$$\alpha = 0.01, z_\alpha = 2.576 \text{ (from Table 3.12)}$$
$$m_2 = m_0 + z_\alpha \sigma/\sqrt{n} = 8.0 + (2.576)(0.26/2)$$
$$= 8.334 \text{ [from Eq. (6.28)]}$$
$$z_\beta = (m - m_2)/(\sigma/\sqrt{n}) = (8.5\text{-}8.334)/(0.26/2)$$
$$= 1.28 \text{ [from Eq. (6.29)]}$$
$$\beta = 0.1003 \text{ (from Table 3.12)}$$
$$1 - \beta = 0.8997, \text{ power} = 90 \text{ per cent}$$

Solution (b):

$$1 - \beta = 0.99, \beta = 0.01, z_\beta = 2.326 \text{ (from Table 3.12)}$$
$$\sqrt{n} = (z_\alpha + z_\beta)\sigma/d = (2.576 + 2.326)(0.26/0.5)$$
$$= 2.56 \text{ [from Table (6.32)]}$$
$$n = 6.5 \sim \text{i.e., 7 samples required.}$$

6.5.2 Application to the t Test. The remarks and illustration just preceding apply to the normal distribution. The values of z_α and z_β are taken as the standard deviates of the normal distribution at the corresponding values of α and β. The more usual situation is to apply the t distribution in which the deviates t_α, and t_β are a function of not only α and β, but also the sample size or number of degrees of freedom. Under ordinary circumstances it would be necessary to proceed by trial and error; however, G. P. Sillitto[1] has published tables, reproduced here with the kind permission of their originator, which give the minimum sample size to be used with t test at fixed values of α and β. The balance of this chapter is a discussion of the use of these tables.

There are two tables, one dealing with the significant difference of the

---

[1] G. P. Sillitto, Research Supplement, *J. Roy. Statis. Soc.* **11**:520 (1948).

mean from some specified value, and the other dealing with the significant difference between two means. When this difference is expressed in units of the estimated standard deviation of the measurements, the tables give the minimum number of measurements necessary to establish this difference with limits set for the probabilities of both types of errors. The tables give the sample size for both symmetrical and unsymmetrical (one-sided) t tests. The tables set the minimum sample size. The t test is carried out in the manner discussed in the earlier sections of this chapter. If the calculated t exceeds the tabulated t at the α probability level, then the initial hypothesis can be rejected with this chance of an error of type I. If the calculated t is less than the tabulated t at the α probability level, then the initial hypothesis can be accepted with β chance of an error of type II. The number of observations needed for the t test for the significance of a mean is given in Table 6.10. The number of observations needed for the t test for the significant difference between two means is given in Table 6.11.

The use of these tables will be illustrated in a series of examples dealing with the different possible symmetrical and unsymmetrical tests. In each case the estimated standard deviation will be assumed to be known. It is not unusual in practice for such an estimate to be available from previous experience or from data on hand which had never been subjected to such a calculation. If no estimate of the standard deviation is available, it may be necessary to make some preliminary measurements in order to obtain such an estimate. After the data have been collected according to the sample size set by Table 6.10 or 6.11, the estimated standard deviation, needed for the subsequent t test, should be compared with the preliminary estimated value to make certain the sample size has been correctly selected.

example 6.8. Sample size for significance of a mean using the symmetrical test (Table 6.10).

A process using anthracite coal has a specification of 8 per cent volatile matter in the coal. Before accepting new shipments of coal, analyses are run to make certain the material meets the volatile matter (V.M.) specification. If the V.M. is higher than the specification, it will require an adjustment of the oxygen feed rate or there will be too much uncombusted carbon on the effluent gases. If the V.M. is too low, it causes difficulties in the ignition rate of the fuel. The t test for the mean value of the volatile matter is therefore a symmetrical test. The acceptable mean range around the specified value of 8 per cent is ± 0.25 per cent. The standard deviation of the volatile matter from previous shipments is known to be 0.177 per cent.

Since this is the buyer running the tests, he takes the stand that he can risk a relatively high number of rejected good shipments and sets the α level at 0.10. On the other hand, he wants to set a fairly low probability level for accepting shipments that do not meet specifications and selects a β value of 0.01.

The minimum deviation between the mean of the samples and the specified value that he wants to determine, in standard-deviation units, is

$$\frac{|\bar{x} - m|}{s(x)} = \frac{0.25}{0.177} = 1.41$$

It is desired to observe this difference when it exists at least 99 times out of 100 and to say the difference exists, when in fact it does not, not more than 1 time in 10. From Table 6.10, at the symmetrical α level of 0.10 and the β level of 0.01, to find a difference of 1.4 standard deviation units requires a minimum sample size of 10. Note that if a sample of 5 is used and the t test is accepted at the 0.10 level, there is a 20 per cent chance of accepting material that has a deviation greater than that specified.

example 6.9. Sample size for significance of a mean using the unsymmetrical test (Table 6.10).

A user of coal has a specification for ash 6 per cent. In sampling shipments, he sets a maximum mean value of 6.75 per cent. He does not care if the ash content is less than control value, he is only interested in rejecting shipments that exceed his specifications. In this case he applies the unsymmetrical, or one-sided, t test to the mean of his samples.

Suppose that the supplier agrees to pay for half the cost of inspecting the shipments if control limits are set so that the probability of rejecting a good batch is the same as the probability of accepting a poor shipment. The buyer wishes to set the β level, that of accepting a bad shipment, at the 0.01 level. Therefore the unsymmetrical α level is also set at the 0.01 level. The standard deviation for ash analyses is about 0.265 per cent. The minimum deviation above the ash specification of 6 per cent that is to be detected at the two specified probability levels in units of the standard deviation is

$$\frac{\bar{x} - m}{s(x)} = \frac{0.75}{0.265} = 2.83$$

Table 6.10 indicated that with a difference of 2.83 standard deviation units, between six and seven determinations are required to detect a mean which exceeds the specified value by 0.75 ninety-nine times out of a hundred and to assume this difference exists when it does not only one time in a hundred.

example 6.10. Sample size for a significant difference between two means, using the symmetrical test (Table 6.11).

Two metal coatings are to be compared in a standard bending test in which the coated-metal sample piece is flexed until cracks appear in the coating. The test is not particularly sensitive and has a standard deviation of about 10 per cent of the flex life of a reasonably good coating. If we desire to detect a difference between the mean flex life of two coatings if it is more than 10 per cent of the observed values, we shall need the number of tests corresponding to $|\bar{x}_1 - \bar{x}_2|/\bar{s}(x)$ equal to 1.0. Since we are interested in detecting a difference between the mean values of either sign, we shall use the symmetrical t test for the difference between means. At the 0.05 level for both types of errors, assumption of a difference when none exists, the α probability, and the assumption of no difference when one does exist, the β probability, the number of samples of *each* coating to be tested is 27. Note that if only nine samples of each coating are tested at the 0.05 level of α, there is a 50 per cent chance of accepting the two coatings as equal when there is a difference between them, the β level.

The use of per cent deviation in both the standard deviation and the allowable difference between means cannot be applied to the actual calculation of t. For this purpose the numerical value of the difference between the means and the numerical value of the standard deviation are employed.

example 6.11. Sample size for a significant difference between two means using the unsymmetrical test (Table 6.11).

When an organization is using a particular commodity and a vendor offers a substitute which is claimed to be better, a comparison of the two materials, from the buyer's standpoint, need be made only to see if the new material is significantly better than the one in use. This is the unsymmetrical test of the difference between means.

A test is to be made on an alternative cement mix to see whether the yield load after 7 days of aging is greater than for the standard mix. The standard deviation for a test of this type is about 22.8 lb, after Youden.[1] If we wish to detect a difference between means for the two mixes of at least 50 lb, we shall need the number of samples correspond-

---

[1] W. J. Youden, "Statistical Methods for Chemists," John Wiley & Sons, Inc., New York, 1951.

TABLE 6.10 The Numbers of Observations Needed in a t Test of the Significance of a Mean, in Order to Control the Probabilities of Errors of Types I and II at α and β, Respectively†

$\|\bar{x}-m\|/s(x)$ $(\bar{x}-m)/s(x)$	Level of t test = 0.10 Symmetrical test: $\alpha = 0.10$ Unsymmetrical test: $\alpha = 0.05$					Level of t test = 0.05 Symmetrical test: $\alpha = 0.05$ Unsymmetrical test: $\alpha = 0.025$					Level of t test = 0.02 Symmetrical test: $\alpha = 0.02$ Unsymmetrical test: $\alpha = 0.01$					Level of t test = 0.01 Symmetrical test: $\alpha = 0.01$ Unsymmetrical test: $\alpha = 0.005$				
	β:0.01	0.05	0.10	0.20	0.50	β:0.01	0.05	0.10	0.20	0.50	β:0.01	0.05	0.10	0.20	0.50	β:0.01	0.05	0.10	0.20	0.50
0.05																				
0.10																				
0.15					122															
0.20					70					99					139					
0.25			139	101	45				128	64					90					110
0.30		122	97	71	32			119	90	45				115	63				134	78
0.35		90	72	52	24		109	88	67	34			109	85	47			125	99	58
0.40	101	70	55	40	19	117	84	68	51	26		101	85	66	37		115	97	77	45
0.45	80	53	44	33	15	93	67	54	41	21	110	81	68	53	30		92	77	62	37
0.50	65	45	36	27	13	76	54	44	34	18	90	66	55	43	25	100	75	63	51	30
0.55	54	38	30	22	11	63	45	37	28	15	75	55	46	36	21	83	63	53	42	26
0.60	46	32	26	19	9	53	38	32	24	13	63	47	39	31	18	71	53	45	36	22
0.65	39	28	22	17	8	46	33	27	21	12	55	41	34	27	16	61	46	39	31	20
0.70	34	24	19	15	8	40	29	24	19	10	47	35	30	24	14	53	40	34	28	17
0.75	30	21	17	13	7	35	26	21	16	9	42	31	27	21	13	47	36	30	25	16
0.80	27	19	15	12	6	31	22	19	15	9	37	28	24	19	12	41	32	27	22	14
0.85	24	17	14	11	6	28	21	17	13	8	33	25	21	17	11	37	29	24	20	13
0.90	21	15	13	10	5	25	19	16	12	7	29	23	19	16	10	34	26	22	18	12
0.95	19	14	11	9	5	23	17	14	11	7	27	21	18	14	9	31	24	20	17	11
1.00	18	13	11	8	5	21	16	13	10	6	25	19	16	13	9	28	22	19	16	10

THE t TEST

† Reprinted from an article by G. P. Sillitto in *J. Roy. Statis. Soc.*, Research 1 (1948), by permission of the author and publisher.

TABLE 6.11 The Numbers of Observations Needed in a *t* Test of the Significance of the Difference of Two Means in Order to Control the Probabilities of Errors of Types I and II at α and β, Respectively

The number tabulated is the number of observations contributing to each mean.

$|\bar{x}_1-\bar{x}_2|/s(x)$ $(\bar{x}_1-\bar{x}_2)/s(\bar{x})$	Level of *t* test = 0.10 Symmetrical test: α = 0.10 Unsymmetrical test: α = 0.05					Level of *t* test = 0.05 Symmetrical test: α = 0.05 Unsymmetrical test: α = 0.025					Level of *t* test = 0.02 Symmetrical test: α = 0.02 Unsymmetrical test: α = 0.01					Level of *t* test = 0.01 Symmetrical test: α = 0.01 Unsymmetrical test: α = 0.005				
	β:0.01	0.05	0.10	0.20	0.50	β:0.01	0.05	0.10	0.20	0.50	β:0.01	0.05	0.10	0.20	0.50	β:0.01	0.05	0.10	0.20	0.50
0.05																				
0.10																				
0.15																				
0.20					137															
0.25					88					124										
0.30					61					87					123					
0.35				102	45					64					90					110
0.40			108	78	35				100	50					70					85
0.45		108	86	62	28			105	79	39				101	55				118	68
0.50		88	70	51	23		106	86	64	32			106	82	45				96	55
0.55	112	73	58	42	19		87	71	53	27		106	88	68	38			101	79	46
0.60	89	61	49	36	16	104	74	60	45	23		90	74	58	32		101	85	67	39
0.65	76	52	42	30	14	88	63	51	39	20	104	77	64	49	27		87	73	57	34
0.70	66	45	36	26	12	76	55	44	34	17	90	66	55	43	24	100	75	63	50	29
0.75	57	40	32	23	11	67	48	39	29	15	79	58	48	38	21	88	66	55	44	26
0.80	50	35	28	21	10	59	42	34	26	14	70	51	43	33	19	77	58	49	39	23
0.85	45	31	25	18	9	52	37	31	23	12	62	46	38	30	17	69	51	43	35	21
0.90	40	28	22	16	8	47	34	27	21	11	55	41	34	27	15	62	46	39	31	19
0.95	36	25	20	15	7	42	30	25	19	10	50	37	31	24	14	55	42	35	28	17
1.00	33	23	18	14	7	38	27	23	17	9	45	33	28	22	13	50	38	32	26	15

144 APPLIED STATISTICS FOR ENGINEERS

THE t TEST

† Reprinted from an article by G. P. Sillitto in *J. Roy. Statis. Soc.*, Research 1 (1948), by permission of the author and publisher.

ing to the difference:

$$\frac{\bar{x}_2 - \bar{x}_1}{\bar{s}(x)} = \frac{50}{22.8} = 2.19$$

where x_2 refers to the new mix and x_1 to the standard mix.

The hypothesis to be tested (see Table 6.9) is that $\mu_2 \leq \mu_1$, that the new mix is only equal to or less than the standard. We are most anxious not to reject the new mix if it is superior; i.e., we do not want to falsely accept the hypothesis. We therefore set the β level at 0.01. We are willing to take a reasonable chance of accepting the new mix when it is not superior, i.e., falsely rejecting the hypothesis. We therefore set the α level at 0.05. From Table 6.11, at these probability levels, for the unsymmetrical test, we require between eight and nine samples of *each* mix for comparison.

The sample sizes given in Tables 6.10 and 6.11 are the minimum samples required for a definitive test at the probability levels selected, assuming the corresponding critical value of t will be obtained. If the difference between means, or between the mean and some theoretical value, is highly significant, it is possible to establish this fact with less samples than the number indicated in the tables. It is often easiest to perform the total number of tests before making the t calculation. On the other hand, it is sometimes possible to make the t calculation after each set of analyses has been obtained, and when a significant t is obtained, to terminate the test. A discussion of this type of sequential testing is given in Chap. 10.

6.6 Problems

1. Select 10 cards at random from a 52-card deck (replacing each card before selecting the next one) and determine whether the mean of your sample is significantly different from the mean of a perfect deck, 7.0.

2. The following data represent sulfur analyses of an oil specified to have not more than 3 per cent sulfur. Do the data, at the 0.01 level, substantiate the claim?

Data

% S = 2.6, 2.4, 2.8, 2.9, 3.0, 2.9, 3.2, 2.6, 3.1, 2.9, 3.2, 2.8
$t = 1.87$, $P(t) > 0.05$

3. The following data are 10 determinations of the molecular weight of sulfur. Does the mean of these determinations differ significantly from the accepted molecular weight of 32.066?

Data

32.05, 32.18, 31.92, 31.95, 32.02, 31.98, 32.04, 32.11, 32.06, 32.08
$t = 1.11$, $P(t) \sim 0.3$

4. Ten samples of low-grade gasoline were divided in half, one-half being treated with a potential upgrading chemical, the other half being untreated. The octane number was run on each half of the 10 samples. Is there evidence from these data of any effect of the added chemical?

Data

Sample no.	Octane number	
	Untreated	Treated
1	63	65
2	68	68
3	68	70
4	60	63
5	65	64
6	60	62
7	72	73
8	75	75
9	73	72
10	64	66

$t = 2.23$, $t_{0.05,9} = 2.26$

5. Assume that the data of example 4 are duplicate determinations. From this series of duplicates, what would be the 95 per cent confidence range on a single result of 67 octane-number run on an entirely independent sample? 64.6 to 69.4.

6. The following data represent inspections of shipments received from two different vendors. Is there a significant difference between the means of the two shipments?

Data

Vendor 1	Vendor 2
75.0	79.0
77.0	77.5
84.0	74.0
84.5	71.5
74.5	79.0
74.0	
85.0	

$t = 1.11$, $P(t) = 0.25$

7. On the basis of the following data, test whether the mean inspections of the second shipment are significantly higher than the mean inspections of the first shipment.

Data

First shipment.......... 76.0, 77.5, 77.0, 75.5, 75.0

Second shipment........ 80.0, 76.0, 80.0, 80.5, 78.5, 79.0, 80.0

$$t = 3.67, P(t) = 0.005$$

8. How many tests need be run to establish a difference of 0.45 between the mean and a theoretical value, if the standard deviation is 0.55 and the acceptable risks are $\alpha = 0.05$ and $\beta = 0.02$?

chapter seven
ANALYSIS OF VARIANCE

The variance is defined as the mean squared deviation from the mean

$$\sigma^2 = \frac{\Sigma(x - \mu)^2}{n} \tag{7.1}$$

In practice μ, the true mean, is usually not known, but is only estimated from the sample mean $\bar{x} = \Sigma x/n$. The variance must therefore be estimated from the deviations from $\bar{x}$ instead of from μ. To counteract the bias that tends to minimize the sum of the squares of deviations when they are calculated from $\bar{x}$ instead of from μ, the variance is estimated by dividing by $n - 1$ instead of by n. The estimated variance $s^2(x)$ is therefore calculated as follows:

$$s^2(x) = \frac{\Sigma(x - \bar{x})^2}{n - 1} \tag{7.2}$$

The sum of squares of deviations from the mean is usually referred to for simplicity as *the sum of squares*, and $\Sigma(x - \bar{x})^2$ is abbreviated as $\Sigma' x^2$. The term $n - 1$ corresponds to the number of degrees of freedom involved in the calculation of $s^2(x)$. The estimate of the variance is the sum of squares divided by the degrees of freedom:

$$s^2(x) = \frac{\Sigma' x^2}{\nu} \tag{7.3}$$

where ν is the degrees of freedom.

The standard deviation, the square root of the variance, which has been used to set confidence limits to a mean or a single determination, has the same units as the measurement under investigation and is therefore a more familiar function for measuring the scatter of the data. However, it is the variance that is amenable to mathematical analysis. If a process has a number of factors contributing to the variance of the final measurement, this total variance is equal to the sum of the variances of the individual factors when these are statistically independent. The standard deviation of the final measurement is *not* the sum of the standard deviations of the individual factors. If the precision, and hence the standard deviation, of a single measurement is desired from a process in which several factors are contributing to the variation of the measurement, it is necessary to analyze the total variance for its specific components and then to calculate the desired standard deviation. The variance and the analysis of variance are therefore very important tools in the statistical interpretation of data.

The calculation of the variance can be made from any of the formulas, (6.2) to (6.6), for the standard deviation if the square-root sign is removed. The simplest method of calculation is to compute $\Sigma x^2 - \bar{x}\Sigma x$ for $\Sigma' x^2$, the sum of squares (of deviation), and then to divide by $n - 1$. Σx^2 and Σx are obtained in one series of operations on a desk computer, and $\bar{x}$ usually required for other purposes, is easily obtained from Σx and n.

7.1 Propagation of Variance

7.1.1 Variance of a Sum.

It can be shown,[1] although we shall not prove it, that the variance of a sum of independently varying functions is equal to the sum of the separate variances. If $X = x_1 + x_2 + x_3 + \cdots$, then

$$\sigma^2(X) = \sigma^2(x_1) + \sigma^2(x_2) + \sigma^2(x_3) + \cdots \tag{7.4}$$

providing x_1, x_2, x_3 vary independently. We have indicated that the confidence range of a single measurement, if normally distributed, is given by the expression $x \pm t_{\alpha,\nu} s(x)$, where α refers to the probability of the true

---

[1] O. L. Davies (ed.), "Statistical Methods in Research and Production," 2d ed., Oliver & Boyd, Ltd., Edinburgh and London, 1954, p. 46.

ANALYSIS OF VARIANCE

measurement being outside this range, and ν, the degrees of freedom, are those associated with the estimate of the standard deviation. The term $\pm t_{\alpha,\nu} s(x)$ was referred to as the precision limits of the measurement. If the standard deviation, or variance, is known, then the degrees of freedom are infinite and t corresponds to z of the normal distribution.

In Eqs. (6.16) and (6.21) the precision has been defined as $l(x) = \pm z \cdot \sigma(x)$ or $\pm t \cdot s(x)$, depending on whether the standard deviation was known or was estimated from the data. If each of the terms of Eq. (7.4) is multiplied by z^2, or if $s^2(x)$ is substituted for $\sigma^2(x)$, each term can be multiplied by t^2, and Eqs. (6.16) and (7.4) can be combined to give an expression for the precision of a measurement in terms of the precision of the variable involved in the statement. In other words, if

$$X = x_1 + x_2 + x_3 + \cdots$$

then

$$l(X) = \sqrt{\Sigma [l(x_i)]^2} \tag{7.4.1}$$

We have said that the variance of a measurement made up of a sum of several separate measurements is equal to the sum of the individual variances. If we know or can measure or have an estimate of the standard deviations or precision limits of the individual measurements, we can calculate the variance, and hence the standard deviation and precision limits, of the total measurement.

TABLE 7.1

Flow stream	Flow rate x_i	$\sigma(x_i)$	Variance $\sigma^2(x_i)$	Precision $\pm (1.96) \cdot \sigma(x_i)$	$\Sigma [l(x_i)]^2$
Fresh feed.....	1,000	50	2,500	±98	9,604
Recycle........	4,000	200	40,000	±392	153,644
Bleeds.........	100	5	25	±9.8	96.04
Total........	5,100	...	42,525	....	163,344.04

Variance of the total flow = total variance = 42,525
Standard deviation of total flow = $\sqrt{42,525}$ = 206.2
Precision of the total flow = (1.96)(206.2) = ±404.2
Precision by Eq. (7.4.1) = $\sqrt{163,344.04}$ = ±404.2

For example, if the total gas stream to a unit is made up of 1,000 scfh of fresh feed, 4,000 scfh of recycle, and 100 scfh of an instrument bleed stream, and if the standard deviation of each measurement is 5 per cent of the individual flow, then the 95 per cent precision limits of each stream, calculated as $\pm 1.96\sigma$ (1.96 being the 0.05 value of t at infinite degrees of freedom) are: fresh feed, ± 98 scfh, recycle, ± 392 scfh, and bleed stream, ± 9.8 scfh. The precision of the total flow can be calculated either by first calculating the variance of the total flow, following Eq. (7.4), and then obtaining the precision from the total variance according to Eq. (6.21); or the precision of the total flow can be obtained directly from the precisions of the individual measurements, following Eq. (7.4.1). Both of these calculations are illustrated in Table 7.1.

7.1.2 Variance of a Difference. The variance of a difference of two quantities is equal to the variance of their sum, and therefore the sum of their variances. If $X = x_1 - x_2 - x_3 - \cdots$, then Eq. (7.4) still holds and

$$\sigma^2(X) = \sigma^2(x_1) + \sigma^2(x_2) + \sigma^2(x_3) + \cdots \qquad (7.5)$$

In general, if $X = x_1 \pm x_2 \pm x_3 \pm \cdots$,

$$\sigma^2(X) = \sigma^2(x_1) + \sigma^2(x_2) + \sigma^2(x_3) + \cdots \qquad (7.6)$$

An example of this relation has already been presented without explanation in the t test for the difference between two means. We combined the two estimates of the standard deviation to give a pooled estimate $\bar{s}(x)$, or a pooled estimated variance of $\bar{s}^2(x)$. The estimated variances of the separate means were $\bar{s}^2(x)/n_1$ and $\bar{s}^2(x)/n_2$. For the denominator of Eq. (6.24) we added these quantities (under the square-root sign) although we were testing the *difference* between two means; i.e., the variance of the difference between two means is equal to the sum of their variances.

The propagation of variance equation as given in Eq. (7.6), or as will be given in the next several paragraphs, applies to the variance of a measurement that is composed of a group of other measurements which are independently determined and which have independent variances. It does not apply to a general algebraic equation. If $X = x_1 - x_2$. then $x_1 = X + x_2$, but it does not follow that both $\sigma^2(X) = \sigma^2(x_1) + \sigma^2(x_2)$ and $\sigma^2(x_1) = \sigma^2(X) + \sigma^2(x_2)$ are true. (Which variance equation is correct depends on which variable is actually calculated from the other two.) If all three terms are separately measured, each with its own independent variance, there is no basis for writing a variance equation since all the variances are independent.

example 7.1. Suppose in a process we measured the total feed to a reactor and a product stream of 90 per cent and vented the other 10 per cent unmeasured. If the standard deviations of the two measured streams were 5 per cent of the flow quantities, what would be the 95 per cent confidence range on the vent stream?

$$\text{Vent} = \text{feed} - \text{product}$$

$$\sigma^2(\text{vent}) = \sigma^2(\text{feed}) + \sigma^2(\text{product})$$

ANALYSIS OF VARIANCE

TABLE 7.2

	Flow rate, scfh	Standard deviation, 5% of flow	Variance
Feed............	10,000	500	250,000
Product........	9,000	450	202,500

Vent-gas flow, by difference = 1,000 scfh

Vent-gas variance = total variance = 452,500

Vent-gas standard deviation = $\sqrt{452{,}500}$ = 672.5 scfh

Vent-gas 95% precision limits = ±(1.96)(672.5) = ±1,318.1 scfh

Vent-gas 95% confidence range = 1,000 ± 1,318.1 scfh

7.1.3 Variance of a Simple Function.

If the standard deviation of a measurement, from which we could calculate the precision limits, was expressed in certain units, say Btu, and we wished to convert to some other units, say foot-pounds, we would multiply the standard deviation (and the precision limits) by the suitable conversion factor, in this case 778.2. Symbolically, if $X = ax$, then

$$\sigma(X) = a\sigma(x) \tag{7.7}$$

and

$$\sigma^2(X) = a^2\sigma^2(x) \tag{7.8}$$

If a function is made up of the sum of several terms, we have shown in Eq. (7.6) that the variance of the sum (or difference) is equal to the sum of the variances. If each term is made up of a constant and a variable, the variance of the sum is the sum of the product of the constant factor squared and the variance of the variable factor of each term. If $X = y_1 + y_2 + y_3 + \cdots$ and $y_1 = ax_1$, $y_2 = bx_2$, $y_3 = cx_3$, etc., then

$$\sigma^2(X) = a^2\sigma^2(x_1) + b^2\sigma^2(x_2) + c^2\sigma^2(x_3) + \cdots \tag{7.9}$$

Equation (7.9) is merely a combination of Eqs. (7.6) and (7.8).

example 7.2. If the total pressure drop in a system is determined by adding a number of separate pressure-drop readings all in different units and the precision of the total pressure drop is desired, the variance of the total pressure will be the sum of the variances of the separate readings, each adjusted by the square of the suitable conversion factor, as in Eq. (7.9). Suppose that the system consisted of a line with some kind of a filter pot and a control valve and that the pressure drops and standard deviations are as in Table 7.3. What is the precision of the calculated total pressure drop in standard atmospheres?

TABLE 7.3

	Pressure drop	Standard deviation
Line............	2.4 psia	0.05 psia
Filter vessel.....	13.0 in. Hg	0.10 in. Hg
Control valve....	10.0 in. H_2O	0.10 in. H_2O

The conversion factors for each of these pressure units to atmospheres are 0.06804, 0.002456, and 0.03342. Table 7.4 illustrates the solution.

TABLE 7.4

	ΔP	ΔP atm	σ	σ^2
Line.........	2.4 psia × 0.06804 =	0.1633	0.05	0.0025
Filter........	13.0 in. Hg × 0.03342 =	0.4345	0.10	0.0100
Valve........	10.0 in. H_2O × 0.002456 =	0.0246	0.10	0.0100
Total.......	....	0.6224	....	0.00002280 †

† σ^2 (total) = $(0.06804)^2(0.0025) + (0.03342)^2(0.01) + (0.002456)^2(0.01)$
   = 0.00002280.

σ (total) = 0.004775.

$l_{0.95,\text{total}} = \pm(1.96)(0.004775) = \pm 0.009359$.

95% confidence range = 0.6224 ± 0.0094 atm.

7.1.4 Variance of a General Function. The variance of a general function of variables involves the square of the partial derivative of the function with respect to each variable. This is the only extension into the field of calculus in this book, and we hope it rings a familiar note with the majority of readers. If not, we can only advise skipping this short section.

If X is calculated from some function of $x_1, x_2, x_3, \cdots$; that is,

$$X = f(x_1, x_2, x_3, \ldots) \tag{7.10}$$

then
$$\sigma^2(X) = \left(\frac{\partial X}{\partial x_1}\right)^2 \sigma^2(x_1) + \left(\frac{\partial X}{\partial x_2}\right)^2 \sigma^2(x_2) + \left(\frac{\partial X}{\partial x_3}\right)^2 \sigma^2(x_3) + \cdots \quad (7.11)$$

This formula, which applies to the simple functions of Eqs. (7.4), (7.5), (7.6), and (7.9), applies to any function with the only restriction that all the measured quantities indicated by x's vary independently. Equation (7.11), as well as the equations for the more simple functions, provides a method for calculating the total variance of a function, and hence the total precision. It also is a method for determining which component of the function makes the largest contribution to the variance of the total, and therefore points the way both to where the maximum improvement can be effected and where it is fruitless to attempt an improvement in the precision.

When applying the propagation of variance formulas involving the derivatives, these are to be evaluated at the mean values of the variable terms, x_i.

Before demonstrating the use of the general equation, some specific formulas will be developed and a table of variances of a number of relations will be given which may facilitate the calculation of the precision of involved measurements in quite a few engineering applications.

If $X = ax_1x_2$ (for example, if x_1 is a measured flow rate and x_2 is the measured concentration of a specific component, and a is some orifice or meter constant, then X would be the quantity of the specific component flowing), to get the variance and precision of X, following Eq. (7.11), we calculate:

$$\frac{\partial X}{\partial x_1} = a\bar{x}_2; \quad \left(\frac{\partial X}{\partial x_1}\right)^2 \sigma^2(x_1) = a^2 \bar{x}_2^2 \sigma^2(x_1) \quad \text{(i)}$$

$$\frac{\partial X}{\partial x_2} = a\bar{x}_1; \quad \left(\frac{\partial X}{\partial x_2}\right)^2 \sigma^2(x_2) = a^2 \bar{x}_1^2 \sigma^2(x_2) \quad \text{(ii)}$$

Multiplying Eq. (i) by $\bar{x}_1^2/\bar{x}_1^2$ and Eq. (ii) by $\bar{x}_2^2/\bar{x}_2^2$, we get

$$\left(\frac{\partial X}{\partial x_1}\right)^2 \sigma^2(x_1) = a^2 \bar{x}_1^2 \bar{x}_2^2 \frac{\sigma^2(x_1)}{\bar{x}_1^2} \quad \text{(iii)}$$

$$\left(\frac{\partial X}{\partial x_2}\right)^2 \sigma^2(x_2) = a^2 \bar{x}_1^2 \bar{x}_2^2 \frac{\sigma^2(x_2)}{\bar{x}_2^2} \quad \text{(iv)}$$

Since $a^2 \bar{x}_1^2 \bar{x}_2^2 = \bar{X}^2$, adding Eqs. (iii) and (iv), we get

$$\sigma^2(X) = \bar{X}^2 \frac{\sigma^2(x_1)}{\bar{x}_1^2} + \bar{X}^2 \frac{\sigma^2(x_2)}{\bar{x}_2^2} \quad \text{(v)}$$

If $X = ax_1^{-n}x_2^m$ (for example, the heat-transfer film coefficient equation for turbulent flow of gases in straight tubes is in this form, where X is the film coefficient, a is a constant times the specific heat, x_1 is the tube diameter to the -0.2 power, and x_2 is the mass velocity to the 0.8 power), to get the variance of X, following Eq. (7.11), we calculate

$$\frac{\partial X}{\partial x_1} = -na\bar{x}_2^m\bar{x}_1^{-n-1} \qquad \left(\frac{\partial X}{\partial x_1}\right)^2 \sigma^2(x_1) = n^2 a^2 \bar{x}_2^{2m}\bar{x}_1^{-2n-2}\sigma^2(x_1) \qquad \text{(vi)}$$

$$\frac{\partial X}{\partial x_2} = ma\bar{x}_1^{-n}\bar{x}_2^{m-1} \qquad \left(\frac{\partial X}{\partial x_2}\right)^2 \sigma^2(x_2) = m^2 a^2 \bar{x}_1^{-2n}\bar{x}_2^{2m-2}\sigma^2(x_2)$$

(vii)

Multiplying Eq. (vi) by $\bar{x}_1^2/\bar{x}_1^2$ and Eq. (vii) by $\bar{x}_2^2/\bar{x}_2^2$, we get

$$\left(\frac{\partial X}{\partial x_1}\right)^2 \sigma^2(x_1) = a^2 \bar{x}_1^{-2n}\bar{x}_2^{2m}\frac{n^2\sigma^2(x_1)}{\bar{x}_1^2} \qquad \text{(viii)}$$

$$\left(\frac{\partial X}{\partial x_2}\right)^2 \sigma^2(x_2) = a^2 \bar{x}_1^{-2n}\bar{x}_2^{2m}\frac{m^2\sigma^2(x_2)}{\bar{x}_2^2} \qquad \text{(ix)}$$

Since $a^2 \bar{x}_1^{-2n}\bar{x}_2^{2m} = \bar{X}^2$, adding Eqs. (viii) and (ix), we get

$$\sigma^2(X) = \bar{X}^2 \frac{n^2\sigma^2(x_1)}{\bar{x}_1^2} + \bar{X}^2 \frac{m^2\sigma^2(x_2)}{\bar{x}_2^2} \qquad \text{(x)}$$

TABLE 7.5 Variances of General Functions

Function	Variance	
$X = x$	$\sigma^2(X) = \bar{X}^2 \dfrac{\sigma^2(x)}{\bar{x}^2}$	$= \bar{X}^2 C^2$
$X = x + a$	$\sigma^2(X) = \bar{X}^2 \dfrac{\sigma^2(x)}{(\bar{x}+a)^2}$	$= \bar{X}^2 C^2$
$X = ax$	$\sigma^2(X) = \bar{X}^2 \dfrac{\sigma^2(x)}{\bar{x}^2}$	$= \bar{X}^2 C^2$
$X = a/x$	$\sigma^2(X) = \bar{X}^2 \dfrac{\sigma^2(x)}{\bar{x}^2}$	$= \bar{X}^2 C^2$
$X = ax_1 x_2$	$\sigma^2(X) = \bar{X}^2 \dfrac{\sigma^2(x_1)}{\bar{x}_1^2} + \bar{X}^2 \dfrac{\sigma^2(x_2)}{\bar{x}_2^2}$	$= \bar{X}^2(C_1^2 + C_2^2)$
$X = \dfrac{ax_1}{x_2}$	$\sigma^2(X) = \bar{X}^2 \dfrac{\sigma^2(x_1)}{\bar{x}_1^2} + \bar{X}^2 \dfrac{\sigma^2(x_2)}{\bar{x}_2^2}$	$= \bar{X}^2(C_1^2 + C_2^2)$
$X = ax^n$	$\sigma^2(X) = \bar{X}^2 \dfrac{n^2\sigma^2(x)}{\bar{x}^2}$	$= \bar{X}^2 n^2 C^2$
$X = ax_1^{-n}x_2^m$	$\sigma^2(X) = \bar{X}^2 \dfrac{n^2\sigma^2(x_1)}{\bar{x}_1^2} + \bar{X}^2 \dfrac{m^2\sigma^2(x_2)}{\bar{x}_2^2}$	$= \bar{X}^2(n^2 C_1^2 + m^2 C_2^2)$

$C = \dfrac{s(x)}{\bar{x}}$

ANALYSIS OF VARIANCE

The variation of a measurement is sometimes reported in terms of the *coefficient of variation*, C. The coefficient of variation is the ratio of the standard deviation to the mean: $C = s(x)/\bar{x}$. In Table 7.5 a number of general functions are listed together with the total variance, calculated from Eq. (7.11), and also are expressed in terms of the coefficient of variation.

example 7.3. The equation for gas flow measured by orifice meter can be written

$$\text{scfh} = C_a \sqrt{\Delta H} \sqrt{1/M.W.} \sqrt{P} \sqrt{520/(t+460)} \qquad (7.12)$$

where C_a = a constant depending on orifice size and units employed
ΔH = measured pressure differential
$M.W.$ = molecular weight of gas flowing
P = absolute pressure
t = temperature, °F

For the mean flow conditions given in the following table, what is the precision of the calculated flow rate and which variable makes the largest contribution to the variance of the calculated flow?

TABLE 7.6

Variable	Mean value	Standard deviation
C_a	38.4	0.00
ΔH	64 in. H_2O	0.50
$M.W.$	16	0.10
P	361 psia	2.00
t	165 °F	0.50

For simplification, we rewrite Eq. (7.12) in terms of x's and constants to get

$$X = ax_1^{1/2} x_2^{-1/2} x_3^{1/2} (x_4 + b)^{-1/2} \qquad (7.13)$$

where X = scfh

$a = C_a \sqrt{520}$
$x_1 = \Delta H$
$x_2 = M.W.$
$x_3 = P$
$x_4 = t$
$b = 460$

From the table of variances of functions, we can get

$$\sigma^2(X) = \bar{X}^2 \left[\frac{\sigma^2(x_1)}{4\bar{x}_1{}^2} + \frac{\sigma^2(x_2)}{4\bar{x}_2{}^2} + \frac{\sigma^2(x_3)}{4\bar{x}_3{}^2} + \frac{\sigma^2(x_4)}{4(\bar{x}_4 + b)^2} \right] \quad (7.14)$$

Substituting the variances and the values of $\bar{x}_i$ in Eq. (7.14) and taking out the common factor of 1/4, we get the following values:

TABLE 7.7

$\sigma^2(x_1)/\bar{x}_1{}^2 = 0.00006103$
$\sigma^2(x_2)/\bar{x}_2{}^2 = 0.00003906$
$\sigma^2(x_3)/\bar{x}_3{}^2 = 0.00003069$
$\sigma^2(x_4)/(\bar{x}_4 + b)^2 = 0.00000064$
= ───────────
 0.00013142

The variance of the total flow rate is therefore $(0.00013142)(\bar{X}^2/4)$. $\bar{X}$ from Eq. (7.13) is 1,330 scfh, which gives a variance of 58.12 and a standard deviation of 7.62 scfh. The precision limits, at the 95 per cent confidence level, are ± 1.96 times this value, or ± 14.9 scfh—about 1 per cent of the total flow.

Table 7.7 indicates the relative contribution of the different variables to the variance of the total flow. Approximately 45 per cent of the variance comes from the x_1 factor, the pressure-differential reading. To decrease this variance contribution we need either to reduce the numerator, the $\sigma^2(x_1)$ term, or to increase the denominator, the $\bar{x}_1{}^2$ term. In this example, it means either to cut down the variation in the readings, i.e., increase their precision, or to increase their absolute value by decreasing the size of the orifice or by using a lighter fluid in the pressure-differential gauge, assuming the variance of the readings will stay the same under these changes. It is obvious from Table 7.7 that nothing will be gained by improving on the temperature measurements, the x_4 term, and that only half as much will be gained toward improving the precision by bettering the molecular weight or the pressure determinations than by improving the pressure-differential readings.

In these examples we have assumed that the standard deviation of the individual measurements were known in all cases. Methods of determining the standard deviation are discussed in Chaps. 4 and 6. The best way to obtain the standard deviation is to calculate it from data under the conditions of operation in which it is desired. If this is not feasible, it is sometimes possible to take a series of readings under similar conditions just for the purpose of determining the standard deviation and precision of the readings. Sometimes previous experience has already established

ANALYSIS OF VARIANCE

the standard deviations. It is good policy to recheck such values from time to time. As has been pointed out in Chap. 6, if some experience with duplicate readings is available, the standard deviation can be obtained from the mean difference between duplicates. The standard deviation of a single determination is equal to 0.8862 times the mean difference between duplicates.

7.2 The F Test for Variances

In discussing the t test for means, it was pointed out that samples drawn from the same batch, or measurements made of the same variable, could not be expected to be identical. The means of several samples will vary over a range which can be approximated for any desired probability level. The same situation holds for variances. If several groups of samples are drawn from the same batch of material, the estimated variances of the parent population calculated from each sample will not be identical. The F test provides a method for determining whether the ratio of two variances is larger than might be expected by chance if they had been drawn from the same population. A test for checking the homogeneity of more than two variances is given in the next section.

The null hypothesis for comparing two variances, like that for comparing two means, is

$$H_0; \sigma^2(x_1) = \sigma^2(x_2) \tag{7.15}$$

In the t test for the difference between two means, we determined the difference between the two observed means, hypothesized to have come from populations with the same means, and compared this difference (divided by the pooled estimate of the standard deviation) with the statistic t. There is no statistic yet discovered that corresponds to the difference between two estimated variances. However, R. A. Fisher discovered the distribution of the statistic corresponding to the function $\frac{1}{2} \log_e [s^2(x_2)/s^2(x_1)]$. Snedecor[1] modified this to give the values corresponding to $s^2(x_2)/s^2(x_1)$ and named the ratio F in honor of Fisher.

$$F = \frac{s^2(x_2)}{s^2(x_1)} \tag{7.16}$$

$s^2(x_1)$ and $s^2(x_2)$ are the estimated variances of the variables x_1 and x_2. These are calculated from n_1 measurements of x_1 and n_2 measurements of x_2. If $s^2(x_1)$ and $s^2(x_2)$ are in fact measures of the same thing, i.e., samples drawn from the same population, or from populations with identical variances, the two estimated variances will not necessarily be the same, no more than the means of the two sets of measurements would be exactly

[1] G. W. Snedecor, "Analysis of Variance and Covariance," Collegiate Press, Inc., Ames, Iowa, 1934.

identical. *F.* measures the difference, in the form of a ratio, of two estimates of the same variance that can be expected to occur, depending on the number of degrees of freedom available for the calculation of each estimate. If $s^2(x_1)$ is calculated with ν_1 degrees of freedom, $n_1 - 1$, and if $s^2(x_2)$ is calculated with ν_2 degrees of freedom, $n_2 - 1$, and the ratio written with the larger estimate in the numerator, the F value at the 2α level (the reason for using 2α will be explained in just a minute) indicates that there is only a 2α probability of observing a ratio this large if the two variance estimates were obtained from the same population or from two populations with the same variance.

The F values have been tabulated to test specifically whether one variance estimate is larger than another, not just whether two variances are significantly different. This corresponds to the one-sided, or unsymmetrical, test used in the comparison of means. The probability levels, $\alpha = 0.05$, and $\alpha = 0.01$, are those corresponding to one side of the F distribution only and are half the 2α probabilities mentioned above for the two-sided test. If the tables are to be used for a two-sided test, i.e., to see if there is significant difference between two variance estimates, then the indicated probability levels must be doubled.

Tables 7.8 and 7.9 give the F values for the usual variance test corresponding to the 0.05 and 0.01 probability or significance levels. These tables are used in the following way: if two variance estimates are at hand and one is suspected of including not only the factors affecting the first but other factors as well, we test whether these other factors are operating by seeing whether the ratio of the second variance estimate to the first is larger than the corresponding F value for the number of degrees of freedom available for the calculation of the two variances. If it is, we take this as evidence that the additional factors are operating on the second variance estimate and accept the indicated probability risk of being in error; i.e., we reject the hypothesis that the variances are equal. If the ratio of the second variance estimate to the first does not exceed the tabulated F value at the significance level we have decided to use, we state there is insufficient evidence that the second variance includes factors not included in the first, and we accept the hypothesis that the variances estimated are equal.

For example, an analytical tool, say a mass spectrometer, is known from long experience to have a variance for a particular component of 0.24. A new analyst runs a series of determinations on the mass spectrometer, and for 25 determinations the estimated variance is 0.326. Has the new analyst contributed another variance factor to that of the instrument, i.e., is the variance estimate of 0.326 significantly larger than the 0.24 value for the number of tests involved? The value of 0.24 is assumed to be accurately known with infinite degrees of freedom. We shall designate this σ^2 (using the symbol σ^2 instead of s^2 since the value is assumed to be the true value). $s^2(x_2)$ is the variance estimate from the 25 new determinations, 0.326, calculated with $25 - 1$, 24 degrees of freedom. We test whether $s^2(x_2)$ with 24 degrees of freedom is significantly larger than σ^2 with infinite degrees of freedom by setting up the hypothesis that they are equal and comparing the ratio of $s^2(x_2)/\sigma^2$ with the tabulated F values at the corresponding degrees of freedom. For our example the variance ratio, $s^2(x_2)/\sigma^2 = 0.326/0.24 = 1.356$. The F value from Table 7.8 for 24 and ∞ degrees of freedom is 1.52. In other words, there is a 0.05 probability, 1 chance in 20, of observing a variance ratio as large as 1.52 when the two variance estimates are in fact estimates of the same thing. Since our variance ratio is less than 1.52, actually 1.356, we cannot reject the hypothesis that the two variances are equal. Symbolically,

$$s^2(x_2) = 0.326, \quad \nu_2 = 24$$
$$\sigma^2 = 0.24, \quad \nu_1 = \infty$$

Test:

$$F = \frac{s^2(x_2)}{\sigma^2}$$

$H_0; \quad s^2(x_2) = \sigma^2$

$$F = \frac{s^2(x_2)}{\sigma^2} = \frac{0.326}{0.24} = 1.356$$

$$F_{0.05,24,\infty} = 1.52$$

Accept H_0.

If the test of variances had been to determine whether there was significant difference between them, the proper probability levels would be double those given in Tables 7.8 and 7.9. Suppose you were testing two different rubber-compounding methods to see if one was more variable than the other and were using the tensile-strength measurements as the criterion. If the variance estimate of one, from 15 samples, was 107 and the variance estimate of the other, from 25 samples, was 41.1, you would divide the larger by the smaller and compare the quotient with the tabulated F value for the corresponding degrees of freedom. In this illustration, $s^2(x_2) = 107$ with 14 degrees of freedom and $s^2(x_1)$ is 41.1 with 24 degrees of freedom. $s^2(x_2)/s^2(x_1) = 107/41.1 = 2.60$. From Table 7.8, at the 0.05 probability level, the F value for 14 and 24 degrees of freedom is 2.13, and from Table 7.9, at the 0.01 probability level, the F value is 2.93. Our observed F ratio was 2.60. If we reject the hypothesis that the variability of the two methods under test is the same with our calculated variance, ratio of 2.60 we have less than 10 but more than 2 per cent chance of being in error, = twice the tabulated probability values because we are using a two-sided probability.

TABLE 7.8 $F_{0.05}$†

0.05 probability of a larger value of F

ν_2 = degrees of freedom for numerator
ν_1 = degrees of freedom for denominator

ν_2\ν_1	1	2	3	4	5	6	7	8	9	10	11	12	14	16	20	24	30	40	50	75	100	200	500	∞
1	161	200	216	225	230	234	237	239	241	242	243	244	245	246	248	249	250	251	252	253	253	254	254	254
2	18.51	19.00	19.16	19.25	19.30	19.33	19.36	19.37	19.38	19.39	19.40	19.41	19.42	19.42	19.44	19.45	19.46	19.47	19.47	19.48	19.49	19.49	19.50	19.50
3	10.13	9.55	9.28	9.12	9.01	8.94	8.88	8.84	8.81	8.78	8.76	8.74	8.71	8.69	8.66	8.64	8.62	8.60	8.58	8.57	8.56	8.54	8.54	8.53
4	7.71	6.94	6.59	6.39	6.26	6.16	6.09	6.04	6.00	5.96	5.93	5.91	5.87	5.84	5.80	5.77	5.74	5.71	5.70	5.68	5.66	5.65	5.64	5.63
5	6.61	5.79	5.41	5.19	5.05	4.95	4.88	4.82	4.78	4.74	4.70	4.68	4.64	4.60	4.56	4.53	4.50	4.46	4.44	4.42	4.40	4.38	4.37	4.36
6	5.99	5.14	4.76	4.53	4.39	4.28	4.21	4.15	4.10	4.06	4.03	4.00	3.96	3.92	3.87	3.84	3.81	3.77	3.75	3.72	3.71	3.69	3.68	3.67
7	5.59	4.74	4.35	4.12	3.97	3.87	3.79	3.73	3.68	3.63	3.60	3.57	3.52	3.49	3.44	3.41	3.38	3.34	3.32	3.29	3.28	3.25	3.24	3.23
8	5.32	4.46	4.07	3.84	3.69	3.58	3.50	3.44	3.39	3.34	3.31	3.28	3.23	3.20	3.15	3.12	3.08	3.05	3.03	3.00	2.98	2.96	2.94	2.93
9	5.12	4.26	3.86	3.63	3.48	3.37	3.29	3.23	3.18	3.13	3.10	3.07	3.02	2.98	2.93	2.90	2.86	2.82	2.80	2.77	2.76	2.73	2.72	2.71
10	4.96	4.10	3.71	3.48	3.33	3.22	3.14	3.07	3.02	2.97	2.94	2.91	2.86	2.82	2.77	2.74	2.70	2.67	2.64	2.61	2.59	2.56	2.55	2.54
11	4.84	3.98	3.59	3.36	3.20	3.09	3.01	2.95	2.90	2.86	2.82	2.79	2.74	2.70	2.65	2.61	2.57	2.53	2.50	2.47	2.45	2.42	2.41	2.40
12	4.75	3.88	3.49	3.26	3.11	3.00	2.92	2.85	2.80	2.76	2.72	2.69	2.64	2.60	2.54	2.50	2.46	2.42	2.40	2.36	2.35	2.32	2.31	2.30
13	4.67	3.80	3.41	3.18	3.02	2.92	2.84	2.77	2.72	2.67	2.63	2.60	2.55	2.51	2.46	2.42	2.38	2.34	2.32	2.28	2.26	2.24	2.22	2.21
14	4.60	3.74	3.34	3.11	2.96	2.85	2.77	2.70	2.65	2.60	2.56	2.53	2.48	2.44	2.39	2.35	2.31	2.27	2.24	2.21	2.19	2.16	2.14	2.13
15	4.54	3.68	3.29	3.06	2.90	2.79	2.70	2.64	2.59	2.55	2.51	2.48	2.43	2.39	2.33	2.29	2.25	2.21	2.18	2.15	2.12	2.10	2.08	2.07
16	4.49	3.63	3.24	3.01	2.85	2.74	2.66	2.59	2.54	2.49	2.45	2.42	2.37	2.33	2.28	2.24	2.20	2.16	2.13	2.09	2.07	2.04	2.02	2.01
17	4.45	3.59	3.20	2.96	2.81	2.70	2.62	2.55	2.50	2.45	2.41	2.38	2.33	2.29	2.23	2.19	2.15	2.11	2.08	2.04	2.02	1.99	1.97	1.96
18	4.41	3.55	3.16	2.93	2.77	2.66	2.58	2.51	2.46	2.41	2.37	2.34	2.29	2.25	2.19	2.15	2.11	2.07	2.04	2.00	1.98	1.95	1.93	1.92
19	4.38	3.52	3.13	2.90	2.74	2.63	2.55	2.48	2.43	2.38	2.34	2.31	2.26	2.21	2.15	2.11	2.07	2.02	2.00	1.96	1.94	1.91	1.90	1.88
20	4.35	3.49	3.10	2.87	2.71	2.60	2.52	2.45	2.40	2.35	2.31	2.28	2.23	2.18	2.12	2.08	2.04	1.99	1.96	1.92	1.90	1.87	1.85	1.84
21	4.32	3.47	3.07	2.84	2.68	2.57	2.49	2.42	2.37	2.32	2.28	2.25	2.20	2.15	2.09	2.05	2.00	1.96	1.93	1.89	1.87	1.84	1.82	1.81
22	4.30	3.44	3.05	2.82	2.66	2.55	2.47	2.40	2.35	2.30	2.26	2.23	2.18	2.13	2.07	2.03	1.98	1.93	1.91	1.87	1.84	1.81	1.80	1.78
23	4.28	3.42	3.03	2.80	2.64	2.53	2.45	2.38	2.32	2.28	2.24	2.20	2.14	2.10	2.04	2.00	1.96	1.91	1.88	1.84	1.82	1.79	1.77	1.76
24	4.26	3.40	3.01	2.78	2.62	2.51	2.43	2.36	2.30	2.26	2.22	2.18	2.13	2.09	2.02	1.98	1.94	1.89	1.86	1.82	1.80	1.76	1.74	1.73
25	4.24	3.38	2.99	2.76	2.60	2.49	2.41	2.34	2.28	2.24	2.20	2.16	2.11	2.06	2.00	1.96	1.92	1.87	1.84	1.80	1.77	1.74	1.72	1.71

ANALYSIS OF VARIANCE

26	4.22	3.37	2.98	2.74	2.59	2.47	2.39	2.32	2.27	2.22	2.18	2.15	2.10	2.05	1.99	1.95	1.90	1.85	1.82	1.78	1.76	1.72	1.70	1.69
27	4.21	3.35	2.96	2.73	2.57	2.46	2.37	2.30	2.25	2.20	2.16	2.13	2.08	2.03	1.97	1.93	1.88	1.84	1.80	1.76	1.74	1.71	1.68	1.67
28	4.20	3.34	2.95	2.71	2.56	2.44	2.36	2.29	2.24	2.19	2.15	2.12	2.06	2.02	1.96	1.91	1.87	1.81	1.78	1.75	1.72	1.69	1.67	1.65
29	4.18	3.33	2.93	2.70	2.54	2.43	2.35	2.28	2.22	2.18	2.14	2.10	2.05	2.00	1.94	1.90	1.85	1.80	1.77	1.73	1.71	1.68	1.65	1.64
30	4.17	3.32	2.92	2.69	2.53	2.42	2.34	2.27	2.21	2.16	2.12	2.09	2.04	1.99	1.93	1.89	1.84	1.79	1.76	1.72	1.69	1.66	1.64	1.62
32	4.15	3.30	2.90	2.67	2.51	2.40	2.32	2.25	2.19	2.14	2.10	2.07	2.02	1.97	1.91	1.86	1.82	1.76	1.74	1.69	1.67	1.64	1.61	1.59
34	4.13	3.28	2.88	2.65	2.49	2.38	2.30	2.23	2.17	2.12	2.08	2.05	2.00	1.95	1.89	1.84	1.80	1.74	1.71	1.67	1.64	1.61	1.59	1.57
36	4.11	3.26	2.86	2.63	2.48	2.36	2.28	2.21	2.15	2.10	2.06	2.03	1.98	1.93	1.87	1.82	1.78	1.72	1.69	1.65	1.62	1.59	1.56	1.55
38	4.10	3.25	2.85	2.62	2.46	2.35	2.26	2.19	2.14	2.09	2.05	2.02	1.96	1.92	1.85	1.80	1.76	1.71	1.67	1.63	1.60	1.57	1.54	1.53
40	4.08	3.23	2.84	2.61	2.45	2.34	2.25	2.18	2.12	2.07	2.04	2.00	1.95	1.90	1.84	1.79	1.74	1.69	1.66	1.61	1.59	1.55	1.53	1.51
42	4.07	3.22	2.83	2.59	2.44	2.32	2.24	2.17	2.11	2.06	2.02	1.99	1.94	1.89	1.82	1.78	1.73	1.68	1.64	1.60	1.57	1.54	1.51	1.49
44	4.06	3.21	2.82	2.58	2.43	2.31	2.23	2.16	2.10	2.05	2.01	1.98	1.92	1.88	1.81	1.76	1.72	1.66	1.63	1.58	1.56	1.52	1.50	1.48
46	4.05	3.20	2.81	2.57	2.42	2.30	2.22	2.14	2.09	2.04	2.00	1.97	1.91	1.87	1.80	1.75	1.71	1.65	1.62	1.57	1.54	1.51	1.48	1.46
48	4.04	3.19	2.80	2.56	2.41	2.30	2.21	2.14	2.08	2.03	1.99	1.96	1.90	1.86	1.79	1.74	1.70	1.64	1.61	1.56	1.53	1.50	1.47	1.45
50	4.03	3.18	2.79	2.56	2.40	2.29	2.20	2.13	2.07	2.02	1.98	1.95	1.90	1.85	1.78	1.74	1.69	1.63	1.60	1.55	1.52	1.48	1.46	1.44
55	4.02	3.17	2.78	2.54	2.38	2.27	2.18	2.11	2.05	2.00	1.97	1.93	1.88	1.83	1.76	1.72	1.67	1.61	1.58	1.52	1.50	1.46	1.43	1.41
60	4.00	3.15	2.76	2.52	2.37	2.25	2.17	2.10	2.04	1.99	1.95	1.92	1.86	1.81	1.75	1.70	1.65	1.59	1.56	1.50	1.48	1.44	1.41	1.39
65	3.99	3.14	2.75	2.51	2.36	2.24	2.15	2.08	2.02	1.98	1.94	1.90	1.85	1.80	1.73	1.68	1.63	1.57	1.54	1.49	1.46	1.42	1.39	1.37
70	3.98	3.13	2.74	2.50	2.35	2.23	2.14	2.07	2.01	1.97	1.93	1.89	1.84	1.79	1.72	1.67	1.62	1.56	1.53	1.47	1.45	1.40	1.37	1.35
80	3.96	3.11	2.72	2.48	2.33	2.21	2.12	2.05	1.99	1.95	1.91	1.88	1.82	1.77	1.70	1.65	1.60	1.54	1.51	1.45	1.42	1.38	1.35	1.32
100	3.94	3.09	2.70	2.46	2.30	2.19	2.10	2.03	1.97	1.92	1.88	1.85	1.79	1.75	1.68	1.63	1.57	1.51	1.48	1.42	1.39	1.34	1.30	1.28
125	3.92	3.07	2.68	2.44	2.29	2.17	2.08	2.01	1.95	1.90	1.86	1.83	1.77	1.72	1.65	1.60	1.55	1.49	1.45	1.39	1.36	1.31	1.27	1.25
150	3.91	3.06	2.67	2.43	2.27	2.16	2.07	2.00	1.94	1.89	1.85	1.82	1.76	1.71	1.64	1.59	1.54	1.47	1.44	1.37	1.34	1.29	1.25	1.22
200	3.89	3.04	2.65	2.41	2.26	2.14	2.05	1.98	1.92	1.87	1.83	1.80	1.74	1.69	1.62	1.57	1.52	1.45	1.42	1.35	1.32	1.26	1.22	1.19
400	3.86	3.02	2.62	2.39	2.23	2.12	2.03	1.96	1.90	1.85	1.81	1.78	1.72	1.67	1.60	1.54	1.49	1.42	1.38	1.32	1.28	1.22	1.16	1.13
1000	3.85	3.00	2.61	2.38	2.22	2.10	2.02	1.95	1.89	1.84	1.80	1.76	1.70	1.65	1.58	1.53	1.47	1.41	1.36	1.30	1.26	1.19	1.13	1.08
∞	3.84	2.99	2.60	2.37	2.21	2.09	2.01	1.94	1.88	1.83	1.79	1.75	1.69	1.64	1.57	1.52	1.46	1.40	1.35	1.28	1.24	1.17	1.11	1.00

† Reprinted from Table 10.5.3 in George W. Snedecor, "Statistical Methods," 5th ed., Iowa State College Press, Ames, Iowa, 1956, by permission of the author and publisher.

TABLE 7.9 $F_{0.01}$†

0.01 probability of a larger value of F

ν_2 = degrees of freedom for numerator
ν_1 = degrees of freedom for denominator

ν_1\\ν_2	1	2	3	4	5	6	7	8	9	10	11	12	14	16	20	24	30	40	50	75	100	200	500	∞
1	4,052	4,999	5,403	5,625	5,764	5,859	5,928	5,981	6,022	6,056	6,082	6,106	6,142	6,169	6,208	6,234	6,258	6,286	6,302	6,323	6,334	6,352	6,361	6,366
2	98.49	99.00	99.17	99.25	99.30	99.33	99.34	99.36	99.38	99.40	99.41	99.42	99.43	99.44	99.45	99.46	99.47	99.48	99.48	99.49	99.49	99.49	99.50	99.50
3	34.12	30.82	29.46	28.71	28.24	27.91	27.67	27.49	27.34	27.23	27.13	27.05	26.92	26.83	26.69	26.60	26.50	26.41	26.35	26.27	26.23	26.18	26.14	26.12
4	21.20	18.00	16.69	15.98	15.52	15.21	14.98	14.80	14.66	14.54	14.45	14.37	14.24	14.15	14.02	13.93	13.83	13.74	13.69	13.61	13.57	13.52	13.48	13.46
5	16.26	13.27	12.06	11.39	10.97	10.67	10.45	10.27	10.15	10.05	9.96	9.89	9.77	9.68	9.55	9.47	9.38	9.29	9.24	9.17	9.13	9.07	9.04	9.02
6	13.74	10.92	9.78	9.15	8.75	8.47	8.26	8.10	7.98	7.87	7.79	7.72	7.60	7.52	7.39	7.31	7.23	7.14	7.09	7.02	6.99	6.94	6.90	6.88
7	12.25	9.55	8.45	7.85	7.46	7.19	7.00	6.84	6.71	6.62	6.54	6.47	6.35	6.27	6.15	6.07	5.98	5.90	5.85	5.78	5.75	5.70	5.67	5.65
8	11.26	8.65	7.59	7.01	6.63	6.37	6.19	6.03	5.91	5.82	5.74	5.67	5.56	5.48	5.36	5.28	5.20	5.11	5.06	5.00	4.96	4.91	4.88	4.86
9	10.56	8.02	6.99	6.42	6.06	5.80	5.62	5.47	5.35	5.26	5.18	5.11	5.00	4.92	4.80	4.73	4.64	4.56	4.51	4.45	4.41	4.36	4.33	4.31
10	10.04	7.56	6.55	5.99	5.64	5.39	5.21	5.06	4.95	4.85	4.78	4.71	4.60	4.52	4.41	4.33	4.25	4.17	4.12	4.05	4.01	3.96	3.93	3.91
11	9.65	7.20	6.22	5.67	5.32	5.07	4.88	4.74	4.63	4.54	4.46	4.40	4.29	4.21	4.10	4.02	3.94	3.86	3.80	3.74	3.70	3.66	3.62	3.60
12	9.33	6.93	5.95	5.41	5.06	4.82	4.65	4.50	4.39	4.30	4.22	4.16	4.05	3.98	3.86	3.78	3.70	3.61	3.56	3.49	3.46	3.41	3.38	3.36
13	9.07	6.70	5.74	5.20	4.86	4.62	4.44	4.30	4.19	4.10	4.02	3.96	3.85	3.78	3.67	3.59	3.51	3.42	3.37	3.30	3.27	3.21	3.18	3.16
14	8.86	6.51	5.56	5.03	4.69	4.46	4.28	4.14	4.03	3.94	3.86	3.80	3.70	3.62	3.51	3.43	3.34	3.26	3.21	3.14	3.11	3.06	3.02	3.00
15	8.68	6.36	5.42	4.89	4.56	4.32	4.14	4.00	3.89	3.80	3.73	3.67	3.56	3.48	3.36	3.29	3.20	3.12	3.07	3.00	2.97	2.92	2.89	2.87
16	8.53	6.23	5.29	4.77	4.44	4.20	4.03	3.89	3.78	3.69	3.61	3.55	3.45	3.37	3.25	3.18	3.10	3.01	2.96	2.89	2.86	2.80	2.77	2.75
17	8.40	6.11	5.18	4.67	4.34	4.10	3.93	3.79	3.68	3.59	3.52	3.45	3.35	3.27	3.16	3.08	3.00	2.92	2.86	2.79	2.76	2.70	2.67	2.65
18	8.28	6.01	5.09	4.58	4.25	4.01	3.85	3.71	3.60	3.51	3.44	3.37	3.27	3.19	3.07	3.00	2.91	2.83	2.78	2.71	2.68	2.62	2.59	2.57
19	8.18	5.93	5.01	4.50	4.17	3.94	3.77	3.63	3.52	3.43	3.36	3.30	3.19	3.12	3.00	2.92	2.84	2.76	2.70	2.63	2.60	2.54	2.51	2.49
20	8.10	5085	4.94	4.43	4.10	3.87	3.71	3.56	3.45	3.37	3.30	3.23	3.13	3.05	2.94	2.86	2.77	2.69	2.63	2.56	2.53	2.47	2.44	2.42
21	8.02	5.78	4.87	4.37	4.04	3.81	3.65	3.51	3.40	3.31	3.24	3.17	3.07	2.99	2.88	2.80	2.72	2.63	2.58	2.51	2.47	2.42	2.38	2.36
22	7.94	5.72	4.82	4.31	3.99	3.76	3.59	3.45	3.35	3.26	3.18	3.12	3.02	2.94	2.83	2.75	2.67	2.58	2.53	2.46	2.42	2.37	2.33	2.31
23	7.88	5.66	4.76	4.26	3.94	3.71	3.54	3.41	3.30	3.21	3.14	3.07	2.97	2.89	2.78	2.70	2.62	2.53	2.48	2.41	2.37	2.32	2.28	2.26
24	7.82	5.61	4.72	4.22	3.90	3.67	3.50	3.36	3.25	3.17	3.09	3.03	2.93	2.85	2.74	2.66	2.58	2.49	2.44	2.36	2.33	2.27	2.23	2.21
25	7.77	5.57	4.68	4.18	3.86	3.63	3.46	3.32	3.21	3.13	3.05	2.99	2.89	2.81	2.70	2.62	2.54	2.45	2.40	2.32	2.29	2.23	2.19	2.17

ANALYSIS OF VARIANCE

df																								
26	7.72	5.53	4.64	4.14	3.82	3.59	3.42	3.29	3.17	3.09	3.02	2.96	2.86	2.77	2.66	2.58	2.50	2.41	2.36	2.28	2.25	2.19	2.15	2.13
27	7.68	5.49	4.60	4.11	3.79	3.56	3.39	3.26	3.14	3.06	2.98	2.93	2.83	2.74	2.63	2.55	2.47	2.38	2.33	2.25	2.21	2.16	2.12	2.10
28	7.64	5.45	4.57	4.07	3.76	3.53	3.36	3.23	3.11	3.03	2.95	2.90	2.80	2.71	2.60	2.52	2.44	2.35	2.30	2.22	2.18	2.13	2.09	2.06
29	7.60	5.42	4.54	4.04	3.73	3.50	3.33	3.20	3.08	3.00	2.92	2.87	2.77	2.68	2.57	2.49	2.41	2.32	2.27	2.19	2.15	2.10	2.06	2.03
30	7.56	5.39	4.51	4.02	3.70	3.47	3.30	3.17	3.06	2.98	2.90	2.84	2.74	2.66	2.55	2.47	2.38	2.29	2.24	2.16	2.13	2.07	2.03	2.01
32	7.50	5.34	4.46	3.97	3.66	3.42	3.25	3.12	3.01	2.94	2.86	2.80	2.70	2.62	2.51	2.42	2.34	2.25	2.20	2.12	2.08	2.02	1.98	1.96
34	7.44	5.29	4.42	3.93	3.61	3.38	3.21	3.08	2.97	2.89	2.82	2.76	2.66	2.58	2.47	2.38	2.30	2.21	2.15	2.08	2.04	1.98	1.94	1.91
36	7.39	5.25	4.38	3.89	3.58	3.35	3.18	3.04	2.94	2.86	2.78	2.72	2.62	2.54	2.43	2.35	2.26	2.17	2.12	2.04	2.00	1.94	1.90	1.87
38	7.35	5.21	4.34	3.86	3.54	3.32	3.15	3.02	2.91	2.82	2.75	2.69	2.59	2.51	2.40	2.32	2.22	2.14	2.08	2.00	1.97	1.90	1.86	1.84
40	7.31	5.18	4.31	3.83	3.51	3.29	3.12	2.99	2.88	2.80	2.73	2.66	2.56	2.49	2.37	2.29	2.20	2.11	2.05	1.97	1.94	1.88	1.84	1.81
42	7.27	5.15	4.29	3.80	3.49	3.26	3.10	2.96	2.86	2.77	2.70	2.64	2.54	2.46	2.35	2.26	2.17	2.08	2.02	1.94	1.91	1.85	1.80	1.78
44	7.24	5.12	4.26	3.78	3.46	3.24	3.07	2.94	2.84	2.75	2.68	2.62	2.52	2.44	2.32	2.24	2.15	2.06	2.00	1.92	1.88	1.82	1.78	1.75
46	7.21	5.10	4.24	3.76	3.44	3.22	3.05	2.92	2.82	2.73	2.66	2.60	2.50	2.42	2.30	2.22	2.13	2.04	1.98	1.90	1.86	1.80	1.76	1.72
48	7.19	5.08	4.22	3.74	3.42	3.20	3.04	2.90	2.80	2.71	2.64	2.58	2.48	2.40	2.28	2.20	2.11	2.02	1.96	1.88	1.84	1.78	1.73	1.70
50	7.17	5.06	4.20	3.72	3.41	3.18	3.02	2.88	2.78	2.70	2.62	2.56	2.46	2.39	2.26	2.18	2.10	2.00	1.94	1.86	1.82	1.76	1.71	1.68
55	7.12	5.01	4.16	3.68	3.37	3.15	2.98	2.85	2.75	2.66	2.59	2.53	2.43	2.35	2.23	2.15	2.06	1.96	1.90	1.82	1.78	1.71	1.66	1.64
60	7.08	4.98	4.13	3.65	3.34	3.12	2.95	2.82	2.72	2.63	2.56	2.50	2.40	2.32	2.20	2.12	2.03	1.93	1.87	1.79	1.74	1.68	1.63	1.60
65	7.04	4.95	4.10	3.62	3.31	3.09	2.93	2.79	2.70	2.61	2.54	2.47	2.37	2.30	2.18	2.09	2.00	1.90	1.84	1.76	1.71	1.64	1.60	1.56
70	7.01	4.92	4.08	3.60	3.29	3.07	2.91	2.77	2.67	2.59	2.51	2.45	2.35	2.28	2.15	2.07	1.98	1.88	1.82	1.74	1.69	1.62	1.56	1.53
80	6.96	4.88	4.04	3.56	3.25	3.04	2.87	2.74	2.64	2.55	2.48	2.41	2.32	2.24	2.11	2.03	1.94	1.84	1.78	1.70	1.65	1.57	1.52	1.49
100	6.90	4.82	3.98	3.51	3.20	2.99	2.82	2.69	2.59	2.51	2.43	2.36	2.26	2.19	2.06	1.98	1.89	1.79	1.73	1.64	1.59	1.51	1.46	1.43
125	6.84	4.78	3.94	3.47	3.17	2.95	2.79	2.65	2.56	2.47	2.40	2.33	2.23	2.15	2.03	1.94	1.85	1.75	1.68	1.59	1.54	1.46	1.40	1.37
150	6.81	4.75	3.91	3.44	3.14	2.92	2.76	2.62	2.53	2.44	2.37	2.30	2.20	2.12	2.00	1.91	1.83	1.72	1.66	1.56	1.51	1.43	1.37	1.33
200	6.76	4.71	3.88	3.41	3.11	2.90	2.73	2.60	2.50	2.41	2.34	2.28	2.17	2.09	1.97	1.88	1.79	1.69	1.62	1.53	1.48	1.39	1.33	1.28
400	6.70	4.66	3.83	3.36	3.06	2.85	2.69	2.55	2.46	2.37	2.29	2.23	2.12	2.04	1.92	1.84	1.74	1.64	1.57	1.47	1.42	1.32	1.28	1.19
1000	6.66	4.62	3.80	3.34	3.04	2.82	2.66	2.53	2.43	2.34	2.26	2.20	2.09	2.01	1.89	1.81	1.71	1.61	1.54	1.44	1.38	1.28	1.19	1.11
∞	6.64	4.60	3.78	3.32	3.02	2.80	2.64	2.51	2.41	2.32	2.24	2.18	2.07	1.99	1.87	1.79	1.69	1.59	1.52	1.41	1.36	1.25	1.15	1.00

† Reprinted from Table 10.5.3 in George W. Snedecor, "Statistical Methods," 5th ed., Iowa State College Press, Ames, Iowa, 1956, by permission of the author and publisher.

Note that nothing is said about the means or the level of the measurements when the variability is being tested. The variance test compares the scatter or variance of a measurement around its mean with the variance of another (or the same) measurement about its mean. It is not necessary that these means be the same. It is not even necessary that the means be known. The individual means are involved only in so far as the variances are measures of deviations from the mean. Although the t test for means includes an assumption of homogeneity or poolability of variances, the F test for variances includes no assumptions as to means.

7.3 Confidence Range of Variances

The last row of values in the F tables, Tables 7.8 and 7.9, corresponds to the variance ratio when the variance in the denominator is the true variance, i.e., calculated from infinite degrees of freedom. A variance estimated with less than infinite degrees of freedom can be expected to exceed the true variance (calculated with infinite degrees of freedom) by the ratio indicated in the last row of the F table with a frequency corresponding to the probability level of the table. If we transpose Eq. (7.16), assuming that the denominator is the true variance σ^2, we can set a lower confidence range on the true variance in terms of the measured variance:

$$\sigma^2 = \frac{s^2(x)}{F_{\alpha,\nu,\infty}} \tag{7.17}$$

where α is the probability level of the F table, and ν is the number of degrees of freedom used in calculating $s^2(x)$.

Following a similar reasoning, and using the right-hand column of the F table, corresponding to the variance of the numerator being known with infinite degrees of freedom, we can set an upper confidence limit to the true variance in terms of the measured variance:

$$\sigma^2 = F_{\alpha,\infty,\nu} s^2(x) \tag{7.18}$$

The total confidence range of the variance, at the $1 - 2\alpha$ level, is the combination of these two expressions.

For example, if a variance is measured with 16 degrees of freedom and a value of 14.7 is obtained, the 90 per cent confidence range $(1 - 2 \times 0.05)$ is from $14.7/1.64$ to $(2.01)(14.7) = 8.96$ to 29.55. The 98 per cent confidence range $(1 - 2 \times 0.01) = 14.7/1.99$ to $(2.75)(14.7) = 7.39$ to 40.42.

Following this procedure it is only possible to obtain confidence ranges corresponding to the levels of the F tables. However, in a manner that is beyond the scope of this book, the F distribution and the χ^2 distribution are related so that the values of F at infinite degrees of freedom in the last row of the F table are identical with the values of χ^2/ν, at the same probability levels, and the values of F at infinite degrees of freedom in the right-hand column of the F table are identical with the values of ν/χ^2 at α levels of F and $1 - \alpha$ levels of χ^2. With this identity it is possible to calculate the variance confidence range corresponding to any probability level for which the χ^2 value is available. The $1 - 2\alpha$ confidence range of the variance is equal to the calculated variance estimate multiplied by the degrees of freedom over the α and $1 - \alpha$ values of χ^2.

$$\frac{\nu}{\chi_\alpha^2} s^2(x) \leqq \sigma^2 \leqq \frac{\nu}{\chi_{1-\alpha}^2} s^2(x) \tag{7.19}$$

Using the numerical values of the previous illustration:

$$s^2(x) = 14.7$$
$$\nu = 16$$

From Table 5.1, at 16 degrees of freedom, we can get the values of χ^2, and from Eq. (7.19) we calculate the corresponding confidence ranges. Table 7.10 shows the results.

TABLE 7.10

	Probability level					
	α 0.01	$1 - \alpha$ 0.99	α 0.02	$1 - \alpha$ 0.98	α 0.05	$1 - \alpha$ 0.95
$\chi^2(\nu = 16)$	32.000	5.812	29.633	6.614	26.296	7.962
Confidence range:						
Probability, $1 - 2\alpha$	0.98		0.96		0.90	
Lower limit	7.35		7.94		8.94	
Upper limit	40.46		35.56		29.54	

7.4 Homogeneity of More Than Two Variances

The test for the homogeneity of several variances is best illustrated by an example. This is Bartlett's[1] χ^2 test. χ^2 is calculated as the difference between the total number of degrees of freedom times the natural logarithm of the pooled estimate of variance and the sum of the product of individual degrees of freedom and the natural logarithm of the individual estimate of variance.

$$\chi^2 = (\ln \bar{s}^2)\Sigma(k_i - 1) - \Sigma[(k_i - 1) \ln s_i^2] \tag{7.20}$$

or with logarithms to the base 10:

$$\chi^2 = 2.3026\{(\log \bar{s}^2)\Sigma(k_i - 1) - \Sigma[(k_i - 1) \log s_i^2]\} \tag{7.21}$$

[1] M. S. Bartlett, *J. Roy. Statis. Soc.*, Suppl., **4**:137 (1937).

where k_i is the number of terms involved in the calculation of s_i^2, and $k_i - 1$ is the number of degrees of freedom associated with s_i^2. When all the variances are estimated with the same number of degrees of freedom, i.e., when all the k_i's are equal, Eq. (7.21) is simplified to

$$\chi^2 = 2.3026(k - 1)(n \log \bar{s}^2 - \Sigma \log s_i^2) \tag{7.22}$$

where n is the number of variances being compared.

The number of degrees of freedom associated with χ^2 is one less than the number of variances being tested for homogeneity; $n - 1$ in Eq. (7.22). The hypothesis being tested is that the variances are equal. The tabulated χ^2 corresponding to the value calculated from Eq. (7.21) or (7.22) for the number of degrees of freedom involved indicates the probability of this large a value if the hypothesis (of equal variances) was true. If the calculated χ^2 value corresponds to a probability of 0.05 or less, then it is reasonable to conclude with this probability of being wrong that the variances are not homogeneous.

The value of χ^2 calculated from these equations is biased slightly high and needs to be adjusted according to the number of degrees of freedom involved. A correction factor C is calculated from the following equation, and a corrected $\chi_c^2 = \chi^2/C$ is computed, and this is the value on which final judgment of the homogeneity of the variances is based.

$$C = 1 + \frac{1}{3(n-1)} \left[\sum \frac{1}{k_i - 1} - \frac{1}{\Sigma(k_i - 1)} \right]$$

With equal k_i's, this equation simplifies to

$$C = 1 + \frac{(n+1)}{3n(k-1)}$$

C is always greater than 1, so that the corrected χ_c^2 is always less than the χ^2 calculated from Eq. (7.21) or (7.22). Therefore, if the initial calculation of χ^2 shows it not to be significant and the variances may be accepted as homogeneous, it is not necessary to correct for the bias. On the other hand, if the initial calculation indicates that the χ^2 value exceeds the tabulated value at some probability level that is deemed significant, then the correction term should be applied to adjust for the bias and perhaps reverse the conclusion of nonhomogeneity.

Table 7.11 shows the general form of the calculation of χ^2, and Example 7.4 gives a numerical illustration.

example 7.4. The following data[1] are from a series of tests on the rate of crack growth of vulcanized rubber. The numbers given in this example are part of the total data and represent the variance of error

[1] J. M. Buist and G. E. Williams, *ASTM Bull.*, vol. 205, p. 35, April, 1955.

ANALYSIS OF VARIANCE

TABLE 7.11 Bartlett's χ^2 Test for Homogeneity of Variances

Sample no.	Sample size, k	Sum of squares	D.F.	Estimated variance $s^2 = \Sigma'x^2/(k-1)$	Log of variance	Product $(k-1)\log s^2$	Reciprocal of D.F.
i	k_i	$\Sigma'x_i^2$	$k_i - 1$	s_i^2	$\log s_i^2$	$(k_i - 1)\log s_i^2$	$1/(k_i - 1)$
.	.	.	.	.	.	.	.
.	.	.	.	.	.	.	.
n	k_n	$\Sigma'x_n^2$	$k_n - 1$	s_n^2	$\log s_n^2$	$(k_n - 1)\log s_n^2$	$1/(k_n - 1)$
Total........		$\Sigma\Sigma'x_i^2$	$\Sigma(k_i - 1)$	...	$\Sigma \log s_i^2$	$\Sigma[(k_i - 1)\log s_i^2]$	$\Sigma[1/(k_i - 1)]$

Pooled estimate of variance, $\bar{s}^2 = \dfrac{\Sigma\Sigma'x_i^2}{\Sigma(k_i - 1)}$

$$\chi^2 = 2.3026\{(\log \bar{s}^2)\Sigma(k_i - 1) - \Sigma[(k_i - 1)\log s_i^2]\}$$

$$c = 1 + \frac{1}{3(n-1)}\left[\sum \frac{1}{(k_i - 1)} - \frac{1}{\Sigma(k_i - 1)}\right]$$

$$\chi_c^2 = \frac{\chi^2}{c}$$

When all the samples are of the same size, i.e., all the k's are equal, the formulas become:

$$\bar{s}^2 = \frac{\Sigma\Sigma'x_i^2}{n(k-1)} = \frac{\Sigma s^2}{n}$$

$$\chi^2 = 2.3026(k - 1)(n \log \bar{s}^2 - \Sigma \log s_i^2)$$

$$c = 1 + \frac{n+1}{3n(k-1)}$$

for five different conditions. The conclusion in the original presentation of the data was that the error variance for the first condition was less than for the other cases. We shall test the variances for homogeneity to see if there are any significant differences among them.

Data

TABLE 7.12

Condition	Degrees of freedom	Variance estimate
1	90	0.0268
2	90	0.0387
3	90	0.0482
4	90	0.0859
5	90	0.0552

Although all the degrees of freedom are the same, we shall illustrate the solution as if they were different, and then demonstrate the use of the simplified formula for equal degrees of freedom.

Solution

TABLE 7.13

n	$k - 1$	$\Sigma' x^2$	s^2	$\log s^2$	$(k - 1) \log s^2$	$1/(k - 1)$
1	90	2.412	0.0268	0.428135 −2	38.532150 − 180	0.01111
2	90	3.483	0.0387	0.587711 −2	52.893990 − 180	0.01111
3	90	4.338	0.0482	0.683047 −2	61.474230 − 180	0.01111
4	90	7.731	0.0859	0.933993 −2	84.059370 − 180	0.01111
5	90	4.968	0.0552	0.741939 −2	66.774510 − 180	0.01111
Total..	450	22.932	......	3.374825 −10	303.734250 − 900	0.05555

$$\bar{s}^2 = \frac{22.932}{450} = 0.050960 \qquad \log \bar{s}^2 = 0.707229 - 2$$

$$\chi^2 = 2.3026[(450)(-1.292771) - (-596.265750)]$$

$$\chi^2 = (2.3026)(14.518800) = 33.431$$

$$C = 1 + (\tfrac{1}{12})(0.05555 - 0.00222)$$

$$C = 1.0044$$

$$\chi_c^2 = \frac{33.431}{1.0044} = 33.284***$$

$$\chi^2_{0.001,4} = 18.465 \qquad \text{from Table 5.1}$$

ANALYSIS OF VARIANCE

Using Eq. (7.22) for equal degrees of freedom, the calculation is:

$$\chi^2 = (2.3026)(90)[(5)(-1.292771) - (-6.625175)]$$
$$\chi^2 = (2.3026)(90)(0.161320) = 33.431$$
$$C = 1 + \frac{(6)}{(15)(90)} = 1.0044$$
$$\chi_c^2 = \frac{33.431}{1.0044} = 33.284 \quad \text{as before}$$

Conclusion. There is less than one chance in a thousand that these variance estimates are from samples from populations with the same variance. (The three asterisks indicate significance at the 0.001 level).

7.5 Analysis of Variance—Single Classification

7.5.1 Comparison of Several Means. One of the uses of t was to test the difference between two means in terms of the standard deviation of the measurements. One of the simplest uses of the analysis of variance is to compare several means in terms of the pooled variance of the measurements. This is in effect an extension of the t test for two means.

If data are obtained at several different temperatures; if several analysts make duplicate determinations of the same material; if several batches of material are made from each of five or six different drums of material from the same shipment; or if several batches of material are made from shipments from several different vendors—in all cases there will be a variance within the replications which is a measure of the "error" or precision or unassignable variation in the system. There will also be a variation in the means of the results obtained under the different conditions. This variation in the means will be affected by the variation in the replicate measurements, and may also be affected by differences between the temperature, or between the analysts, or between the batches of material, or whatever the classifications of groups are. It is the purpose of the analysis of variance to separate the effects of these factors if they exist.

The variation within one group of k measurements of the same thing is usually accepted as the "error variance" or the minimum, unassignable variance. With n groups of measurements there are n estimates of the variance of the measurement. If these estimates are pooled to give an over-all estimate of the measurement variance, this estimate can be used as a sort of least common denominator to test the existence of other factors. This pooled estimate, $\bar{s}^2(x)$, is an estimate of the residual or error variance of the measurements, σ^2. A variance can be calculated from the means of the different groups of measurements. If there is no other factor operating to cause a variation of the data than that included in the pooled estimate of variance, then the variance of the means is another estimate of the residual variance, in the form of Eq. (6.8), $\sigma^2(\bar{x}) = \sigma^2/k$. Or $ks^2(\bar{x}) \cong \sigma^2$. If the variance estimate of the means, $s^2(\bar{x})$, includes some factor other than that affecting the residual variance, $ks^2(\bar{x})$ includes this variance in addition to the residual variance, or $ks^2(\bar{x}) \cong \sigma^2 + k\sigma^2(G)$, where G is the new factor and k is the number of replicates at each condition of G.

We have two estimates of σ^2: $\bar{s}^2(x)$ and $ks^2(\bar{x})$. We suspect that the second estimate may contain the estimate of the variance of another factor, $s^2(G)$. We set up the hypothesis that the two estimates are equal and test by the F test whether $ks^2(\bar{x})$ is significantly larger than $\bar{s}^2(x)$. $ks^2(\bar{x})$ cannot be less than $\bar{s}^2(x)$ since it is at least an estimate of σ^2. We therefore use the one-sided probability values of Tables 7.8 and 7.9. If the ratio of $ks^2(\bar{x})/\bar{s}^2(x)$ exceeds the tabulated F value for the degrees of freedom corresponding to those for the numerator and denominator, then the hypothesis of equal variances is rejected and $ks^2(\bar{x})$ is accepted as including some additional variance estimate, $s^2(G)$. It is because of the frequency of this type of variance test that the F values are tabulated as a one-sided distribution rather than for a symmetrical distribution as given for the t values.

One of the appealing features (to statisticians) of the analysis of variance is its elegance of calculation. The engineer would be well advised not to be misled by the ease of the computations into simply putting data into the formulas and "grinding" out answers. The analysis of variance must be interpreted with understanding in the same way that the application of any fluid-flow or heat-transfer formula is interpreted. We shall endeavor to point out some of the different interpretations that can be applied to the results, but no set of rules can be given that will cover all the data that will be encountered.

Table 7.14 illustrates symbolically a series of replicate data taken at different conditions of some factor G. Each condition has k_i replicates or observations, so that the mean of each condition $\bar{x}_i = \sum_{j=1}^{k} x_{ji}/k_i$. To test whether the factor G has a significant effect on the variation of x, we compare the variance estimates within the replicates and the variance estimate of the sample means. The variance estimates are obtained by dividing the sums of squares (of deviations from the means) by the degrees of freedom. The total sum of squares for all the data is equal to the sum of squares of the means, multiplied by the number of items making up the mean, plus the sum of squares within each group. This equality is evident from the three sum-of-squares equations at the bottom of Table 7.14.

ANALYSIS OF VARIANCE

TABLE 7.14

Groups	G_1	G_2	G_3	$\cdots$	G_i	$\cdots$	G_n
Item in group	x_{11}	x_{12}	x_{13}	$\cdots$	x_{1i}	$\cdots$	x_{1n}
	x_{21}	x_{22}	x_{23}	$\cdots$	x_{2i}	$\cdots$	x_{2n}
	x_{31}	x_{32}	x_{33}	$\cdots$	x_{3i}	$\cdots$	x_{3n}
	$\cdot$	$\cdot$	$\cdot$		$\cdot$		$\cdot$
	x_{j1}	x_{j2}	x_{j3}	$\cdots$	x_{ji}	$\cdots$	x_{jn}
	$\cdot$	$\cdot$	$\cdot$		$\cdot$		$\cdot$
No. of observations	k_1	k_2	k_3	$\cdots$	k_i	$\cdots$	k_n

Total number of observations $N = \sum_{i=1}^{n} k_i$

Over-all mean $\bar{x} = \sum_{i=1}^{n} \sum_{j=1}^{k} \dfrac{x_{ji}}{N} = \dfrac{\Sigma x}{N}$

	Sum of squares	Degrees of freedom
Total	$\sum_{i=1}^{n} \sum_{j=1}^{k} (x_{ji} - \bar{x})^2 = \Sigma x^2 - \dfrac{(\Sigma x)^2}{N}$	$N - 1$
Among groups	$\sum_{i=1}^{n} k_i(\bar{x}_i - \bar{x})^2 = \sum_{i=1}^{n} \dfrac{\left(\sum_{j=1}^{k} x_{ji}\right)^2}{k_i} - \dfrac{(\Sigma x)^2}{N}$	$n - 1$
Within groups	$\sum_{i=1}^{n} \sum_{j=1}^{k} (x_{ji} - \bar{x}_i)^2 = \Sigma x^2 - \sum_{i=1}^{n} \dfrac{\left(\sum_{j=1}^{k} x_{ji}\right)^2}{k_i}$	$\sum_{i=1}^{n} (k_i - 1)$

When all k's are equal:

$$N = kn \quad \text{and} \quad \sum_{i=1}^{n} (k_i - 1) = n(k - 1)$$

The total degrees of freedom (one less than the total number of items) is equal to the degrees of freedom for the means (one less than the number of means) plus the degrees of freedom for the total within sample sum of squares (the sum of the number of observations in each sample, less 1). This equality is shown in the last column at the bottom of Table 7.14.

The analysis of variance can be computed by calculating two of the sums of squares and two of the degrees of freedom and obtaining the third by difference. The advantage of this procedure is not so evident with the simple one-factor example, but it will be more apparent when we come to discuss more involved situations. Table 7.15 shows the analysis-of-vari-

TABLE 7.15

Source	Sum of squares	D.F.	Mean square or variance estimate
Total............	$\Sigma x^2 - \dfrac{(\Sigma x)^2}{N}$	$N - 1$	
Among groups....	$\Sigma \dfrac{\text{(group total)}^2}{\text{no. in group}} - \dfrac{(\Sigma x)^2}{N}$	$n - 1$	$\dfrac{\text{Group sum of squares}}{n - 1}$
Within groups......	Difference	Difference	$\dfrac{\text{Difference sum of squares}}{\text{Difference D.F.} = N - n}$

ance table calculated from two sums of squares, using the symbols of Table 7.14. Note that $\sum_{j=1}^{k} x_{ji}$ is the total of the ith group so that the among-group sum of squares is calculated from the group sums and the number in each group. The term $(\Sigma x)^2/N$, which is common for both calculated sums of squares, is sometimes referred to as the "correction term" inasmuch as it is common to all the calculated sums of squares in a multifactor analysis of variance.

The quotient of the sum of squares by the degrees of freedom is a variance estimate, either of a single factor or of several factors, depending on the problem. Until the significance of the variance factors is established by the F test, this quotient is referred to as the "mean square." In Table 7.15, the difference mean square, or the within-group mean square, is an estimate of the error or residual variance, σ^2. The among-group mean square is an estimate of the residual variance and k times the variance due to the group factor, $\sigma^2 + k\sigma^2(G)$. We test for the existence of $\sigma^2(G)$ by the ratio of the among-group mean square and the within-group mean square. If this ratio exceeds the tabulated F value for the degrees of freedom corresponding to $n - 1$ and $N - n$, we can reject the hypothesis that two mean squares estimate the same variance, σ^2, and can accept the existence of a contributing factor G.

example 7.5. The following data are from five replicate runs on the time of passage of aluminum powder through six different tests units.[1] We shall endeavor to test whether there is a significant difference among the six test units when compared with the variation within the units.

[1] Metal Powder Association, New York, mimeographed supplement to members.

ANALYSIS OF VARIANCE

Data

TABLE 7.16

Unit	1	2	3	4	5	6
Run 1	52.9	54.0	52.6	50.5	54.6	54.0
Run 2	52.3	53.8	53.2	50.8	54.6	53.3
Run 3	52.2	53.8	53.4	50.7	54.4	53.7
Run 4	52.5	53.6	53.4	50.8	54.4	53.5
Run 5	52.7	53.6	53.0	50.5	54.4	53.7
Unit mean	52.5	53.8	53.1	50.7	54.5	53.6

Solution. All the data are coded by subtracting 50 from each value and multiplying the remainder by 10 to remove the decimal. (We have already demonstrated that coding by adding or subtracting a constant has no effect on the calculation of the sum of squares or the variance. Coding by multiplying by a constant will change the resulting variance by the square of this constant. However, since we shall test the mean squares by a ratio of two calculations, the coding factor will cancel. If we ultimately intend to use the calculated variances, they will have to be decoded by dividing by 100.) The coded data and the calculation of the terms for the analysis of variance are given in Table 7.17. The analysis is shown in Table 7.18.

TABLE 7.17

	Coded data					
Unit	1	2	3	4	5	6
Run 1	29	40	26	5	46	40
Run 2	23	38	32	8	46	33
Run 3	22	38	34	7	44	37
Run 4	25	36	34	8	44	35
Run 5	27	36	30	5	44	37
Total T	126	188	156	33	224	182

The F test, $889.26/5.417$, is obviously significant.
The F value from Table 7.9 for 5 and 24 degrees of freedom at the 0.01 level is only 3.90.

TABLE 7.18   Analysis of Variance

Source	Sum of squares	D.F.	Mean square
Among units	31,989 − 27,542.7 = 4,446.3	5	889.26**
Within units	4,576.3 − 4,446.3 = 130.0	24	5.417
Total	32,119 − 27,542.7 = 4,576.3	29	

There is strong evidence of a factor among units which causes a variation in the results greater than can be accounted for by the variation within units.

Calculations

$$\Sigma x = 29 + 23 + \cdots + 35 + 37 = 909$$

$$\Sigma x^2 = (29)^2 + (23)^2 + \cdots + (35)^2 + (37)^2 = 32{,}119$$

$$\Sigma T^2 = (126)^2 + (188)^2 + \cdots = 159{,}945$$

$$(\Sigma x)^2 = (909)^2 = 826{,}281$$

$$N = 30,\, n = 6,\, k = 5$$

$$\frac{(\Sigma x)^2}{N} = \frac{826{,}281}{30} = 27{,}542.7$$

$$\frac{\Sigma T^2}{k} = \frac{159{,}945}{5} = 31{,}989\dagger$$

The coded variance within units is 5.417. We uncode this value by dividing by 100 to get $\bar{s}^2(x)$ of 0.05417, or the standard deviation of a single measurement for a single test unit is 0.2327.

7.5.2   Interpretation of the Analysis of Variance—Model 1. The differences among the means might be due to real differences in levels of the several groups. If the groups are different sources of supply of a raw material, the purpose of the analysis of variance would be to see whether one or more suppliers furnished material that was significantly higher or lower in some measurable attribute than the others. Or the groups might be different lots of shipments from the same supplier, in which case the

---

† In this example all the groups have the same number of replicate determinations; all the k's are equal. If the number varied for the different groups, we would calculate

$$\frac{\Sigma T_i^2}{k_i} = \frac{(126)^2}{5} + \frac{(188)^2}{5} + \cdots = 31{,}989 \quad \text{as before}$$

With equal k's, it is simpler to calculate ΣT^2 and then to divide by k, as in the numerical illustration.

analysis of variance would be performed to determine how much of the variation in the end product was due to manufacturing procedures and how much was due to variation in the raw material. For the analysis of variance of these two cases with a single factor, the calculation procedures are identical but the interpretation of the results is different.

The analysis of the first type of variation, due to different fixed levels of the group factor, is called a Model 1 type analysis. For the second type of variation, due to random effect of the group factors, the term Model 2 is usually applied. There is a third type of variation, in which the means have some determinable correlation with the levels of the group factor. This relation may be a simple linear one or one of a higher order. The nomenclature is not yet firmly established for this type of variation, but it is sometimes referred to as Model 3. With more than one factor, there can be different combinations of variations, all referred to as Mixed Models.

When the levels of the factors under study represent the total population of interest, if we are studying the results obtained by different analysts in a laboratory, not as samples of all analysts but merely to determine what effect these specific analysts have on the laboratory's results, the variance contribution of the analyst factor is not of interest. What is useful is the difference in the means between analysts, i.e., a Model 1 analysis. When the factor levels are only a sample from a whole population of this factor, if the few analysts studied are taken as a representative sample of all the analysts in a large laboratory, the variation of the results will contain a variance contribution due to different analysts operating on the product; then the number of interest is the variance contribution of the analyst factor under study, i.e., a Model 2 analysis. If the factor is a controllable variable such as temperature, pressure, catalyst concentration, then what may be of interest is the relation between the measurement and the variation of this factor, a Model 3 analysis. The point of all this is not to establish the nomenclature, but to emphasize that the variation among means must be interpreted in terms of the problem at hand, which requires an understanding of the problem as well as a familiarity with the statistics.

If the analysis of variance shows no significant difference among means, the work is finished as far as that particular analysis is concerned. The data may be inspected for some other factor which causes the "within-group" variance to be so large as to mask the among-group effect. Or more data may be taken at the same conditions to decrease the residual variance and to decrease the critical F value by increasing the degrees of freedom. But as it stands, if the computed F value does not exceed the tabulated F value for the corresponding degrees of freedom, we cannot conclude without too great a chance of error that the variation among means is larger than can be accounted for from the variation within groups. If there is ground for rejecting the hypothesis of equal estimates of variance, we may wish to compare all possible pairs of means, or we may wish to group the means into various homogeneous sets. These two procedures are discussed in the next two sections.

Model 1—Differences between All Means. We have completed an analysis of variance and found a significant effect of a set of fixed factors. As with the data of example 7.5, we wish to find which of the means or how many of them differ significantly from the others to cause the variation among means to be significant. The residual variance gives us a measure of the precision of a single measurement. This measure is identical with the pooled estimate of the standard deviation, $\bar{s}(x)$. In our study of the t test we saw that the difference between two means was tested by

$$t = \frac{|\bar{x}_1 - \bar{x}_2|}{\bar{s}(x)\sqrt{1/k_1 + 1/k_2}} \tag{6.24}$$

When $k_1 = k_2$, this equation reduces to

$$t = \frac{|\bar{x}_1 - \bar{x}_2|}{\bar{s}(x)\sqrt{2/k}} \tag{7.23}$$

Since the standard deviation of the mean is equal to $1/\sqrt{k}$ times the standard deviation of a single measurement, Eq. (7.23) may be written

$$t = \frac{|\bar{x}_1 - \bar{x}_2|}{\bar{s}(\bar{x})\sqrt{2}} \tag{7.24}$$

We can transpose the denominator of the right-hand member and obtain an expression for the absolute difference between two means:

$$|\bar{x}_1 - \bar{x}_2| = t\bar{s}(\bar{x})\sqrt{2} \tag{7.25}$$

For any significance level of t at the degrees of freedom corresponding to the calculation of $\bar{s}(x)$, we have an expression for the least significant difference (LSD) between two means. At the 0.05 level of t we can expect a difference between means as large as that of Eq. (7.25) 1 time in 20 when there is no real difference between the factors. Since with n means, there are $(n/2)(n - 1)$ comparisons (the number of combinations of n things taken two at a time), it is sometimes possible to obtain a least significant difference between two means simply because of the large number of comparisons even when no difference exists. Tukey[1] has developed a term which he calls a "wholly significant difference" (WSD), for testing the multiple differences among a number of means. Table 7.19, which

---

[1] John W. Tukey, "The Problem of Multiple Comparison," mimeographed publication of limited circulation, 1953.

ANALYSIS OF VARIANCE

TABLE 7.19 WSD/$s(\bar{x})$

ν \ n	2	3	4	5	6	7	8	9	10	12	15	20	30	60
1	21.96	28.80	34.56	39.60	44.28	48.78	53.10	55.62	57.96	62.46	69.12	77.58	91.98	116.64
2	6.83	8.54	9.88	10.98	11.95	12.87	13.73	14.27	14.76	15.68	17.02	18.67	21.41	26.11
3	4.89	6.00	6.84	7.50	8.07	8.60	9.08	9.41	9.69	10.22	10.98	11.89	13.40	15.96
4	4.19	5.10	5.76	6.27	6.70	7.10	7.45	7.70	7.92	8.31	8.88	9.55	10.62	12.47
5	3.86	4.66	5.24	5.68	6.04	6.37	6.66	6.88	7.06	7.39	7.86	8.41	9.28	10.77
6	3.64	4.38	4.91	5.31	5.63	5.92	6.17	6.37	6.53	6.82	7.23	7.70	8.45	9.73
7	3.50	4.20	4.70	5.06	5.35	5.62	5.85	6.03	6.17	6.44	6.81	7.24	7.91	9.05
8	3.41	4.08	4.55	4.89	5.17	5.41	5.62	5.79	6.00	6.18	6.53	6.92	7.53	8.57
9	3.34	3.98	4.44	4.76	5.03	5.26	5.46	5.62	5.75	5.98	6.31	6.68	7.25	7.99
10	3.28	3.91	4.35	4.66	4.91	5.13	5.32	5.48	5.61	5.83	6.14	6.49	7.02	7.94
12	3.19	3.80	4.22	4.52	4.75	4.96	5.13	5.28	5.40	5.61	5.90	6.22	6.70	7.54
15	3.11	3.69	4.09	4.37	4.60	4.79	4.95	5.09	5.20	5.39	5.66	5.95	6.39	7.14
20	3.04	3.60	3.98	4.25	4.45	4.63	4.78	4.91	5.02	5.19	5.44	5.71	6.11	6.79
30	2.97	3.51	3.87	4.12	4.32	4.48	4.61	4.74	4.84	5.00	5.23	5.47	5.83	6.44
60	2.90	3.41	3.76	4.00	4.18	4.33	4.45	4.57	4.66	4.81	5.02	5.24	5.56	6.09
∞	2.82	3.32	3.66	3.88	4.04	4.18	4.29	4.40	4.49	4.63	4.82	5.01	5.29	5.76

WSD = wholly significant difference between means at 0.05 level.
$s(x)$ = estimate of error variance.
$s(\bar{x})$ = estimate of variance of the means = $s(x)/\sqrt{k}$.
n = number of means under test.
ν = degrees of freedom associated with $s(x)$, the pooled estimate of error variance.
k = number of terms included in the mean.

is a modification of Tukey's tables, gives the values of the WSD/$\bar{s}(\bar{x})$ at the 0.05 significance level for various degrees of freedom for $\bar{s}(x)$, the pooled estimate of error variance, and for different numbers of means, n. $\bar{s}(\bar{x})$ to be used with Table 7.19 is calculated from the residual variance and k, the number of replicates included in the means being tested, $\bar{s}(\bar{x}) = \bar{s}(x)/\sqrt{k}$.

In using Table 7.19, all pairs of means are compared. Any pairs with differences larger than the wholly significant difference are accepted as being different with the 0.05 probability of error. This test is illustrated in the following example with the data from Example 7.5.

example 7.6. The means from Table 7.17, coded to conform with the analysis of variance of Table 7.18, are 25.2, 37.6, 31.2, 6.6, 44.8, 36.4.

$$\bar{s}(x) = 2.327$$
$$k = 5$$
$$\bar{s}(\bar{x}) = 2.327/\sqrt{5} = 1.0406$$
$$n = 6$$
$$\nu = 24$$
$$\frac{\text{WSD}}{\bar{s}(\bar{x})} = 4.39 \quad \text{(by interpolation)}$$
$$\text{WSD} = (4.39)(1.0406) = 4.57$$

To compare the several means, they are arranged in rank order, and the difference between succeeding pairs is compared with the WSD.

TABLE 7.20

Rank	Unit no.	Mean	Difference
1	4	6.6	
			18.6
2	1	25.2	
			6.0
3	3	31.2	
			5.2
4	6	36.4	
			1.2
5	2	37.6	
			7.2
6	5	44.8	

Only the difference between unit 6 and unit 2 is less than the WSD; therefore we may conclude that all the other units give results significantly different from each other.

If more than two consecutively ranked values differ by less than the WSD, these must all be compared with each other and may conveniently be so compared in a table as follows. If there were six means between 31.2 and 44.8, with the following values, none of which differs from the succeeding value by more than the WSD of 4.57,

31.2, 33.3, 36.4, 37.6, 41.2, 44.8

these could be arranged in a table of differences, and those differences exceeding the WSD could be readily selected. They are indicated by the asterisk.

TABLE 7.21 Table of Differences

Mean	44.8	41.2	37.6	36.6	33.3
31.2	13.6*	10.0*	6.4*	5.2*	2.1
33.3	11.3*	7.9*	4.3	3.1	
36.4	8.4*	4.8*	1.2		
37.6	7.2*	3.6			
41.2	3.6				

If, when the means are ranked, the largest and the smallest value do not differ by more than the WSD, it is obvious that none of the intervening differences can be greater than the WSD.

Confidence limits may be set on any of the means in the analysis-of-variance table for any probability level by establishing the range $\pm t\bar{s}(x)/\sqrt{k}$, where t is selected for the desired probability level at the degrees of freedom corresponding to the calculation of $\bar{s}^2(x)$, and k is the number of terms included in the calculation of the mean. This formulation is identical with that for the confidence range of a single mean calculated from a series of replicate measurements.

Model 1—Differences among Several Groups of Means. If instead of comparing the means in pairs we desire to group the means into homogeneous sets, we can proceed by an alternative method also devised by Tukey.[1] This method involves a series of steps which are described below and illustrated in example 7.7:

[1] John W. Tukey, *Biometrics*, **5**:99 (1949).

1. Select a significance level, α.
2. Arrange the means in rank order.
3. From the residual variance $\bar{s}^2(x)$ calculate (a) the variance of the means, $\bar{s}^2(\bar{x}) = \bar{s}^2(x)/k$; (b) the standard deviation of the means, $\bar{s}(\bar{x})$.
4. From Table 6.1 select the t value corresponding to the significance level and degrees of freedom associated with $\bar{s}^2(x)$. With this t value, calculate the least significant difference between means, LSD = $t\bar{s}(\bar{x})\sqrt{2}$.
5. Separate the ranked means into groups wherever the difference between two successive means is greater than LSD.
6. Any group containing one or two means is considered a separate group.
7. With any group containing more than two means, determine the group mean $\bar{x}_m$ and determine the largest difference for the group, $d_L = |\bar{x}_i - \bar{x}_m|$.
8. Calculate t from one of the two equations which applies to the group:
For groups of three means:

$$t = \frac{d_L/\bar{s}(\bar{x}) - 0.5}{3(0.25 + 1/\nu)}$$

For groups of greater than three means:

$$t = \frac{d_L/\bar{s}(\bar{x}) - 1.2 \log n'}{3(0.25 + 1/\nu)}$$

where n' = number of means in group
ν = number of degrees of freedom associated with $\bar{s}(x)$

9. If t is greater than the tabulated t at the significance level selected and infinite degrees of freedom, separate from the group the mean corresponding to $\bar{x}_i$ and repeat the calculation from step 8 using a new group mean. If the t value is less than the tabulated t, accept the group as being homogeneous at that selected significance level.

example 7.7. The following results were obtained from data from another series of runs with different tests units similar to those of example 7.5. The analysis of variance followed the same form as in the earlier example. Only the results are given.

The F ratio for the two estimates of variance, 267.91/5.56, is obviously significant. Because of the number of units tested, we are more interested in grouping them into homogeneous sets than in examining the differences between all pairs. We proceed as outlined in the preceding section.

TABLE 7.22

Source	Sum of squares	D.F.	Mean square
Between groups.....	2,679.10	10	267.91***
Within groups.......	244.56	44	5.56
Total............	2,923.66	54	

$k = 5$ Means: 13.1, 12.3, 8.4, 27.6, 6.8, 18.6,
$n = 11$ 11.0, 17.4, 20.1, 21.0, 29.0
$N = 55$

1. Select an α level, say 0.05.
2. Arrange the means in rank order: 6.8, 8.4, 11.0, 12.3, 13.1, 17.4, 18.6, 20.1, 27.6, 29.0.
3. Calculate $\bar{s}(\bar{x})$: $= \sqrt{5.56/5} = 1.058$.
4. Calculate LSD (least significant difference):

$$= t_{0.05,44}\bar{s}(\bar{x}) \sqrt{2} = (2.000)(1.058)(1.414)$$
$$= 2.992$$

5. Separate means into preliminary groups where successive values differ by 2.992 or more:

6.8	17.4	27.6
8.4	18.6	29.0
11.0	20.1	
12.3	21.0	
13.1		

6. Test the first group:

$n' = 5$ Mean $= 10.32$ $d_L = 10.32 - 6.8 = 3.52$

$\log n' = 0.69897$

$1.2 \log n' = 0.8388$

$$t = \frac{(3.52/1.058) - 0.8388}{3(0.25 + 1/44)}$$

$$= \frac{3.327 - 0.8388}{0.8181} = 3.041$$

$t_{0.05,\infty} = 1.96$ (therefore remove 6.8 value from group 1)

7. Test truncated group:

$n' = 4$ Mean = 11.20 $d_L = 11.20 - 8.4 = 2.80$

$\log n' = 0.60206$

$1.2 \log n' = 0.7225$

$$t = \frac{(2.80/1.058) - 0.7225}{0.8181}$$

$$= 2.351$$

Again remove 8.4 from group since $t > 1.96$.

8. Test remaining three-member group:

Mean = 12.13 $d_L = 12.13 - 11.0 = 1.13$

$$t = \frac{(1.13/1.058) - 0.5}{0.8181}$$

$$= \frac{0.568}{0.8181} = 0.694$$

Accept groups as homogeneous since $t < 1.96$.

9. Test the second group:

$n' = 4$ Mean = 19.3 $d_L = 19.3 - 17.4 = 1.9$

$1.2 \log n' = 0.7225$

$$t = \frac{(1.90/1.058) - 0.7225}{0.8181}$$

$$= \frac{1.104}{0.8181} = 1.35$$

Accept group as homogeneous since $t < 1.96$.

10. Final grouping:

6.8	11.0	17.4	27.6
8.4	12.3	18.6	29.0
	13.1	20.1	
		21.0	

7.5.3 Interpretation of the Analysis of Variance—Model 2. For the single-factor analysis of variance, model 2 offers very few problems. When the variation in the means represents the action of some random factor, it is the variance of this factor and the contribution of this variance to the total variance that is of interest. The residual mean square is an estimate of the error or unassignable variance, σ^2. The among-group mean square is an estimate of the error variance plus k times the variance due to the particular factor if it exists, where k is the number of items in the groups, $\sigma^2 + k\sigma^2(G)$. The F test to establish the existence of $\sigma^2(G)$ is the ratio of these terms: (group mean square)$/\bar{s}^2(x)$, where $\bar{s}^2(x)$ is the error mean square, or estimate of σ^2. If the F term is significant, the calculation of $s^2(G)$, the estimate of $\sigma^2(G)$, follows directly from these relations and is equal to [group mean square $- \bar{s}^2(x)]/k$.

This relation is shown schematically in Table 7.23.

TABLE 7.23   Analysis of Variance, Model 2

For n groups with k observations in each group

Source	Sum of squares	D.F.	Mean square	Variance estimated
Groups.......	$\Sigma \dfrac{(\text{group sum})^2}{\text{no. in group}} - \dfrac{(\Sigma x)^2}{N}$	$n - 1$	$\dfrac{\text{Group S.S.}}{\text{Group D.F.}}$	$\sigma^2 + k\sigma^2(G)$
Within groups or residual	Difference	Difference	$\dfrac{\text{Residual S.S.}}{\text{Residual D.F.}}$	σ^2
Total.......	$\Sigma x^2 - \dfrac{(\Sigma x)^2}{N}$	$N - 1$		

$$F \text{ test: } \frac{\sigma^2 + k\sigma^2(G)}{\sigma^2}$$

If each of the means is not calculated from the same number of terms, an average k value, $\bar{k}$, is used in the calculation of the final variance estimate for groups, $\sigma^2(G)$. This average $\bar{k}$ value cannot be used in the calculation of the group sum of squares. The square of each group sum must be divided by the k value corresponding to the number of terms in that group. The $\bar{k}$ term in the group-mean-square variance estimate, $\sigma^2 + \bar{k}\sigma^2(G)$, is an average of the terms in the groups. This average value is calculated from the following equation:

$$\bar{k} = \frac{1}{n - 1}\left(N - \frac{\Sigma k^2}{N}\right) \tag{7.26}$$

When all the k's are equal, this equation for $\bar{k}$ is identical with $\Sigma k/n$.

The total variance of the individual measurements is the sum of the variances of the factors involved in the measurements; in our example with a single factor G, the total variance is the sum of the error variance and the variance of G.

$$\sigma^2(X) = \sigma^2 + \sigma^2(G) \tag{7.27}$$

The confidence range of an individual measurement can be established by using the standard deviation and t at the significance level desired. The number of degrees of freedom associated with the total variance is the weighted mean of the degrees of freedom of the separate variances, weighted in proportion to the magnitude of the variances.

$$\nu_X = \frac{\nu_r \sigma^2 + \nu_G \sigma^2(G)}{\sigma^2 + \sigma^2(G)} \tag{7.28}$$

where ν_r is the degrees of freedom associated with the residual mean square.

7.5.4 Interpretation of the Analysis of Variance—Model 3. There are often cases when the factor under examination is a variable at several levels, such as temperature, pressure, etc. When the analysis of variance shows this factor to make a significant contribution to the variance of the measurements, what is of interest is not the effect of specific levels of the variable or the over-all variance of the variable, but the manner in which the variation of the measurement is related to the variation of the temperature or pressure. This is a regression or correlation problem. When the levels of the factor are such that they can be coded into a series of integers, 1, 2, 3, . . . , etc., and there are at least three levels of the factor, then the variance due to this factor can be readily partitioned into linear, quadratic, cubic, etc., components, depending on the number of levels available for analysis.

If data are obtained at several levels of catalyst concentration, and if the levels can be coded into a series of regular integers, the analysis of variance is carried out in the usual manner and the sum of squares due to "catalyst level" is partitioned into a mean square of linear regression, a mean square of quadratic regression, etc., and each individually tested for significance against the "error" mean square. (Note that the levels of the variable need not actually be coded into integers. It is only necessary that it be possible to code them. As long as they are in some systematic order: 0.005, 0.05, 0.5, 5.0 per cent; or 100, 150, 200, 250, 300 degrees; or $\frac{1}{2}$, $\frac{1}{4}$, $\frac{1}{8}$, $\frac{1}{16}$, $\frac{1}{32}$ min; the partitioning into difference orders of regression can be performed by the method to be outlined. This fact should be borne in mind when planning the conditions of an experiment.)

The partitioning of the sum of squares into various orders of regression is equivalent to determining the incremental decrease in the sum of squares of deviation attributable to the successive terms of an equation of the form

$$x = a + bG + cG^2 + dG^3 + \cdots \tag{7.29}$$

where x is the measurement and G is the factor which is varied in some regular manner.

The sum of squares of means may be looked upon as the sum of the squares of the vertical distances of the individual means from a horizontal line drawn through the means at the value of the over-all mean and in the units of the particular measurement involved. This is the case when just the first two terms of Eq. (7.29) are involved and $a = \bar{x}$; $\bar{x}$ is substituted for each x, and the sum of the squares of the differences between $\bar{x}$ and x is the sum of squares. Sometimes a line of slope b drawn through the means will have a smaller sum of squares of vertical distances to the individual means than does the horizontal line. This is represented by the first three terms of Eq. (7.29): $x = a + bG$. Perhaps a parabolic equation of the second order might give even less squared deviation. In each case the addition of another term in the line from which the deviations are measured involves the calculation of one constant, a, b, c, etc., and hence a corresponding decrease in the number of degrees of freedom associated with the calculation of the mean square—the sum of squares divided by the degrees of freedom. The partitioning of the sum of squares into the various regression components is a method for determining the decrease in the sum of squares of deviation as each higher-order term is added to the regression equation. Each incremental decrease in sum of squares has 1 degree of freedom and can be tested for significance against the estimate-of-error variance (the residual mean square) by the F test. When the number of terms in the right-hand side of Eq. (7.29) is the same as the number of levels of the variable being tested, a perfect fit of the means should result and the total sum of squares of deviation should be accounted for.

We shall leave the discussion of the method of calculating the actual constants of the equation until the chapter dealing with regression and correlation. We deal here only with the sum of squares of deviation attributable to each term.

Table 7.24 shows the formulas for the total sum of squares of means and the sum of squares for the various regression terms. The total sum of squares of means is identical with the formula given in Table 7.15 first illustrating the analysis of variance for the difference between means.

When the number of degrees of freedom for the terms of the regression equation is equal to the degrees of freedom for the sum of squares of means, i.e., when the order of the regression equation is $n - 1$, the total sum of squares accounted for by the successive terms must equal the total sum of squares attributed to the means. This relationship provides a check to the calculations. However, it is not usual for a regression above the third order to be of much interest in ordinary engineering problems.

The calculation of the sum of squares removed by each term of the regression equation can be made one at a time, and when the major por-

TABLE 7.24

	Sum of squares of deviation of means	D.F.
Total S.S. of means.............	$\Sigma(T_i^2/k_i) - [(\Sigma x)^2/N]$	$n-1$
Removed by linear term........	$[\Sigma(h_{1i}T_i)]^2/k\Sigma h_1^2$	1
Addition removed by:		
Quadratic term..............	$[\Sigma(h_{2i}T_i)]^2/k\Sigma h_2^2$	1
Cubic term.................	$[\Sigma(h_{3i}T_i)]^2/k\Sigma h_3^2$	1
Quartic term...............	$[\Sigma(h_{4i}T_i)]^2/k\Sigma h_4^2$	1
.		
.		
.		

where h_1 = coefficient for first-order, or linear, regression
        h_2 = coefficient for second-order, or quadratic, regression
h_3, h_4, etc. = cubic and quartic regressions
        T_1 = total of first group of data
        T_2 = total of second group of data
        T_i = total of ith group of data, etc., when groups are arranged in codable order corresponding to sequence of integers
       h_{11} = linear coefficient for first total
       h_{12} = linear coefficient for second total
         .
         .
         .
       h_{23} = quadratic coefficient for third total
         .
         .
         .
       h_{4i} = quartic coefficient for ith total
        n = number of groups of data; also number of means
        k = number of items making up a group, i.e., number of replicates

Table 7.25 gives the values of h for several different sets of means. A more extensive table is given in Chap. 9. See Table 9.33.

$\Sigma(h_{1i}T_i)$ indicates the sum of the product of the linear partitioning coefficient times the group total: $h_{11}T_1 + h_{12}T_2 + h_{13}T_3 + \cdots$.

tion of the sum of squares has been accounted for, or when the remaining sum of squares divided by the remaining degrees of freedom is no longer significantly larger than the error mean square, the computation can be stopped. If a regression relation is customarily accepted as linear, or of some higher order, it is possible to calculate the additional sum of squares that would be removed by the addition of a higher-order term without calculating the sum of squares removed by the lower-order terms for the particular data involved.

ANALYSIS OF VARIANCE

TABLE 7.25 Partitioning Coefficients, h

No. of means n	Order of h term	Sequence of means, i						Σh^2
		1	2	3	4	5	6	
3	h_1	−1	0	1				2
	h_2	1	−2	1				6
4	h_1	−3	−1	1	3			20
	h_2	1	−1	−1	1			4
	h_3	−1	3	−3	1			20
5	h_1	−2	−1	0	1	2		10
	h_2	2	−1	−2	−1	2		14
	h_3	−1	2	0	−2	1		10
	h_4	1	−4	6	−4	1		70
6	h_1	−5	−3	−1	1	3	5	70
	h_2	5	−1	−4	−4	−1	5	84
	h_3	−5	7	4	−4	−7	5	180
	h_4	1	−3	2	2	−3	1	28
	h_5	−1	5	−10	10	−5	1	252

TABLE 7.26

	550°	600°	650°	700°	750°
	6	32	45	63	87
	4	26	45	62	85
	5	24	44	44	72
	5	22	34	39	80
Totals:	20	104	168	208	324

example 7.8. Table 7.26 represents the results of a series of runs made at increasing temperatures. The numbers in the table are the per cent of equilibrium of the water-gas shift reaction for a particular catalyst and set of operating conditions.

Solution

$$\Sigma x = 824 \quad (\Sigma x^2) = 678{,}976 \quad k = 4$$
$$\Sigma x^2 = 47{,}652 \quad n = 5 \quad N = 20$$

$$\Sigma T^2 = (20)^2 + (104)^2 + \cdots = 187{,}680$$

$$\frac{(\Sigma x)^2}{20} = 33{,}948.8$$

$$\frac{\Sigma T^2}{4} = 46{,}920$$

Since the mean square for temperatures is highly significant, and since the temperature factor can be coded into the sequence of integers by subtracting 500 from each term and dividing the remainder by 50, we shall partition the mean sum of squares into the various components of a polynomial regression to find the simplest type of equation to account for the variation of the approach to equilibrium as a function of temperature. Following the formulas of Table 7.24, the sum of squares accounted for by each order of the regression equation is calculated as follows. Note that with the fourth-order equation the total sum of squares is equal to the sum of squares of means in Table 7.27, since $n - 1 = 4$.

TABLE 7.27   Analysis of Variance

Source	Sum of squares	D.F.	Mean square
Mean	46,920 − 33,948.8 = 12,971.2	4	3,242.8***
Error	Difference = 732.0	15	48.8
Total	47,652 − 33,948.8 = 13,703.2	19	

$$F = \frac{3{,}242.8}{48.8} = 66.4^{***}$$

$$F_{0.001,4,15} = 8.25$$

Linear:

$$\frac{[(-2)(20) + (-1)(104) + (0)(168) + (1)(208) + (2)(324)]^2}{(4)(10)}$$

$$= \frac{(712)^2}{40} = 12673.6$$

Quadratic:

$$\frac{[(2)(20) + (-1)(104) + (-2)(168) + (-1)(208) + (2)(324)]^2}{(4)(14)}$$

$$= \frac{(40)^2}{56} = 28.6$$

Cubic:

$$\frac{[(-1)(20) + (2)(104) + (0)(168) + (-2)(208) + (1)(324)]^2}{(4)(10)}$$

$$= \frac{(96)^2}{40} = 230.4$$

Quartic:

$$\frac{[(1)(20) + (-4)(104) + (6)(168) + (-4)(208) + (1)(324)]^2}{(4)(70)}$$

$$= \frac{(104)^2}{280} = 38.6$$

Summarizing:

TABLE 7.28

Regression term	Sum of squares	D.F.
Linear	12,673.6	1
Quadratic	28.6	1
Cubic	230.4	1
Quartic	38.6	1
Total	12,971.2	4

The quadratic and the quartic terms are not significant when compared to the error mean square of 48.8. The cubic term is just significant at the 0.05 level, $F = 230.4/48.8 = 4.72$. $F_{0.05,1,15} = 4.54$. The linear term is highly significant. The variation in the mean approach to equilibrium can be well represented by a function $a + bT$ where T is the temperature; or it might be slightly better represented by a function $a + bT + cT^3$. The calculation of the constants a, b, and c is discussed in Chap. 9 dealing with regression and correlation. The regression equation, when it is calculated, is the equation relating the mean values of x at each temperature to the temperature.

7.6 Single Classification with Subgrouping

It is often the case in collecting data that the principal classifications can be divided into subclassifications which do not cut across the main classes. Several analysts might draw two, or three, or four specimens (not necessarily the same number for each analyst) from a batch of material and run several replicate analyses on each specimen. The replicate analyses give an estimate of the error variance. The specimens run by the same analyst give a measure of the variation within the batch plus the error. The differences among the results of the several analysts include not only the variation among analysts but also both the variation between specimens and the error variation. The same situation holds if several runs are made on each of several reactors; or if several test pieces are cut from each of several bolts of fabric made wholly from several different machines from each of several plants. As long as there is no relation between the corresponding members of the different groupings the analysis of variance by subgroups, or nesting as it is sometimes called, applies. When there is relation between the members of the subgroups, so that the corresponding division of each subgroup can be considered as a separate class, we have an analysis of variance for more than one main classification. This situation is discussed in the next section.

The general arrangement of data for a single classification with several hierarchies of subgroupings is shown in Fig. 7.1. The analysis of variance is shown symbolically in Table 7.29. The mean-square column has been omitted from Table 7.29 as this is the quotient of the sum of squares by the degrees of freedom. The last column in Table 7.29 indicates the variance that is estimated by the corresponding mean square. The significance of the effect of each subgrouping and of the main classification is

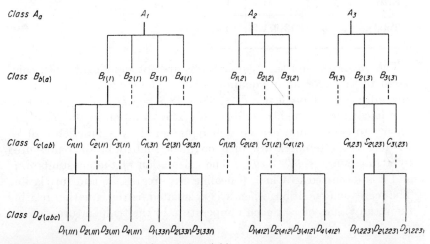

Fig. 7.1 System of nested classes of variables.

measured by testing each mean square against the next lower one in the table by the F test. The "variance estimated" column shows that each mean square contains the variances estimated by the preceding value plus one additional factor for the particular subgroups involved. If the

ANALYSIS OF VARIANCE

TABLE 7.29 Analysis of Variance for Single Classification of Nested Groups

- a Main classes, factor A
- b Subgroups in A, factor $B(A)$
- c Subgroups in B, factor $C(AB)$
- d Subgroups in C, factor $D(ABC)$
- k replicates in D

Total data $= abcdk = N$

Source	Sum of squares	D.F.	Variance estimated
Main class, A	$\sum \dfrac{(A \text{ totals})^2}{\text{no. in class} = kdcb} - \dfrac{(\Sigma x)^2}{N}$	$a-1$	$\sigma^2 + k\sigma^2(D) + kd\sigma^2(C) + kdc\sigma^2(B) + kdcb\sigma^2(A)$
Among B's within A's	$\sum \dfrac{(B \text{ totals})^2}{kdc} - \sum \dfrac{(A \text{ totals})^2}{kdcb}$	$a(b-1)$	$\sigma^2 + k\sigma^2(D) + kd\sigma^2(C) + kdc\sigma^2(B)$
Among C's within B's	$\sum \dfrac{(C \text{ totals})^2}{kd} - \sum \dfrac{(B \text{ totals})^2}{kdc}$	$ab(c-1)$	$\sigma^2 + k\sigma^2(D) + kd\sigma^2(C)$
Among D's within C's	$\sum \dfrac{(D \text{ totals})^2}{k} - \sum \dfrac{(C \text{ totals})^2}{kd}$	$abc(d-1)$	$\sigma^2 + k\sigma^2(D)$
Error	$\Sigma x^2 - \sum \dfrac{(D \text{ totals})^2}{k}$	$abcd(k-1)$	σ^2
Total	$\Sigma x^2 - (\Sigma x)^2/N$	$N-1$	

F test indicates the mean square to be significant, the additional variance estimate can be readily calculated from the difference between the two mean squares.

For example, in Table 7.29, the mean square for "among C factors" is an estimate of $\sigma^2 + k\sigma^2(D) + kd\sigma^2(C)$, and the mean square for "among D factors" is an estimate of $\sigma^2 + k\sigma^2(D)$. The ratio of these, if found to be significant by the F test at the corresponding degrees of freedom, $ab(c-1)$ and $abc(d-1)$, indicates that the term $kd\sigma^2(C)$ is real and effective. The value of $s^2(C)$, which is the estimate of $\sigma^2(C)$ is calculated by subtracting the mean square for the D factors from the mean square for the C factors and dividing by kd.

If any variance estimate is found to be not significant, that term may be ignored in testing the next higher mean square for the significance of the next variance factor.

example 7.9. The data in Table 7.30 represent a series of runs made by three operators to locate a source of variation in a chemical analysis. The procedure consisted of taking a specimen, treating the specimen in a combustion-tube furnace, and performing a chemical analysis. In the test, three operators each took two specimens and made three

TABLE 7.30

A	1						2						3					
$B(A)$	11			12			21			22			31			32		
$C(AB)$	111	112	113	121	122	123	211	212	213	221	222	223	311	312	313	321	322	333
Duplicate results	156	151	154	148	154	147	125	94	98	118	112	98	184	172	181	172	181	175
	154	154	160	150	157	149	125	95	102	124	117	110	184	186	191	176	184	177

combustion runs on each specimen and titrated each run in duplicate. The single letter A represents the operator factor, the double letter B within A, that is $B(A)$, indicates the first or second specimen taken by the operator, the triple letter C within A and B, that is, $C(AB)$, indicates the combustion run on each specimen by each operator. The results are shown in duplicate. In the nomenclature of Table 7.29, $a = 3$, $b = 2$, $c = 3$, and $k = 2$.

Solution

$$\Sigma x = 156 + 151 + 154 + \cdots + 176 + 184 + 177 = 5{,}315$$

$$\Sigma x^2 = (156)^2 + (151)^2 + \cdots + (184)^2 + (177)^2 = 817{,}085$$

ANALYSIS OF VARIANCE

$$\Sigma(\Sigma C)^2 = (310)^2 + (305)^2 + \cdots + (365)^2 + (352)^2 = 1{,}633{,}556$$
$$\Sigma(\Sigma B)^2 = (929)^2 + (905)^2 + \cdots + (1{,}098)^2 + (1{,}065)^2$$
$$= 4{,}891{,}254$$
$$\Sigma(\Sigma A)^2 = (1{,}834)^2 + (1{,}318)^2 + (2{,}163)^2 = 9{,}779{,}249$$
$$\frac{\Sigma(\Sigma C)^2}{k} = \frac{1{,}633{,}556}{2} = 816{,}778$$
$$\frac{\Sigma(\Sigma B)^2}{kc} = \frac{4{,}891{,}254}{6} = 815{,}209$$
$$\frac{\Sigma(\Sigma A)^2}{kcb} = \frac{9{,}779{,}249}{12} = 814{,}937$$
$$\frac{(\Sigma x)^2}{N} = \frac{(5{,}315)^2}{36} = 784{,}700$$

Discussion. By far the largest source of variation in results is operators. Whether this is a model 1 or a model 2 type of difference is a matter of interpretation of the problem. There is no evidence of variation between specimens. There is a definite (probability of error < 0.001) source of variation in the combustion step in the analysis. The estimated variances for the two significant factors, calculated from the mean squares, are:

$$s^2(\text{operators}) = \frac{15{,}618 - 131}{(2)(3)(2)} = 1{,}288$$

$$s^2(\text{combustion}) = \frac{131 - 17}{2} = 57$$

$$s^2(\text{error}) = 17$$

The discussion for the simple single-factor analysis with regard to model 1, 2, or 3 type of variation applies to the analysis with subgrouping. The procedure for analysis with respect to the subgroups is the same whether the factors are fixed or random, model 1 or model 2, or whether there is regression, model 3. The interpretation, however, will be different, and that is a matter of understanding the problem under investigation.

In the formulas given in Table 7.29 the number of data points in each similar type of classification was the same, so that values representing the number in a class or subgroup were a constant. It is not necessary that the groups or subgroups be of the same size, and the analysis of variance will be rigorously correct if instead of dividing the sum of the group totals squared by the number in all the groups, each group total squared is divided by the particular number for that group.

$$\sum \frac{(\text{group total})^2}{\text{no. in group}}$$

However, the exact calculation of the variances of the different factors is difficult because of the necessity of using an average value for the number of measurements in a group in the estimate-of-variance formulas. A

TABLE 7.31 Analysis of Variance

Source	Sum of squares	D.F.	Mean square
Operators, A............	$814{,}937 - 784{,}700 = 30{,}237$	2	15,618***
Specimen within operators, $B(A)$................	$815{,}209 - 814{,}937 = 272$	3	91
Combustion within operators and specimens, $C(AB)$................	$816{,}778 - 815{,}209 = 1{,}569$	12	131***
Replicates, error.........	$817{,}085 - 816{,}778 = 307$	18	17
Total................	$817{,}085 - 784{,}700 = 32{,}385$	35	

$$F \text{ test for } C\text{'s: } \frac{131}{17} = 7.70; \quad F_{0.001, 12, 18} = 5.13$$

$$F \text{ test for } A\text{'s: } \frac{15{,}618}{131} = 119; \quad F_{0.001, 2, 12} = 12.97$$

working approximation for the average value can be obtained by using a formula similar to Eq. (7.26):

$$\bar{k} = \frac{1}{m-1}\left(N - \frac{\Sigma k_i^2}{N}\right) \tag{7.30}$$

where $\bar{k}$ = average number in a subgroup
m = number of subgroups of the particular factor involved
k_i = number in a subgroup
N = total number of measurements

7.7 Analysis of Variance—Two-factor Classification of Data

If a series of experiments is run at several different temperatures and at several different pressures, or if material from several sources of supply is tested under a variety of conditions, or if a group of operators makes a

ANALYSIS OF VARIANCE

series of runs in a number of pilot plants, the data can be arranged as in Fig. 7.2. The analysis of variance for this arrangement of data is similar to that for the single-factor arrangement: the column-factor effect is calculated from the squares of the column totals, and the row-factor effect

C factor	C_1	C_2	$\cdots$	C_j	$\cdots$	C_c
R factor			Data			
R_1	X_{11}	X_{12}	$\cdots$	X_{1j}	$\cdots$	X_{1c}
R_2	X_{21}	X_{22}	$\cdots$	X_{2j}	$\cdots$	X_{2c}
.	.					
.	.					
.	.					
R_i	X_{i1}	X_{i2}	$\cdots$	X_{ij}	$\cdots$	X_{ic}
.	.					
.	.					
.	.					
R_r	X_{r1}	X_{r2}	$\cdots$	X_{rj}	$\cdots$	X_{rc}

Fig. 7.2 Two-factor classification of data.

is calculated from the squares of the row totals. Since each column total includes one run at each row condition, the effect of the row factor is canceled out in comparing column totals or means. The same argument applies to the row totals in which one set of data for each column condition is included in the total, so that in the comparison of these totals the effect of the column factor is removed.

The analysis of variance for the two-classification system is shown in Table 7.32.

The number of degrees of freedom associated with the column effect is one less than the number of columns, $c - 1$, and the number of degrees

TABLE 7.32

Source	Sum of squares		D.F.
Column effect......	$\Sigma \dfrac{\text{(column totals)}^2}{\text{no. in column}}$	$- \dfrac{(\Sigma x)^2}{N}$	No. of columns $- 1$
Row effect.........	$\Sigma \dfrac{\text{(row totals)}^2}{\text{no. in row}}$	$- \dfrac{(\Sigma x)^2}{N}$	No. of rows $- 1$
Error.............	Difference		Difference
Total...........	Σx^2	$- \dfrac{(\Sigma x)^2}{N}$	$N - 1$

of freedom for the row effect is one less than the number of rows, $r - 1$. The total degrees of freedom is one less than the total number of data points, $N - 1$, or $rc - 1$. This leaves $(c - 1)(r - 1)$ degrees of freedom for estimating the error or residual variance. Without any replication of experimental conditions it is therefore possible to obtain an estimate of the error variance by subtracting from the total sum of squares the sum of squares due to the row and column factors in an experiment which has been prepared in this manner.

The calculation of the mean square for each effect and for the error is, as always, the quotient of the sum of squares by the degrees of freedom. The column mean square estimates the error variance plus the number of items in a column times the column-factor variance. The row mean square estimates the same thing for the row factor.

$$\text{Column mean square} = \sigma^2 + (\text{no. in column}) \sigma^2(C) \tag{7.31}$$

$$\text{Row mean square} = \sigma^2 + (\text{no. in row}) \sigma^2(R) \tag{7.32}$$

$$\text{Error mean square} = \sigma^2 \tag{7.33}$$

The F test for each effect is the ratio of the mean square for that effect to the error mean square. If this ratio is significant, i.e., larger than the tabulated F value for the corresponding degrees of freedom, then the effect can be accepted as being real, and the magnitude of the effect can be calculated by subtracting the error variance and dividing by the constant term in Eq. (7.31) or (7.32). This is the same procedure followed for the single-factor analysis.

example 7.10. The following data[1] give in coded form the effect of different-size fractions of graphite powder on the shrinkage of powdered iron, compacted at two different pressures. The analysis of variance is carried out to see whether the size fractions and the pressure significantly affect the shrinkage of the finished iron; a large number indicates a greater shrinkage, which is undesirable.

Analysis of Variance Solution

$c = 6$ $\Sigma x = 19.32$ $\Sigma(\text{col. totals})^2 = 63.7598$

$r = 2$ $\Sigma x^2 = 32.1948$ $\Sigma(\text{row totals})^2 = 188.2874$

$N = 12$ $\dfrac{(\Sigma x)^2}{N} = 31.1052$ $\dfrac{\Sigma(\text{col. totals})^2}{2} = 31.8799$

$\dfrac{\Sigma(\text{row totals})^2}{6} = 31.3812$

[1] Arthur Lesser, Jr., *Iron Age*, Aug. 22, 1946, p. 50.

ANALYSIS OF VARIANCE

TABLE 7.33 Shrinkage of Sintered Specimen

Compacting pressure, tons/sq in.	Powder fraction						
	1	2	3	4	5	6	Total
25	1.47	1.59	1.78	1.51	1.90	2.32	10.57
50	1.15	1.38	1.50	1.38	1.44	1.90	8.75
Total	2.52	2.97	3.28	2.89	3.34	4.22	19.32

Discussion. The difference between the means at the two pressures is obviously a model 1 effect. The 25-ton pressure mean is 1.76, and the 50-ton pressure mean is 1.46. The difference between the means at the two pressures is 0.30. The least significant difference, as given in Eq. (7.25), is t times the standard deviation of the mean (the square-root of the error variance divided by the number of measurements included in the mean) times the square-root of two. t at the 0.05 level for five degrees of freedom is 2.571. The LSD is therefore $(2.571) \sqrt{(0.0078/6)} \, (1.414) = 0.13$. The effect of pressure is therefore at least $0.30 - 0.13 = 0.17$ units.

The effect of powder fraction is probably also a model 1 effect, but it might be interpreted as a model 3 effect. If the sequence of powder sizes is in some systematic order in Table 7.33, there appears to be evidence of a linear regression of mean shrinkage with increase in powder fraction. However, since no units are given for this factor, it is not possible to establish this regression with any certainty. The

TABLE 7.34 Analysis of Variance

Source	Sum of squares	D.F.	Mean square
Powder fractions	$31.8799 - 31.1052 = 0.7747$	5	0.1549**
Pressure	$31.3812 - 31.1052 = 0.2760$	1	0.2760**
Error	Difference $= 0.0389$	5	0.0078
Total	$32.1948 - 31.1052 = 1.0896$	11	

F tests: $\dfrac{1549}{78} = 19.8, \; F_{0.01,5,5} = 10.97$

$\dfrac{2760}{78} = 35.4, \; F_{0.01,1,5} = 16.26$

means can be tested for wholly significant differences between pairs, or they may be grouped into homogeneous sets according to the method outlined in Sec. 7.5.2. under Model 2—Differences between Several Groups of Means.

If either factor had been a random effect, the variance of that factor could be calculated by the difference between the mean square and the error mean square as in Eq. (7.31).

The interpretation of the analysis of variance depends on the specific problem, and it is important to understand the type of variable operating on the factor under study. The computation of the analysis of variance for the simple row times column type of data is the same regardless of the model involved.

7.8 Multiple Balanced Blocks

An extension of the simple two-factor arrangment of data is one in which the number of rows and columns is the same. This is the balanced-block design or arrangment of data. For example, if data were taken at five different temperatures and at five different pressures, making a total of 25 data points, this would be a 5×5 balanced square. Each set of runs at one temperature would contain one made at each pressure, so that in comparing the means at different temperatures the effect of pressure would be canceled. So far this is what has already been said about the two-factor arrangement. But with the balanced block it is possible to superimpose on the 25 combinations of temperature and pressure another variable, say five different catalyst concentrations, in such a manner that the five runs at each level of each variable include one run at each level of the other two variables. In this way the comparison of totals or means for each variable is not affected by the levels of the other variables. In some cases it is possible to include additional variables with the same number of levels and not duplicate any combination of two levels of two variables.

The balanced block with three variables is called a Latin square because the third variable is usually identified by letters of the Latin alphabet. The balanced block with four variables is called a Greco-Latin square for a similar reason. An example of a 5×5 Greco-Latin square is illustrated in Fig. 7.3. Higher-order balanced blocks are called orthogonal squares of whatever order they are. There are very few published examples of actual use of balanced blocks of higher than the fourth order, or the Greco-Latin type. Fisher and Yates[1] have published complete sets of orthogo-

---

[1] R. A. Fisher and Frank Yates, "Statistical Tables," Hafner Publishing Company, Inc., New York, 1953.

nal squares up to 9 × 9. Textbooks dealing with experimental design[1,2] also include several examples of orthogonal squares for use in laying out experimental programs. We merely illustrate the application of the analysis of variance to the separation of the different factor effects for data arranged in this manner.

	Column variable level				
Row variable level	1	2	3	4	5
1	A_α	B_γ	C_ϵ	D_β	E_δ
2	B_β	C_δ	D_α	E_γ	A_ϵ
3	C_γ	D_ϵ	E_β	A_δ	B_α
4	D_δ	E_α	A_γ	B_ϵ	C_β
5	E_ϵ	A_β	B_δ	C_α	D_γ

Fig. 7.3 Greco–Latin square.

With a 5 × 5 Greco–Latin square, there are four factors at five levels each and 25 data points. The sum of squares for each factor is calculated from the totals for the five levels of that factor, and therefore with 4 degrees of freedom. The four factors utilize 16 degrees of freedom, leaving 8 for an estimate of error. Therefore, with 25 data points properly taken, it is possible to find the effect of five levels of four factors and still have 8 degrees of freedom for estimating error without any replication of measurements.

The calculation of the analysis of variance follows the same formulation for that of a two-factor arrangement. The sum of squares for each factor is the sum of the squares of the totals for each level of the factor divided by the number of levels, less the correction term, $(\Sigma x)^2/N$. The sum of squares for error is equal to the difference between the sum of the factor sums of squares and the total sum of squares calculated from $\Sigma x^2 - (\Sigma x)^2/N$. The degrees of freedom for each factor is one less than the number of levels, and the number of degrees of freedom for error is the difference between the degrees of freedom for factors and the total degrees

---

[1] W. G. Cochran and Gertrude M. Cox, "Experimental Designs," John Wiley & Sons, Inc., New York, 1956.

[2] Owen L. Davies (ed.), "The Design and Analysis of Industrial Experiments," Hafner Publishing Company, Inc., New York, 1954.

of freedom, $N - 1$. The solution is illustrated for a 5×5 Greco-Latin square with a numerical problem in example 7.11.

The mean squares are calculated as before, and each is an estimate of the error variance plus the number of levels times the variance for the particular factor. Each mean square is tested against the error mean square for significance of the factor effect. And the interpretation of the result, as with the other examples mentioned thus far, depends on whether the factor is a fixed or random effect or whether there might be regression between the factor and the measurement.

example 7.11. The data in Table 7.35 represent a series of 25 runs made at five temperatures and five durations to find the effect of these variables on the extent of conversion. The numbers in Table 7.35 are

TABLE 7.35 Per Cent Conversion

Temperature, °F	Time, min					
	30	60	90	120	150	Total
100	$16A\alpha$	$40B\gamma$	$50C\epsilon$	$20D\beta$	$15E\delta$	141
125	$30B\beta$	$25C\delta$	$62D\alpha$	$67E\gamma$	$30A\epsilon$	214
150	$50C\gamma$	$50D\epsilon$	$83E\beta$	$85A\delta$	$45B\alpha$	313
175	$80D\delta$	$80E\alpha$	$95A\gamma$	$98B\epsilon$	$70C\beta$	423
200	$90E\epsilon$	$92A\beta$	$98B\delta$	$100C\alpha$	$88D\gamma$	468
Total.......	266	287	388	370	248	1,559

Reactor totals Operator totals
A 318 α 303
B 311 β 295
C 295 γ 340
D 300 δ 303
E 335 ϵ 318

per cent conversion. In order to see whether the different reactors and different operators had an effect on the results, the 25 runs were made in five reactors by five operators with the runs distributed among the reactors and operators in a balanced arrangement so that each operator used each reactor only once at each temperature and pressure. The arrangement is indicated in Table 7.35, where the Latin letters indicate the reactors and the Greek letters indicate the operators.

Calculations

$$\Sigma x = 1{,}559 \qquad \Sigma(\text{time totals})^2/5 = 100{,}414.6$$
$$\Sigma x^2 = 117{,}119 \qquad \Sigma(\text{temperature totals})^2/5 = 112{,}319.8$$
$$(\Sigma x)^2/N = 97{,}219.24 \qquad \Sigma(\text{reactors totals})^2/5 = 97{,}419.0$$
$$\Sigma(\text{operator totals})^2/5 = 97{,}473.2$$

Discussion. Both the time and the temperature effects are significant, the temperature making the larger contribution to the total variance. Both of these factors appear to be of the model 3 type, and the evidence of regression of conversion with time and temperature should be checked by the method of Sec. 7.5.4. This is one of the suggested problems at the end of the chapter.

TABLE 7.36 Analysis of Variance

Source	Sum of squares	D.F.	Mean square
Time...........	100,414.6 − 97,219.24 = 3,195.36	4	798.84*
Temperature....	112,319.8 − 97,219.24 = 15,100.56	4	3,775.14***
Reactors........	97,414.0 − 97,219.24 = 194.76	4	48.69
Operators.......	97,473.2 − 97,219.24 = 253.96	4	63.49
Error..........	Difference = 1,155.12	8	144.39
Total.........	117,119.0 − 97,219.24 = 19,899.76	24	

$$F \text{ tests: Time} \quad \frac{798.84}{144.39} = 5.53, \quad F_{0.05,4,8} = 3.84$$

$$\text{Temp.} \quad \frac{3{,}775.14}{144.39} = 26.14, \quad F_{0.001,4,8} = 14.37$$

Neither the reactors nor the operators make a significant contribution to the variation of the measured variable. The fact that the mean squares for these two factors are both smaller than the error mean squares has no significance and is merely the result of the variation in the estimates of the error variance. Since these three mean squares are estimates of the error variance, they may be pooled to give a better estimate. They are pooled by adding the three sums of squares and dividing by the 3 degrees of freedom: (194.76 + 253.96 + 1,155.12)/(4 + 4 + 8) = 1,603.84/16 = 100.24. This pooled error variance increases the significance of the time mean square to the 0.01 level. To pool or not to pool depends on what is to be done with the error variance and the number of degrees of freedom that are involved. If the error variance is to be used in a model 1 or model 2 type problem to establish the confidence range on the means or to calculate the variance contribution of some factor, then the best estimate available of the error variance should be used. If the original degrees of freedom for the error variance are small, less than 10, then pooling will usually be of significant help both in obtaining a better estimate and in establishing F tests of the other factors. If the original degrees of freedom are more than 25, then pooling will seldom be of much value.

7.9 Estimating Missing Data

Sometimes in an experiment one piece of data is lost, and this missing datum point leaves a blank in an analysis-of-variance table. It is possible to supply a missing number with which to complete the calculation of the analysis of variance. This number is *not* a substitute for the lost datum point in the sense that experimental information can be obtained from the value. The number substituted in the analysis-of-variance table from the following equations is a number which permits a calculation of the variances from the remaining data. The formulas supply a value which minimizes the sums of squares of deviation from the means for the known data. The substituted value adds nothing in the way of information to the calculation, but merely completes the symmetry of the data so that the analysis-of-variance formulas may be used.

For a two-factor system of data with r rows and c columns the formula for one missing datum point is

$$X = \frac{rR + cC - S}{(r-1)(c-1)} \tag{7.34}$$

where R = total of row with missing datum point
C = total of column with missing datum point
S = total of all known data
X = value to be substituted for missing point

The estimation of the missing value decreases by 1 the total degrees of freedom and the degrees of freedom available for calculating the error mean square.

example 7.12. Suppose that in the data for example 7.10, the value of 1.38 for the fourth powder fraction and 50 tons pressure was missing. This value would be estimated from Eq. (7.34) as

$$X = \frac{(2)(7.37) + (6)(1.51) - 17.84}{(1)(5)} = 1.192$$

If two data points are missing, it is necessary to proceed by trial and error, first estimating one and calculating the other by Eq. (7.34). Then on the basis of the calculated point, check the first estimated value by Eq. (7.34). If the second calculation gives a value much different from the estimated value, the first calculation must be done over on the basis of the modified value of the first point. When two data points are estimated in this manner, 2 degrees of freedom are lost to the error mean square.

TABLE 7.37   Modified Analysis of Variance

Source	Sum of squares	D.F.	Mean square
Powder fractions...	31.3488 − 30.4964 = 0.8574	5	0.1705***
Pressure............	30.8331 − 30.4964 = 0.3367	1	0.3367***
Error...............	Difference = 0.0210	4	0.0052
Total...............	31.7065 − 30.4964 = 1.2101	10	

In a Latin-square arrangement of data, the formula for replacing a missing datum point is

$$X = \frac{r(R + C + T) - 2S}{(r - 1)(r - 2)} \qquad (7.35)$$

where r = number of rows or columns in square
R = total of row with missing datum point
C = total of column with missing datum point
T = total of level of superimposed factor with missing datum point
S = total of known data points
X = value to be substituted in analysis of variance

Again, the estimation of a single datum point involves the loss of 1 degree of freedom for the total analysis and 1 degree of freedom for the estimation of the error variance.

example 7.13. In the data of example 7.11, assume that the value of 67 per cent conversion were missing from the run at 125°F and 120 min reaction time. We shall treat the data as a Latin square and make the calculation for the missing point using "reactors" as the internal factor. A similar result will be obtained if any of the possible combinations of three factors are used.

$$X = \frac{5(147 + 303 + 268) - (2)(1492)}{(4)(3)} = 50.5$$

TABLE 7.38 Modified Analysis of Variance

Source	Sum of squares	D.F.	Mean square
Time.........	98,027.05 − 95,172.25 = 2,854.80	4	718.70
Temperature...	110,961.85 − 95,172.25 = 15,789.60	4	3,947.40***
Reactors.......	95,262.45 − 95,172.25 = 90.20	4	22.55
Operators......	95,183.85 − 95,172.25 = 11.60	4	2.90
Error..........	Difference = 1,261.80	7	180.25
Total........	115,180.25 − 95,172.25 = 20,008.00	23	

7.10 Two Factors with Replication

In the two-factor type of experiment with one observation at each condition, the difference between the two main effect sums of squares and the total sum of squares is taken as the error sum of squares. If a significant effect of either or both of the classification factors is found to exist, it must be assumed that the effect of one factor is the same at all levels of the other factor. When replicate determinations are made at the different conditions of the experiment, the comparison of the replicates provides a measure of the error. A sum of squares can be calculated for all the replicates by pooling the sum of squares for each set. This sum of squares, plus the sum of squares for the main classifications, subtracted from the total sum of squares, leaves a remainder to be accounted for in some new manner. This remainder sum of squares represents a measure of the different effect of one factor at different levels of the other and is usually referred to as *interaction*.

The interaction sum of squares has different interpretations for different models of analysis of variance. We shall first show the method of calculation and then discuss the meaning.

Figure 7.4 illustrates symbolically a two-factor arrangement of data with replication. There are c levels of the column factor, r levels of the row factor, and k replicates at each condition, making $crk = N$ data points. The analysis-of-variance calculation is shown in Table 7.39. The sums of squares for each main factor and for the total are calculated in the usual way: the sum of the squares of the totals of each level of each factor divided by the number of data points included in the total, less the correction term, $(\Sigma x)^2/N$. The sum of squares for interaction is obtained by first obtaining a sum of squares for replicates from the squares of the replicate totals divided by the number of each replication, less the correction term, and subtracting from this the two main factor sums of squares. The sum of squares for error is the difference between the total sum of

ANALYSIS OF VARIANCE

Column factor	C_1	C_2	$\cdots$	C_c
Row factor				
	X_{111}	X_{121}	$\cdots$	X_{1c1}
R_1	X_{112}	X_{122}	$\cdots$	X_{1c2}
	.	.		.
	.	.		.
	.	.		.
	X_{11k}	X_{12k}	$\cdots$	X_{1ck}
	X_{211}	X_{221}	$\cdots$	X_{2c1}
R_2	X_{212}	X_{222}	$\cdots$	X_{2c2}
	.	.		.
	.	.		.
	.	.		.
	X_{21k}	X_{22k}	$\cdots$	X_{2ck}
.				
.				
.				
	X_{r11}	X_{r21}	$\cdots$	X_{rc1}
R_r	X_{r12}	X_{r22}	$\cdots$	X_{rc2}
	.	.		.
	.	.		.
	.	.		.
	X_{r1k}	X_{r2k}	$\cdots$	X_{rck}

c Column levels   r Row levels   k Replicates

Fig. 7.4 Two classes of data with replication.

TABLE 7.39 Analysis of Variance for Two Classes of Data with Replications

C factor at c levels   total data $= crk = N$
R factor at r levels   (See Fig. 7.4 for arrangement
k replicates at each   of data.)
     condition

Source	Sum of squares		Degrees of freedom
C factor	$\sum \dfrac{(C \text{ totals})^2}{\text{no. in totals} = rk} - \dfrac{(\Sigma x)^2}{N}$	(1)	$c - 1$
R factor	$\sum \dfrac{(R \text{ totals})^2}{\text{no. in totals} = ck} - \dfrac{(\Sigma x)^2}{N}$	(2)	$r - 1$
Interaction: $R \times C$	Difference $= (3) - (2) - (1)$		$(c-1)(r-1)$
Subtotal	$\sum \dfrac{(\text{replicate totals})^2}{k} - \dfrac{(\Sigma x)^2}{N}$	(3)	$rc - 1$
Error	Difference $= (4) - (3)$		$rc(k-1)$
Total	$\Sigma x^2 - \dfrac{(\Sigma x)^2}{N}$	(4)	$rck - 1 = N - 1$

squares and the replicate sum of squares. The degrees of freedom for the interaction and the error are calculated by similar differences. The mean squares in each case are obtained by dividing the sum of squares by the corresponding degrees of freedom. The subtotal and the total mean squares are not calculated as they do not serve any function in the ordinary analysis of variance. The subtotal and total sums of squares are only computed to obtain by suitable differences the sums of squares for interaction and error.

The error mean square estimates the residual variance—the measure of the unassignable variation in the results when the effects, if any, of the main factors and their interaction have been removed. A series of repeat runs at ostensibly the same conditions will have a variance equivalent to that estimated by the error mean square.

7.10.1 Model 1. The interpretation of the interaction mean square depends on the type of factors under study in the analysis. To take the simplest case first: when both main factors are model 1, we are determining the mean effect of specific levels of two factors on some measurement. If the interaction mean square is significantly larger than the error mean square, it indicates that the effects of the factors are different at different levels of each other and that the individual results should be compared rather than the means for the different levels, to find the optimum effect. If the interaction mean square is not significant, i.e., not larger than the error mean square, then the means can be judged according to the procedure outlined in Sec. 7.5 dealing with a single factor.

For example, if a plant has three reaction vessels, a wooden tank, a glass-lined Pfaudler, and a stainless steel kettle, and makes two batches of product from three different sources of raw material in each vessel, we have a simple example of a two-factor experiment at three levels of each with duplication at each set of conditions. The factors are strictly fixed level—model 1—since we are testing only these particular reaction vessels and these sources of raw material and not samples from a population of kettles or vendors. The analysis of variance would be as shown in Table 7.40.

The "variance estimate" column of Table 7.40 gives the term corresponding to the mean square for each factor. The terms (I), (V), and (K) refer to the interaction effect, vendor effect, and kettle effect and are measures of the effect of the different levels of these factors on the measured variable. The interaction mean square is tested against the error mean square by the F test with 4 and 9 degrees of freedom. If it is significant, it indicates there is an (I) effect, i.e., the materials from different vendors respond differently in the kettles of different materials, and to find the best combination, each of the nine pairs of results must be examined separately to see whether the mean of the duplicate is significantly larger than the mean of other duplicates when judged by the error variance as outlined in the discussion for comparing several means. If the interaction mean square is not significant, the kettle and vendor mean squares are tested against the error mean square in the same way, and if either or both are found significant, the means for that factor are investigated to find which ones differ significantly, in terms of the error variance, from the others. If the kettle factor is significant and the vendor factor is not, it indicates that the difference between kettles is independent of the source of supply. Likewise, if the vendor factor is

TABLE 7.40 Analysis of Variance for Three Reactors and Three Vendors with Duplicate Batches

Source	Sum of squares	D.F.	Variance estimate
Kettles..........	$\Sigma \dfrac{(\text{kettle totals})^2}{6} - \dfrac{(\Sigma x)^2}{N}$	2	$\sigma^2 + 6(K)$
Vendors..........	$\Sigma \dfrac{(\text{vendor totals})^2}{6} - \dfrac{(\Sigma x)^2}{N}$	2	$\sigma^2 + 6(V)$
Interaction: $K \times V$	Difference	4	$\sigma^2 + 2(I)$
Subtotal.........	$\Sigma \dfrac{(\text{duplicate totals})^2}{2} - \dfrac{(\Sigma x)^2}{N}$	8	
Error...........	Difference	9	σ^2
Total...........	$\Sigma x^2 - \dfrac{(\Sigma x)^2}{N}$	17	

significant and the kettle factor is not, it indicates that the difference between sources of supply is real, and which kettle is used is immaterial on the basis of the measurements of this test. If both factors are significant and there is no interaction, some type of orderly relationship is indicated, and the best source of supply and the best kettle can be paired for optimum results. This is not necessarily the same result that would be obtained from examining all nine pairs if the interaction had been significant. The existence of the interaction factor indicates that best results in one kettle were with one source of supply while the best results in another kettle were with another source of material.

This might appear to be a somewhat lengthy description of an analysis of data when the best combination of vendor and kettle would be "obvious" if the actual yields were given. It is true that some results are obviously better than others and may be selected without mathematical analysis. On the other hand, it is also true that some "obviously better" results are often statistically not significant when compared with the unassignable variation in the data. In each case when a factor is said to be significant or not in an analysis of variance, it is in comparison with some other variance estimate and with some specific degrees of freedom associated with both estimates. If the comparison variance estimate is relatively large or the degrees of freedom are few, a fairly large effect will have to be present to be significant. If intuition (or desire) establishes the presence of an effect which statistical tests do not support, to obtain statistical substantiation, it is necessary either to decrease by improvement of technique the variance used as a basis for comparison, or to increase the degrees of freedom associated with the test by additional levels of the factor so as to make the F test more sensitive to differences in the variance estimates.

7.10.2 Model 2. With factors in which strictly random effects are active, the contribution of each factor to the total variance of the measured results is a variance contribution as indicated in Table 7.42. A comparison of the means at the different levels of the factors tested is not of interest inasmuch as these levels are only random selections from a whole population of possible levels. What is of interest is the variance component of the factor. If the experiment is arranged so that an interaction effect is determinable, the measurement estimates a variance contribution due to the random effects of the factors acting upon each other. For example, if two raw materials are mixed in a batch operation, the

TABLE 7.41

Raw material B	Raw material A		
	Lot 1	Lot 2	Lot 3
		Batches	
Lot 1..........	1, 2	3, 4	5, 6
Lot 2..........	7, 8	9, 10	11, 12
Lot 3..........	13, 14	15, 16	17, 18

random variations in the two raw materials might have a combined effect which would result in a greater variation in the end product than could be accounted for from the ascertainable variations in the materials. This would be an interaction effect. Table 7.41 illustrates an arrangement that would permit the measurement of this effect. The arrangement is identical with any other two-factor test with replicates.

The analysis of variance is calculated from the measurements of the different batches in the same manner as indicated in Table 7.40. The variance estimate column, however, is different. The mean squares for the model 2 analysis estimate the following variances:

TABLE 7.42   Variance Estimates for Model 2 Analysis of Variance

Mean square	Variance estimate
Factor A	$\sigma^2 + k\sigma^2(AB) + kb\sigma^2(A)$
Factor B	$\sigma^2 + k\sigma^2(AB) + ka\sigma^2(B)$
Interaction $A \times B$	$\sigma^2 + k\sigma^2(AB)$
Error	σ^2

where k = number of replications

a = levels of A factor

b = levels of B factor

The significance tests are obvious from the variance estimate terms. Each mean square is compared with the mean square containing all but the one variance term being tested. If a factor is found to be significant, the variance estimate can be calculated by subtracting the mean squares and dividing by the constant coefficient as demonstrated for the single-factor problem.

7.10.3   Mixed Models. With a set of data in which two factors were varied, one with random effect and one with fixed-level effects, the analysis of variance would be for a mixed model. If lots of raw material were treated in different reactors, the variation in the different lots of raw material would represent a model 2 effect and the different levels of performance of the reactors would be a model 1 effect. The calculation of the analysis of variance would be the same as for the cases of either model 1 or model 2, but again the interpretation of the mean squares would be different.

An interaction effect would indicate that the fixed-level factors of the different reactors were different for different lots of raw materials. The fixed-level-factor mean square would therefore contain an interaction effect, while the random-factor mean square would not. This relation is indicated in Table 7.43.

The significance tests for the terms of Table 7.43 follow from the components in the variance estimate. The "interaction" and "factor B" mean squares are tested against the "error" mean square, while the "factor A" mean square is tested against the "interaction" mean square. A general rule for determining the terms in the variance estimate is given in Sec. 7.12.2 dealing with a more involved factorial type of problem. For a simple two-factor mixed model, the form of Table 7.43 may be followed.

TABLE 7.43 Variance Estimates for Mixed-model Analysis of Variance

Mean square	Factor type	Variance estimate
Factor A	Model 1	$\sigma^2 + k\sigma^2(AB) + kb\sigma^2(A)$
Factor B	Model 2	$\sigma^2 + ka\sigma^2(B)$
Interaction $A \times B$		$\sigma^2 + k\sigma^2(AB)$
Error		σ^2

where k = number of replications

a = levels of A factor

b = levels of B factor

7.11 Two Classifications with Subgrouping and Replication

The subgrouping, or nesting, of data within a main classification can occur with data that are collected under two main classifications. If several reactors are producing on a batch basis, and each makes several batches at two or more conditions, we have a situation illustrated in Table 7.44.

The batches from each reactor are not related to the batches from other reactors and for analysis of variance purposes are simply subgroups of the reactor classification. Any variation among reactors would include the variation among batches within reactors. The "condition" factor is an independent classification, and its effect on the variation of the results is reflected in the difference among reactors only in so far as there is "interaction" between the condition and reactor factors or the condition and batch factor.

Subgrouping could exist under either factor or under both factors in the same set of data. Secondary subgrouping could exist within the first subgroups as in example 7.9. In fact, there could be any hierarchy of subgroupings under both main classes of factors. In this section we shall illustrate the solution of an analysis of variance for an arrangement of data as in Table 7.44. The pattern of solution in Table 7.45 can be modified to obtain the solution for similar problems of different arrangements of factors, subgroupings, and replications.

Referring to Table 7.44, let us establish the following nomenclature: There are r levels of the reactor factor, b batches from each reactor, c condition levels, and k replicates at each different set of levels, making a total of $rbck$ data points. We shall designate the total from each reactor as ΣR, the total from each batch as $\Sigma B(R)$, and the total at each condition as ΣC. The data for each reactor at each condition, disregarding the different batches, constitute a block of bk data points. We shall designate the total of each such block as ΣRC. The replicates at each condition constitute a block for each batch at each condition, and the total of each of these blocks is designated $\Sigma B(R) \cdot C$. The main factors are designated by the capital letters R or C. The subgrouping is shown by a parenthesis, $B(R)$, indicating batches within reactors. Interactions

TABLE 7.44

Reactor	Batch	Condition		
		1	2	3
1	1	x x x	x x x	x x x
	2	x x x	x x x	x x x
	3	x x x	x x x	x x x
	4	x x x	x x x	x x x
2	1	x x x	x x x	x x x
	2	x x x	x x x	x x x
	3	x x x	x x x	x x x
	4	x x x	x x x	x x x
3	1	x x x	x x x	x x x
	2	x x x	x x x	x x x
⋮	⋮			

are indicated by a multiplication sign, $R \times C$, the interaction of reactors and conditions; $B(R) \times C$, the interaction of batches within reactors and conditions. The subscript ss designates the sum of squares, R_{ss}, C_{ss}, etc.

The sums of squares for each block of conditions is obtained by subtracting from the sum of the squares of each block total divided by the number of observations in the block the correction factor $(\Sigma x)^2/N$, where N in this case is equal to $rbck$. As in the analysis of variance in the simple two-factor case, some of the sums of squares are obtained by differences. In Table 7.45 illustrating the analysis of variance for the arrangement of Table 7.44, the auxiliary sums of squares that are used for calculation purposes only are included in square brackets. The reason for giving the analysis in this manner is to maintain the systematic method of calculation.

TABLE 7.45 Analysis of Variance of Two Factor Arrangement with Subgrouping in One Factor and with Replication

r levels of main factor, R,—reactors
b levels of subfactor B within R, $B(R)$,—batches in reactors
c levels of main factor, C,—condition of experiments
k replicates at each combination of factor levels
(See Table 7.44 for arrangement of data.)

Source	Sum of squares	Degrees of freedom
R	$\sum \dfrac{(\Sigma R)^2}{cbk} - \dfrac{(\Sigma x)^2}{N} = R_{ss}$	$r - 1$
C	$\sum \dfrac{(\Sigma C)^2}{rbk} - \dfrac{(\Sigma x)^2}{N} = C_{ss}$	$c - 1$
$B(R)$	$B_{ss} - R_{ss} = B(R)_{ss}$	$r(b - 1)$
$\quad [B$	$\sum \dfrac{[\Sigma B(R)]^2}{ck} - \dfrac{(\Sigma x)^2}{N} = B_{ss}$	$rb - 1 \,]$
$R \times C$	$(RC)_{ss} - R_{ss} - C_{ss} = (R \times C)_{ss}$	$(r - 1)(c - 1)$
$\quad [RC$	$\sum \dfrac{(\Sigma RC)^2}{bk} - \dfrac{(\Sigma x)^2}{N} = (RC)_{ss}$	$rc - 1 \,]$
$B(R) \times C$	$(BC)_{ss} - B(R)_{ss} - (RC)_{ss} = [B(R) \times C]_{ss}$	$r(b - 1)(c - 1)$
$\quad [BC$	$\sum \dfrac{[B(R) \times C]^2}{k} - \dfrac{(\Sigma x)^2}{N} = (BC)_{ss}$	$rbc - 1 \,]$
Error	$\text{Total}_{ss} - (BC)_{ss} = \text{Error}_{ss}$	$rbc(k - 1)$
$\quad [\,$Total	$\Sigma x^2 - \dfrac{(\Sigma x)^2}{N} = \text{Total}_{ss}$	$rbck - 1 \,]$

Note that the calculation consists of summing all the different blocks, and for each type of block obtaining the sum of the squares of the totals divided by the number of data points in a block. From each total is subtracted the correction term $(\Sigma x)^2/N$, to give the sum of squares for the particular block. The sum of squares for subgroups or for interactions is obtained by the differences between the sums of squares for the blocks that are involved. The error sum of squares is the residual or difference between the total sum of squares and the sums of squares of all the main factors, subgroups, and interactions.

The degrees of freedom associated with each sum of squares calculated from the block totals is equal to one less than the number of blocks. If the number of levels of the different factors is indicated by lower-case letters as in Table 7.45, the degrees of freedom for each sum of squares is one less than the product of the letters *not* included in the denominator of the sum of squares of the group totals. The degrees of freedom for the sums of squares obtained by the difference between block sums of squares is equal to the difference between the same block degrees of freedom.

ANALYSIS OF VARIANCE

The mean square for each factor is calculated from the sum of squares and degrees of freedom for that factor as usual. Only the mean squares for the main factors, the subgroups, the interactions, and the error have significance in the analysis of variance. The mean squares corresponding to the block sums of squares that are used for computation purposes only are usually not calculated as they serve no useful purpose. The testing of the mean squares and the interpretation of the variances they estimate depend on the type of variables represented by the factors involved. If all the factors involve random effects, model 2 factors, the variances estimated by each mean square include all those terms having the same symbols as the mean square. This is illustrated in the left side of Table 7.46 for symbols of the analysis of variance shown in Table 7.45. If the factors are of the fixed-level type, model 1, or the factors are of both types, a mixed model, then variances estimated differ for different arrangements and can be obtained by the method detailed in Sec. 7.12.2. For the factors of Table 7.45, where "reactors" and "conditions" can be assumed to be of fixed level, model 1, and the "batches within reactors" is probably a random factor, model 2, the variances estimated by the mean squares are shown in the right side of Table 7.46. The coefficients for the different

TABLE 7.46 Variances Estimated by Analysis of Table 7.45

Mean square	Model 2	Mixed model R and C factors model 1 $B(R)$ factor model 2
	Variance estimated	Variance estimated
R	$\sigma^2 + k\sigma^2[B(R) \times C] + kb\sigma^2(R \times C) + kc\sigma^2[B(R)] + kcb\sigma^2(R)$	$\sigma^2 + kc\sigma^2[B(R)] + kcb\sigma^2(R)$
C	$\sigma^2 + k\sigma^2[B(R) \times C] + kb\sigma^2(R \times C) + kbr\sigma^2(C)$	$\sigma^2 + k\sigma^2[B(R) \times C] + kbr\sigma^2(C)$
$B(R)$	$\sigma^2 + k\sigma^2[B(R) \times C] + kc\sigma^2[B(R)]$	$\sigma^2 + kc\sigma^2[B(R)]$
$R \times C$	$\sigma^2 + k\sigma^2[B(R) \times C] + kb\sigma^2(R \times C)$	$\sigma^2 + k\sigma^2[B(R) \times C] + kb\sigma^2(R \times C)$
$B(R) \times C$	$\sigma^2 + k\sigma^2[B(R) \times C]$	$\sigma^2 + k\sigma^2[B(R) \times C]$
Error	σ^2	σ^2

variance-estimate terms are the same as the denominators used for calculating the sums of squares for the different blocks. Or looked at another way, the coefficients of the variance estimates, indicated by lower-case letters, and the variance factors, indicated by upper-case letters, are constant. If a factor is not included as a term in the variance estimate, the levels of this factor are included in the coefficient of the term.

The F test for each mean square follows from the variance estimated by the mean square. The significance of the variance of a factor is determined by comparing the mean square that includes the variance estimate of that factor with the mean square that includes the same variance estimates as the first mean square excepting the factor under test. For example, in Table 7.46 with the model 2 factors, the "batch–condition" interaction variance is tested by the ratio of the $B(R) \times C$ mean square to the error mean square. The "reactor–condition" interaction variance is tested by the ratio of the $R \times C$ mean square to the $B(R) \times C$ mean square. Likewise, the "batches within reactors" variance is tested by the ratio of the $B(R)$ mean square to the $B(R) \times C$ mean square, and the "condition" variance is tested by the ratio of the C mean square to the $R \times C$ mean square. For the mixed-model factors in the right of Table 7.46, the "batches within reactors" and the "batches–condition" interaction variances are tested against the error mean square, as no other variance estimates are included in the mean squares for these factors. The test for the "reactor–condition" interaction is the same for both models. Each test, of course, is made by comparing the calculated F value with the tabulated value at the degrees of freedom for numerator and denominator mean square in the usual manner.

The variance test for the "reactor" factor in the mixed-model example can be seen from the formulas for the estimated variances. For the model 2 example, however, there is no exact test inasmuch as there is no mean square that contains the identical terms of the R mean square minus the variance-of-R term. In this case it is necessary to resort to an approximate test in order to establish the significance of the variance in question. The mean squares of the $B(R)$ and $R \times C$ terms together are equal to the mean squares of the R and $B(R) \times C$ terms except for the $\sigma^2(R)$ term, so that the ratio of the sum of the second pair to the sum of the first pair gives a test for the existence of $\sigma^2(R)$. The degrees of freedom to use for the F test are those of the mean squares making up the two totals, weighted inversely as the squares of the mean squares. If the mean square for the numerator is the sum of mean square 1 and mean square 2, then the degrees of freedom for the numerator are

$$\nu_n = \frac{(MS)_n{}^2}{(MS)_1{}^2/\nu_1 + (MS)_2{}^2/\nu_2} \tag{7.36}$$

Likewise, if the denominator mean square is made up of mean squares 3 and 4, the degrees of freedom to use with the denominators are

$$\nu_d = \frac{(MS)_d{}^2}{(MS)_3{}^2/\nu_3 + (MS)_4{}^2/\nu_4} \tag{7.37}$$

When a mean square is found by F test not to be significantly larger than the mean square against which it is tested, indicating that the additional variance estimate in the numerator may not exist, then both mean squares may be taken to be estimates of the same factors. The estimates may therefore be pooled to give a better estimate by adding the two sums of squares, adding the two degrees of freedom, and calculating a new mean square. Rules about when to combine estimates and when not to cannot be rigorously laid down since we are dealing with probability values. If an F ratio was just slightly less than the 0.05 value we would be less inclined to combine the estimates than if the ratio was close to 1.0. Somewhat informally, it is suggested that if the F ratio is not significant at the 0.05 level *and* if the ratio is less than 2.0, the estimates may be pooled.

If an estimated variance term is found not to be significant, this term may be disregarded in testing further variance estimates. It is sometimes possible to apply an exact F test rather than the approximate test just described if some of the terms in the variance estimate may be disregarded as a result of earlier tests. Also the discarding of nonsignificant terms improves the sensitivity of the F test by increasing the degrees of freedom involved. In the model 2 side of Table 7.46, if the R and C factors are at two levels, the degrees of freedom associated with the mean squares for R, C, and $R \times C$ interaction are 1 each. Therefore the test of the C mean square against the $R \times C$ mean square is at 1 degree of freedom for numerator and denominator and is not very sensitive. If the $R \times C$ variance factor should prove to be nonsignificant, the C mean square could be tested against the $B(R) \times C$ mean square, or against the pooled value of the $R \times C$ and the $B(R) \times C$ terms, giving more degrees of freedom in the denominator. Also if the $R \times C$ variance estimate is found nonsignificant, the R mean square can be tested against the $B(R)$ mean square without resorting to the approximate test required if all the terms are significant.

example 7.14. The operation of a particular type of mass spectrometer for gas analyses involves an operator getting the gas sample into the machine, the printing of a spectrogram by the machine, and the calculation of the analysis from the spectrogram by a calculator. Two operators will not get a sample into the machine in exactly the same way, and the machine will not exactly reproduce the same spectrogram on the same sample, and two calculators will not get exactly the same answer from the same spectrogram. In 1950 the National Bureau of Standards[1] submitted samples of the same gas to 27 laboratories with the request that they have two operators each make two spectrograms

[1] Martin Shepherd, *Natl. Bur. Standards (U.S.) Research Paper* 2098, May, 1950.

and that two calculators each work up the four spectrograms to give eight analyses of each sample. The resulting data supplied information on the precision[1] of the mass spectrometer for the various components of the gas mixture and also the contributions of the various steps in the analysis to the over-all precision. The following data, for one component of one analysis, with the added feature of duplicate calculations by each calculator, serve as an illustration of the analysis of variance. O_1 refers to operator 1, S_1 refers to spectrogram 1, and C_1 refers to calculator 1.

Data

TABLE 7.47

$O_1S_1C_1$	6.6%	$O_1S_2C_2$	6.4%	$O_2S_3C_2$	5.9%
$O_1S_1C_1$	6.4%	$O_1S_2C_2$	6.0%	$O_2S_4C_1$	6.1%
$O_1S_1C_2$	6.8%	$O_2S_3C_1$	6.1%	$O_2S_4C_1$	6.1%
$O_1S_1C_2$	6.6%	$O_2S_3C_1$	5.7%	$O_2S_4C_2$	6.2%
$O_1S_2C_1$	6.1%	$O_2S_3C_2$	5.7%	$O_2S_4C_2$	6.0%
$O_1S_2C_1$	6.1%				

Solution. The data are arranged in a table under the different classifications to facilitate the calculation of the group totals. A constant of 6.2 has been subtracted from each value to simplify the arithmetic. The estimated variances in Table 7.49 are calculated on the basis of operators and calculators being model 1 factors and spectrograms being model 2. The procedure followed is outlined in 7.12.2.

[1] W. Volk, *Anal. Chem.*, **26**:1771 (1954).

TABLE 7.48

	O_1				O_2				
	S_1		S_2		S_3		S_4		C total
C_1	0.4	0.2	−0.1	−0.1	−0.1	−0.5	−0.1	−0.1	−0.4
C_2	0.6	0.4	0.2	−0.2	−0.5	−0.3	−0.0	−0.2	0.0
S Totals..	1.6		−0.2		−1.4		−0.4		
O Totals..	1.4				−1.8				−0.4

$\Sigma x = -0.4$ $\Sigma(\Sigma O)^2/8 = 0.65$ $\Sigma(\Sigma OC)^2/4 = 0.70$

$(\Sigma x)^2/16 = 0.01$ $\Sigma(\Sigma S)^2/4 = 1.18$ $\Sigma(\Sigma SC)^2/2 = 1.24$

$\Sigma x^2 = 1.48$ $\Sigma(\Sigma C)^2/8 = 0.02$

ANALYSIS OF VARIANCE

TABLE 7.49 Analysis of Variance

Source	Sum of squares		D.F.	Mean square	Estimated variance
O	$0.65 - 0.01$	$= 0.64$	1	0.640	$\sigma^2 + 4\sigma^2[S(O)]$ $+ 8\sigma^2(O)$
C	$0.02 - 0.01$	$= 0.01$	1	0.010	$\sigma^2 + 2\sigma^2[S(O) \times C]$
$S(O)$	$1.17 - 0.64$	$= 0.53$	2	0.265	$\sigma^2 + 4\sigma^2[S(O)]$
$[S$	$1.18 - 0.01$	$= 1.17$	3]		
$O \times C$	$0.69 - 0.64 - 0.01$	$= 0.04$	1	0.040	$\sigma^2 + 2\sigma^2[S(O) \times C]$ $+ 4\sigma^2(O \times C)$
$[OC$	$0.70 - 0.01$	$= 0.69$	3]		
$S(O) \times C$	$1.23 - 0.53 - 0.69$	$= 0.01$	2	0.005	$\sigma^2 + 2\sigma^2[S(O) \times C]$
$[SC$	$1.24 - 0.01$	$= 1.23$	7]		
Error	$1.47 - 1.23$	$= 0.24$	8	0.030	σ^2
Total	$1.48 - 0.01$	$= 1.47$	15		

Discussion. The two interaction mean squares and the "calculator" mean square are obviously not significantly different from the error mean square, so that the sums of squares and the degrees of freedom for these factors may be pooled with the error sum of squares to give a revised estimate of the error variance. With this pooling we obtain the following analysis of variance.

TABLE 7.50 Revised Analysis of Variance

Source	Sum of squares	D.F.	Mean square	Estimated variance
O	0.64	1	0.640	$\sigma^2 + 4\sigma^2 S(O) + 8\sigma^2(O)$
$S(O)$	0.53	2	0.265**	$\sigma^2 + 4\sigma^2 S(O)$
Error	0.30	12	0.025	σ^2

The $S(O)$ mean square is highly significant, $F_{0.01, 2, 12} = 6.93$, but the O mean square, compared with the $S(O)$ mean square, is not significant, $F_{0.20, 1, 2} = 3.65$. On the basis of this one set of analyses and one gas component, we would be justified in attributing the variation of results to the error variance and the difference between spectrograms. There is no significant contribution of either calculators or operators. The spectrogram variance estimate is calculated from the error and $S(O)$ mean squares.

$$s^2[S(O)] = \frac{0.265 - 0.025}{4} = 0.060$$

The question of what to do with the "operator" variance estimate needs to be answered. On the basis of this one set of data there is very little basis for assuming an operator effect, while with the original data, where the operator effect of all the gas components could be pooled, the operator effect was significant. We shall see that on the over-all precision it makes little difference whether we identify this factor. If we ignore the fact that the O mean square is not significantly larger than the $S(O)$ mean square and calculate the operator variance estimate, we obtain

$$s^2(O) = \frac{0.640 - 0.265}{8} = 0.047$$

This gives a total variance for a single analysis of

$$0.025 + 0.060 + 0.047 = 0.112$$

If we accept the nonsignificance of $\sigma^2(O)$ and pool the operator and spectrogram mean squares to get a new sum of squares of 1.17 and a new degrees of freedom of 3, we have a new $S(O)$ mean square of 0.390, and hence a new spectrogram variance estimate of

$$s^2[S(O)] = \frac{0.390 - 0.025}{4} = 0.091$$

This gives a total variance for a single analysis of $0.091 + 0.025 = 0.116$.

With a total variance for a single determination of 0.116, the standard deviation is about 0.34. We might therefore expect a mean difference between duplicate determinations of $0.34/0.8862$ (see Sec. 6.3.3) $= 0.38$ per cent, on an analysis of about 6.0 per cent from the data of this example.

7.12 Factorial Analyses

7.12.1 Analysis-of-variance Calculation. There is no limit to the number of main factors that may be examined in an experimental program: temperature, pressure, catalyst concentration, contact time, particle size, etc. In a completely balanced experiment, each level of each factor is tested at all the levels of all the other factors so that the total number of experiments required is the product of all the levels of all the factors. If two factors are to be examined at two levels each and two other factors are to be examined at three levels each, we have a $2 \times 2 \times 3 \times 3$ experiment, and to be completely balanced it requires 36 determinations. When more than two factors are involved, all the possible combinations of interactions may exist, and a complete analysis of variance provides mean squares attributable to all the individual interactions. It is possible by different arrangements of the levels of the different factors to decrease the number of experiments required at the expense of not being able to determine some of the higher-order interactions but still to be able to calculate the main effects and lower-order interactions.

Experimental design is the subject of several books[1,2] which are readily available to engineers who need to plan involved experimental tests. It is beyond the planned scope of this particular book to discuss the method of handling the various combinations of unbalanced and incompletely balanced arrangements of data. And it would be unlikely that a few random examples would fit the specific requirements of many readers. We limit our discussion in the remainder of this chapter to the completely balanced factorial arrangement. In the next chapter we discuss the special case of the factorial design with each factor at two levels, the 2^n factorial.

The procedure for analysis of the multifactor experiment is similar to that described for the two-factor case. For the factorial analysis of A factor at a levels, B factor at b levels, C factor at c levels, D factor at d levels, and with k replications at each condition, the sum of squares for any combination of factors is equal to the sum of the squares of totals for the blocks represented by the upper-case letters for the factors divided by the lower-case letters for the factors not included, less the correction term, $(\Sigma x)^2/N$, and less the sums of squares associated with all possible combinations of the factors included. For example, the $A \times B \times C$ interaction sum of squares is equal to the sum of the squares of the ABC block totals divide by dk, less the correction term and less the A, the B, the C, the AB, the AC, and the BC sums of squares. The degrees of freedom corresponding to each sum of squares is equal to the product of the number of levels less one for each factor involved in the sum of squares. The number of degrees of freedom for the $A \times B \times C$ interaction is $(a-1)(b-1)(c-1)$. The residual or error sum of squares, when there is replication, is the difference between the total and all the other sums of squares. When there is no replication, the highest-order interaction sum of squares is usually taken for the error.

Table 7.51 shows the symbolic calculation of a three-factor analysis of variance with replication. The various block totals can most readily be obtained by systematically arranging the data into the groups indicated

---

[1] W. G. Cochran and Gertrude M. Cox, "Experimental Designs," John Wiley & Sons, Inc., New York, 1956.

[2] Owen L. Davies (ed.), "The Design and Analysis of Industrial Experiments," Hafner Publishing Company, Inc., New York, 1954.

TABLE 7.51 Analysis of Variance of Three-factor Experiments with Replication

A factor at a levels k replicates
B factor at b levels $N = abck$
C factor at c levels

Source	Sum of squares	D.F.
A factor	$\sum \dfrac{(\Sigma A)^2}{bck} - \dfrac{(\Sigma x)^2}{N}$	$a - 1$
B factor	$\sum \dfrac{(\Sigma B)^2}{ack} - \dfrac{(\Sigma x)^2}{N}$	$b - 1$
C factor	$\sum \dfrac{(\Sigma C)^2}{abk} - \dfrac{(\Sigma x)^2}{N}$	$c - 1$
$A \times B$ interaction	$\sum \dfrac{(\Sigma AB)^2}{ck} - \dfrac{(\Sigma x)^2}{N} - A_{ss} - B_{ss}$	$(a-1)(b-1)$
$A \times C$ interaction	$\sum \dfrac{(\Sigma AC)^2}{bk} - \dfrac{(\Sigma x)^2}{N} - A_{ss} - C_{ss}$	$(a-1)(c-1)$
$B \times C$ interaction	$\sum \dfrac{(\Sigma BC)^2}{ak} - \dfrac{(\Sigma x)^2}{N} - B_{ss} - C_{ss}$	$(b-1)(c-1)$
$A \times B \times C$ interaction	$\sum \dfrac{(\Sigma ABC)^2}{k} - \dfrac{(\Sigma x)^2}{N} - A_{ss} - B_{ss} - C_{ss} - AB_{ss} - AC_{ss} - BC_{ss}$	$(a-1)(b-1)(c-1)$
Error	Difference	Difference
Total	$\Sigma x^2 - \dfrac{(\Sigma x)^2}{N}$	$N - 1 = abck - 1$

NOTE: The subscript ss signifies sum of squares.

by the factor symbols. The block totals for the main factors are obviously the totals for each level of each factor, disregarding the other factors. The block totals for the first-order interactions are the totals that occur in the blocks formed when the data are arranged in a simple two-factor arrangement. The second-order interactions are calculated from the totals in the blocks formed by arranging the data under the three factors involved, ignoring any other factors. If there are four main factors, there will be six first-order interactions, four second-order interactions, and one third-order interaction: the number of combinations of four things taken two at a time, taken three at a time, and taken four at a time. A numerical example of a four-factor analysis with two levels of each factor and with replication is given in example 7.15, following the next section.

The mean square for each factor and interaction is calculated in the usual way: the quotient of the sum of squares divided by the degrees of freedom. The variances estimated by the mean squares depend on the type of variables, model 1 or model 2. This subject is dealt with in the next section.

7.12.2 Variance Estimate Calculation. If all the factors investigated in the analysis of variance are model 2, i.e., random effects, the variances estimated by the mean squares follow the rule laid down in Sec. 7.10.2: each variance estimate term includes the error variance, the variance for the particular mean-square term calculated, and for all terms which include all of the factor symbols included in the particular effect being calculated. The coefficients of each variance estimate term include all the symbols representing the number of levels of the factors *not* included in the variance estimate term. For example, with a four-factor analysis involving factors A, B, C, D, at levels a, b, c, d, respectively, and with k replicates at each combination of factor levels, the ABD mean square estimates the sum of the error variance, kc times the ABD interaction variance and k times the $ABCD$ interaction variance. The CD mean square estimates the sum of the error variance, kab times the CD interaction variance, k times the $ABCD$ interaction variance, ka times the BCD interaction variance, and kb times the ACD interaction variance. The F test for any particular mean square is the ratio of the mean square for that term to the mean square for some other term that is a variance estimate of all but one of the terms included in the numerator mean square estimate.

If all the factors are model 1, i.e., fixed effects, the variances estimated by the mean square in each case are the error variance and the square of the effect or combination of effects represented by the symbols for that particular term. The coefficient of each estimated effect is the product of the number of levels of the factors not included in the effect being measured. All mean squares, in this case, are tested against the error mean square for significance by the F test.

For a mixed model combination of factors, each arrangement must be analyzed separately. The following steps describe the method, and two illustrations are given: one for the results previously listed in Table 7.46 without explanation, and one for the complete factorial analysis shown in Table 7.51. For this second case, factors A and B are assumed to be model 1, and factor C is assumed to be model 2.

Step 1. Make a table with a column for each main factor, indicated in our nomenclature by an uppercase letter, and a column for replicates. Provide a row for each mean square from the analysis of variance. Table 7.52 shows the arrangement for the two cases illustrated. The numbers in the body of the table are explained in the next step.

Step 2. In each factor column, if the factor heading that column is *not* included in the mean square term, put the symbol representing the number of levels of that factor. (If the factor *is* included in the mean square term and the factor is model 2, or is the host factor for a nested classification—R, for example, where B is nested in R, $B(R)$—put the number 1 in the table; otherwise put a zero.) In the replicate column, place the symbol indicating the number of replications. A single experiment has one replication.

Step 3. The mean square term estimates the sum of the error variance and the variance of every combination of factors that includes all of the factor terms in the mean square, each multiplied by a coefficient obtained from the constants in the table made according to steps 1 and 2. The coefficients for the variances estimated by the mean square term are obtained by disregarding those values in columns headed by the factor symbols that are included in the mean square terms, and taking the product of the remaining constants.

For example, the $B(R)$ mean square in the left column of Table 7.52 estimates the error variance, the $B(R)$ variance and the $B(R) \times C$ variance. The coefficient of the error variance is 1. The coefficient of the $B(R)$ variance is $c \cdot k$, disregarding the two unity constants under the column headings R and B. The coefficient of the $B(R) \times C$ variance is $0 \cdot k$, again disregarding the constants in the columns headed by R and B, the two terms in the mean square factor under consideration. The C mean square in this same table estimates the error variance, the C variance and the $R \times C$, and the $B(R) \times C$ variances. The coefficients for these terms are 1 for the error variance, and disregarding the constants in the C column, are $r \cdot b \cdot k$, $0 \cdot b \cdot k$, and $1 \cdot 1 \cdot k$ for the other three variance terms.

Table 7.52 illustrates steps 1 and 2 for preparing the table of coefficients, and Table 7.53 illustrates the calculation of the variance estimate terms for two cases of analysis of variance.

ANALYSIS OF VARIANCE

TABLE 7.52 Illustration of the Calculation of Variance Estimate Coefficients

Data from Table 7.44					Data from Table 7.51				
Mean square term	Main factors and model type				Mean square term	Main factors and model type			
	R	C	B	rep.		A	B	C	rep.
	1	1	2	—		1	1	2	—
R	0	c	b	k	A	0	b	c	k
C	r	0	b	k	B	a	0	c	k
$B(R)$	1	c	1	k	C	a	b	1	k
$R \times C$	0	0	b	k	$A \times B$	0	0	c	k
$B(R) \times C$	1	0	1	k	$A \times C$	0	b	1	k
					$B \times C$	a	0	1	k
					$A \times B \times C$	0	0	1	k

TABLE 7.53 Illustration of Variance Estimate Terms from Constants in Table 7.52

Mean square term	Estimated variance

Data from Table 7.44

R	$\sigma_0^2 + (0)(1)(k)\sigma^2(B(R) \times C) + (0)(b)(k)\sigma^2(R \times C)$ $+ (c)(1)(k)\sigma^2(B(R)) + (c)(b)(k)\sigma^2(R)$
C	$\sigma_0^2 + (1)(1)(k)\sigma^2(B(R) \times C) + (0)(b)(k)\sigma^2(R \times C)$ $+ (r)(b)(k)\sigma^2(C)$
$B(R)$	$\sigma_0^2 + (0)(k)\sigma^2(B(R) \times C) + (c)(k)\sigma^2(B(R))$
$R \times C$	$\sigma_0^2 + (1)(k)\sigma^2(B(R) \times C) + (b)(k)\sigma^2(R \times C)$
$B(R) \times C$	$\sigma_0^2 + (k)\sigma^2(B(R) \times C)$

Data from Table 7.51

A	$\sigma_0^2 + (0)(1)(k)\sigma^2(ABC) + (b)(1)(k)\sigma^2(AC)$ $+ (0)(c)(k)\sigma^2(AB) + (b)(c)(k)\sigma^2(A)$
B	$\sigma_0^2 + (0)(1)(k)\sigma^2(ABC) + (a)(1)(k)\sigma^2(BC)$ $+ (0)(c)(k)\sigma^2(AB) + (a)(c)(k)\sigma^2(B)$
C	$\sigma_0^2 + (0)(0)(k)\sigma^2(ABC) + (a)(0)(k)\sigma^2(BC)$ $+ (0)(b)(k)\sigma^2(AC) + (a)(b)(k)\sigma^2(C)$
$A \times B$	$\sigma_0^2 + (1)(k)\sigma^2(ABC) + (c)(k)\sigma^2(AB)$
$A \times C$	$\sigma_0^2 + (0)(k)\sigma^2(ABC) + (b)(k)\sigma^2(AC)$
$B \times C$	$\sigma_0^2 + (0)(k)\sigma^2(ABC) + (a)(k)\sigma^2(BC)$
$A \times B \times C$	$\sigma_0^2 + (k)\sigma^2(ABC)$

TABLE 7.54 Power Consumption in Electric-furnace Heats, Kwhr/ton

Repli-cate	High roof								Low roof							
	High power				Low power				High power				Low power			
	Tube		Plate		Tube		Plate		Tube		Plate		Tube		Plate	
	700 lb	1,000 lb	700 lb	1,000 lb	700 lb	1,000 lb	700 lb	1,000 lb	700 lb	1,000 lb	700 lb	1,000 lb	700 lb	1,000 lb	700 lb	1,000 lb
1	709	784	829	752	946	966	1,028	1,056	774	702	817	798	866	988	1,017	922
2	789	700	806	714	800	976	906	870	834	658	783	726	862	808	990	808
3	646	596	691	714	840	876	977	908	746	650	771	700	800	650	954	868
Total...	2,144	2,080	2,326	2,180	2,586	2,818	2,911	2,834	2,354	2,010	2,371	2,224	2,528	2,446	2,961	2,598

ANALYSIS OF VARIANCE

The F test to establish the significance of any particular variance depends on the specific analysis. In any case, it is a comparison of the mean square estimating a number of variances to a mean square estimating all but one of the same group of variances. The one in the numerator that is not in the denominator is the one whose significance is tested by the F ratio.

The following example illustrates a four-factor completely balanced analysis of variance with three replicates at each condition. The data are for each factor at two levels, and in a case of this kind, the calculation techniques illustrated in the next chapter are simpler. However, the data are from some actual experiments and the calculation in this example are for illustrative purposes only. The same method would apply if some of the factors had been tested at more than two levels, but the arithmetic for illustration purposes becomes difficult to follow. The formulation of Table 7.51 can be rigorously followed for a completely balanced factorial of any size.

example 7.15. The data in Table 7.54 are from a series of tests run on an electric furnace[1] with a high roof and low roof, with two different power settings, with tube scrap and with plate scrap, and with 700- and 1,000-pound charges. Each test is reported in triplicate, and the numbers in the table are in kilowatthours of power consumed per ton of melted product. Table 7.54 gives the experimental data.

Step 1. Calculate Σx, Σx^2, and $(\Sigma x)^2/N$.

$\Sigma x = 39{,}371$

$\Sigma x^2 = 32{,}917{,}817$

$$\frac{(\Sigma x)^2}{N} = \frac{(39{,}371)^2}{48} = 32{,}293{,}242.5$$

Step 2. Calculate all the main-factor totals. For each main-factor sum of squares, obtain the sum of the squares of that main-factor total divided by the number of measurements in each total, less the correction term.

Roof totals: high = 19,879
                low   = 19,492
Power totals: high = 17,689
                low   = 21,682

[1] W. H. Glaisher et al., *J. Iron Steel Inst.* (*London*), **183**:22 (1956).

Scrap totals: tube = 18,966
plate = 20,405
Charge totals: 700 lb = 20,181
1,000 lb = 19,190

Roof sum of squares $= \dfrac{(19,879)^2 + (19,492)^2}{24} - 32,293,242.5$

$= 32,296,362.7 - 32,293,242.5 = 3,120.2$

Power sum of squares $= 32,625,410.2 - 32,293,242.5 = 332,167.7$
Scrap sum of squares $= 32,336,382.5 - 32,293,242.5 = 43,140.0$
Charge sum of squares $= 32,313,702.5 - 32,293,242.5 = 20,460.0$

Step 3. Calculate all the first-order interactions by first obtaining the totals for the different pairs of factors, then calculating the sum of the squares of these totals divided by the number of measurements in each total, less the correction term and less the sums of squares for the two main factors involved. Tables 7.55 to 7.60 show the totals for the various pairs.

TABLE 7.55

	High roof	Low roof
High power.	8,730	8,959
Low power.	11,149	10,533

TABLE 7.56

	High roof	Low roof
Tube......	9,628	9,338
Plate......	10,251	10,154

TABLE 7.57

	High roof	Low roof
700 lb......	9,967	10,214
1,000 lb....	9,912	9,278

TABLE 7.58

	High power	Low power
Tube......	8,588	10,378
Plate......	9,101	11,304

TABLE 7.59

	High power	Low power
700 lb......	9,195	10,986
1,000 lb....	8,494	10,696

TABLE 7.60

	Tube	Plate
700 lb......	9,612	10,569
1,000 lb....	9,354	9,836

ANALYSIS OF VARIANCE

Roof × power sum of squares

$$= \frac{(8,730)^2 + (8,959)^2 + (11,149)^2 + (10,533)^2}{12} - 32,293,242.5$$

$- 3,120.2 - 332,167.7 = 32,643,405.8 - 32,628,530.4 = 14,875.4$

Roof × scrap sum of squares
$= 32,340,278.7 - 32,293,242.5 - 3,120.2 - 43,140.0 = 776.0$

Roof × charge sum of squares
$= 32,332,992.7 - 32,293,242.5 - 3,120.2 - 20,460.0 = 16,170.0$

Power × scrap sum of squares
$= 32,672,103.7 - 32,293,242.5 - 332,167.7 - 43,140 = 3,553.5$

Power × charge sum of squares
$= 32,649,389.4 - 32,293,242.5 - 332,167.7 - 20,460.0 = 3,519.2$

Scrap × charge sum of squares
$= 32,361,543.0 - 32,293,242.5 - 43,140.0 - 20,560.0 = 4,700.5$

Step 4. Calculate the second-order interactions by obtaining the totals for the different combinations of three factors, squaring these totals for each combination, and from the sum of these squared totals divided by the number of measurements in each total subtract the correction term, the sums of squares for each of the main factors involved,

TABLE 7.61

	High roof		Low roof	
	High power	Low power	High power	Low power
Tube.........	4,224	5,404	4,364	4,974
Plate.........	4,506	5,745	4,595	5,559

TABLE 7.62

	High roof		Low roof	
	High power	Low power	High power	Low power
700 lb.........	4,470	5,497	4,725	5,489
1,000 lb.......	4,260	5,652	4,234	5,044

TABLE 7.63

	High roof		Low roof	
	Tube	Plate	Tube	Plate
700 lb	4,730	5,237	4,882	5,332
1,000 lb	4,898	5,014	4,456	4,822

TABLE 7.64

	High power		Low power	
	Tube	Plate	Tube	Plate
700 lb	4,498	4,697	5,114	5,872
1,000 lb	4,090	4,404	5,264	5,432

and the sums of squares for each pair of factors included. Tables 7.61 to 7.64 show the totals for the four combinations of three factors.

Roof × power × scrap sum of squares

$$= \frac{(4,224)^2 + (4,506)^2 + \cdots + (4,974)^2 + (5,559)^2}{6} - 32,293,242.5$$
$$- 3,120.2 - 332,167.7 - 43,140.0 - 14,875.4 - 776.0 - 3,553.5$$
$$= 32,692,688.5 - 32,690,875.3 = 1,813.2$$

Roof × power × charge sum of squares
$$= 32,685,675.1 - 32,293,242.5 - 3,120.2 - 332,167.7$$
$$- 20,460.0 - 14,875.4 - 16,170.0 - 3,519.2 = 2,120.1$$

Roof × scrap × charge sum of squares
$$= 32,383,572.8 - 32,293,242.5 - 3,120.2 - 43,140.0$$
$$- 20,460.0 - 776.0 - 16,170.0 - 4,700.5 = 1,963.6$$

Power × scrap × charge sum of squares
$$= 32,711,138.1 - 32,293,242.5 - 332,167.7 - 43,140.0$$
$$- 20,460.0 - 3,553.5 - 3,519.2 - 4,700.5 = 10,354.7$$

Step 5. The third-order interaction of all the factors is calculated from the sum of the squares of the totals from Table 7.54 divided by the

ANALYSIS OF VARIANCE

number of replicates less all the previous sums of squares. Thus

Roof × power × scrap × charge sum of squares

$$= \frac{(2{,}144)^2 + (2{,}080)^2 + \cdots + (2{,}961)^2 + (2{,}598)^2}{3}$$
$$- 32{,}293{,}242.5 - 3{,}120.2 - 332{,}167.7 - 43{,}140.0$$
$$- 20{,}460.0 - 14{,}875.4 - 776.0 - 16{,}170.0 - 3{,}553.5$$
$$- 3{,}519.2 - 4{,}700.5 - 1{,}813.2 - 2{,}120.1 - 1{,}963.6$$
$$- 10{,}354.7 = 32{,}753{,}289.0 - 32{,}751{,}976.6 = 1{,}312.4$$

Step 6. The error sum of squares is the difference between the total sum of squares, $\Sigma x^2 - (\Sigma x)^2/N$, less all the factor sums of squares:

$$(32{,}917{,}817.0 - 32{,}293{,}242.5) - 460{,}046.5 = 164{,}528.0$$

All the sums of squares are summarized in Table 7.65 with the corresponding degrees of freedom and mean squares. In this particular example, with all factors at two levels, the degree of freedom for all terms except error is 1.

TABLE 7.65

Source	Sum of squares	D.F.	Mean squares
Roof..................	3,120.2	1	3,120.2
Power.................	332,167.7	1	332,167.7**
Scrap.................	43,140.0	1	43,140.0**
Charge................	20,460.0	1	20,460.0
$R \times P$..............	14,875.4	1	14,875.4
$R \times S$..............	776.0	1	776.0
$R \times C$..............	16,170.0	1	16,170.0
$P \times S$..............	3,553.5	1	3,553.5
$P \times C$..............	3,519.2	1	3,519.2
$S \times C$..............	4,700.5	1	4,700.5
$R \times P \times S$..........	1,813.2	1	1,813.2
$R \times P \times C$..........	2,120.1	1	2,120.1
$R \times S \times C$..........	1,963.6	1	1,963.6
$P \times S \times C$..........	10,354.7	1	10,354.7
$R \times P \times S \times C$......	1,312.4	1	1,312.4
Error.................	164,528.0	32	5,141.5
Total...............	624,574.5	47	

$F_{0.05,1,32} = 5.14$
$F_{0.01,1,32} = 7.50$

From Table 7.65 it can be seen that only the power-factor and the scrap-factor effects are significant. Since all the factors in this example are model 1, the significant mean squares indicate an effect of the different levels of the two factors on the power consumption. Referring to step 2 in the calculation, the mean values, obtained by dividing the totals by 24, for high and low power are 737 and 903 kwhr/ton, and the mean values for the two different types of scrap are 790 for tubes and 850 for plate. The differences between the means in these two pairs are a measure of the increased power consumption, in the first case when operating at a low-power level, and in the second case when using the plate scrap. The error mean square of 5,141.5 indicates a standard deviation for a single measurement of about 71.8 kwhr/ton and for a mean of 24 measurements of $\sqrt{5{,}141.5/24} = 14.6$ kwhr/ton.

7.13 Problems

1. Calculate the 95 per cent precision limits for the molecular weight of a flue gas made up of 13.1 per cent CO_2, 7.7 per cent O_2, and 79.2 per cent N_2, if the variances of the analyses (in fractions, *not* per cents) are $\sigma^2(CO_2) = 0.000625$, $\sigma^2(O_2) = 0.000625$, $\sigma^2(N_2) = 0.0025$. Ans.: ± 3.82.

2. Calculate the 95 per cent precision limits for the amount of ingredient A fed to a process if the total feed is made up of two streams, one with 100 gph containing 4 per cent A, and the other with 150 gph containing 7 per cent A. The variances of the four measurements are: first stream 0.5, second stream 0.7, first analysis (fraction units) 0.00002, second analysis (fraction units) 0.00002. Ans.: ± 1.58 gph.

3. Calculate the 95 per cent precision limits for a Reynolds number assuming reasonable precisions for the diameter, velocity, density, and viscosity terms.

4. Check the following variance estimates for homogeneity.

s^2	Degrees of Freedom
3,975	4
1,107	4
1,442	4
4,720	4
2,500	4
962	4

$\chi^2 = 4.02$.

5. The tabulated data[1] represent tests run by five different operators on three different types of equipment. Establish by analysis of variance whether there is significant difference between operators and between equipment types.

[1] Charles R. Hicks, *Ind. Quality Control*, August, 1956.

ANALYSIS OF VARIANCE

Operator	Equipment		
	1	2	3
1	97.6	96.6	97.0
2	97.2	96.4	96.0
3	96.4	97.0	95.0
4	97.4	.96.2	95.8
5	97.8	96.8	97.0

6. The tabulated data represent residual sulfur in a treated oil after different catalyst ages and with different types of reactor operation. Determine whether the sulfur varies with catalyst age and with different type of operation. If there is significant variation with catalyst age, determine whether there is linear or higher-order correlation.

Type of operation	Catalyst age, days			
	20	40	60	80
A	0.61	0.66	0.98	0.91
B	0.72	0.90	1.00	1.13
C	0.72	0.93	1.07	1.04
D	0.80	0.97	1.02	1.16
E	0.82	0.88	0.93	1.00

7. The tabulated data[1] represent percentage moisture absorbed by different water-repellent cottons tested by four different laundries under four different test conditions, each test run in duplicate.

Test	Laundry							
	A		B		C		D	
1	7.20	9.06	2.40	2.14	2.19	2.69	1.22	2.43
2	11.70	11.79	7.76	7.76	4.92	1.86	2.62	3.90
3	15.12	14.38	6.13	6.89	5.34	4.88	5.50	5.27
4	8.10	8.12	2.64	3.17	2.47	1.86	2.74	2.31

[1] Norman R. Garner, *Ind. Quality Control*, May, 1956.

Assuming both factors are model 1, find which tests and which laundries give similar results and which are significantly different, if any.

8. Assume that the tests run by each laundry in Prob. 7 are not related, so that there are 32 different tests, eight for each laundry, and determine whether there is significant difference between laundries.

9. Calculate the wholly significant difference (WSD) for the means of the different powder fractions of example 7.10, and group the results into homogeneous sets.

10. Assume the powder fractions of example 7.10 are in order of integers, and calculate the linear, quadratic, and cubic components of the powder effect.

11. Calculate the regression sum of squares for the time and/or the temperature effect of example 7.11.

12. Following example 7.13, assume the categories in the example are reactors, time, and operators, and calculate the missing datum point.

13. Alter the symbols in example 7.16 so that the first and second levels of some of the factors are reversed, and recalculate the analysis of variance.

chapter eight
THE 2^n FACTORIAL

The factorial experiment with each factor at two levels is a special case of the general balanced factorial analysis discussed in Sec. 7.12. The completely balanced factorial with n factors at two levels each will require 2^n experiments—hence the classification of these designs as the "2^n factorials."

The factorial experiment with each factor at two levels can be analyzed by the formal method for factorial designs as illustrated in example 7.15. However, there are features about the 2^n factorial which make its use in experimental work particularly attractive. First, the analysis of variance can be carried out in a much simpler manner than that employed for the general balanced factorial design. Second, when large numbers of experiments are involved, the sequence of experiments can be arranged so that differences among groups of experiments (blocks) can be equated to higher-order interactions and the main effects and lower-order interactions can be determined without any interference (this feature is discussed in detail in Sec. 8.3). Third, the 2^n factorial lends itself to fractionation so that results can be obtained from a half or a quarter, or some other $1/2^p$ fraction, of the total factorial experiment (this feature is discussed in Sec. 8.4).

The 3^n factorial—all factors at three levels—and mixed factorials with some factors at two levels and others at three levels can also be calculated in a somewhat simpler manner than outlined in Chap. 7 for the general factorial. However, the use of special factorials other than the 2^n is beyond the planned scope of this general statistics text. Readers interested in further factorial design discussion are referred to books dealing specifically with experimental design.[1,2]

8.1 Nomenclature

Uppercase (capital) letters are used to designate the factors being tested: temperature, pressure, catalyst concentration, etc. Uppercase letters are also used to designate the particular effects being determined. A is the effect of the A factor, i.e., the difference between doing the experiment at one level of A and doing it at the other level of A. AB designates the interaction of the A and B factors, i.e., the difference in results between experiments with both A and B at their higher level or with both at their lower levels, and with experiments with either A or B at the higher level and the other factor at the low level. Although the uppercase letters are used for both the factors and for the effects, there will be no confusion in interpreting the use of the symbols.

Lowercase letters, a, b, c, ab, . . . , are used both to designate the experiments to be run and also the results obtained from the experiment. For example, ab in an experiment involving factors A, B, C, and D designates the experiment with factors A and B at their upper levels and factors C and D at their lower levels. It also designates the response obtained in this experiment. c designates the experiment carried out at the upper level of factor C and at the lower level of all other factors. The symbol (1) is reserved for the experiment at the low level of all factors.

A complete factorial involving four factors at two levels each, a 2^4 factorial, would have 16 experiments to be run, and if the factor designations were A, B, C, and D, the experiments would be (1), a, b, ab, c, ac, bc, abc, d, ad, bd, abd, cd, acd, bcd, $abcd$. These same lowercase combinations of letters designate the responses obtained for each of the experiments.

"A" designates the A effect—the difference in response when the experiments are run at the upper level of the A factor and when they are run at the lower level.

$$A = (a + ab + ac + abc) - ([1] + b + c + bc) \qquad (8.1)$$

Although we refer to one level as the "upper" and the other as the "lower," this designation is merely descriptive and it does not necessarily

---

[1] W. G. Cochran and G. M. Cox, "Experimental Designs," John Wiley & Sons, Inc., New York, 1957.

[2] W. L. Davies, "Design and Analysis of Industrial Experiments," Hafner Publishing Company, Inc., New York, 1956.

indicate a quantitative difference in the factor levels or a difference in the direction indicated by "upper" and "lower." If an experimental program involved determining the difference between two reactors or two sources of supply, experiments in one reactor or with one source of supply would be designated by the presence of the symbol and the other by the absence. If temperature was one of the factors being investigated, one temperature would be indicated by the presence of the symbol designated and one by the absence of the symbol, and the presence or absence would not necessarily be connected with the higher and lower temperatures. It is only necessary to know how the symbols are used so that when the effects are determined from the analysis, the signs of the effects are properly applied. Positive effects indicate that the experiments designated by the presence of the symbol give higher responses than the experiments designated by the absence of the symbol.

8.2 Analysis of the 2^n Factorial

8.2.1 Table of Signs. Equation (8.1) shows the A effect calculated from the differences between the responses at the upper level of A and the responses at the lower level of A. A similar equation can be written for each of the effects, each effect being calculated from differences involving all the experiments.

$$B = (b + ab + bc + abc) - ([1] + a + c + ac)$$
$$C = (c + ac + bc + abc) - ([1] + a + b + ab)$$

The interaction was defined as the difference between results when the factors were held at the same levels and the results when the effects were at different levels.

$$AB = ([1] + c + ab + abc) - (a + b + ac + bc)$$
$$AC = ([1] + b + ac + abc) - (a + c + ab + bc)$$

In the left parentheses of the above equations, the first two terms represent the responses when the interacting factors are both at their lower levels. The second two terms are the responses when the factors are both at their upper levels. In the second parentheses each term represents the response for the experiment when one of the interacting factors is at the upper level and the other is at the lower.

There are several ways in which the equations given above can be obtained. The table of signs illustrated below is presented as the preferred method because the table will be used for other purposes as will be explained in later sections.

The table of signs is constructed in the following manner: the columns are designated by the effect symbols A, B, AB, C, etc., and the rows are designated by the response symbols (1), a, b, ab, etc. Table 8.1 illustrates the arrangement for a three factor experiment. After the column and row headings are put in, the signs under the main effect symbols are entered, a plus sign in the rows indicating experiments at the upper level of that factor and a minus sign in rows indicating experiments at the lower levels. The signs in the columns headed by interaction effect symbols (combinations of main effect symbols) are obtained by multiplying the signs in the corresponding main effect columns of the same row. For example, in row "ac" the signs in the A and C column are $+$ and in the B column are $-$. The sign in the AB interaction column for row "ac" would be $(+)(-) = -$, and in the AC interaction column would be $(+)(+) = +$.

TABLE 8.1 Table of Signs

	A	B	AB	C	AC	BC	ABC
(1)	−	−	+	−	+	+	−
a	+	−	−	−	−	+	+
b	−	+	−	−	+	−	+
ab	+	+	+	−	−	−	−
c	−	−	+	+	−	−	+
ac	+	−	−	+	+	−	−
bc	−	+	−	+	−	+	−
abc	+	+	+	+	+	+	+

Each effect, indicated by the uppercase letters at the top of the columns, is calculated by adding those responses (indicated by the lowercase letters designating the rows), carrying a plus sign and subtracting those carrying the minus sign. Thus the total BC effect equals $(1) + a - b - ab - c - ac + bc + abc$. This is the same as $([1] + a + bc + abc) - (b + ab + c + ac)$, which is the difference between those responses with B and C at the same levels and those with B and C at different levels.

Although the order of the rows in the table of signs is not important for the formation of the table, it is important for further use of the table as will be described. We will therefore establish a "standard order" for the arrangement of the row symbols. The first symbol in the standard order is for all experiments at their lower level, (1). Successive symbols are obtained by multiplying the existing symbols by one lowercase letter symbol. For example, we start with (1) and multiply it by a to give (1), a. We then multiply these two symbols by b to give (1), a, b, ab. We then multiply these four by c and get (1), a, b, ab, c, ac, bc, abc. If there were a fourth factor D, we would then multiply by d and get (1), a, b, ab,

c, ac, bc, abc, d, ad, bd, abd, cd, acd, bcd, $abcd$. The symbols used, of course, have nothing to do with the order. They could just as well be q, r, t, z. It is the method of developing the sequence that establishes the standard order.

It is also important to understand that the "standard order" applies to the formation of the table of signs and later to the calculation of the effects, and *not* to the order in which the experiments are to be carried out. The actual order of doing the experiments depends on the experimental situation. If possible, the order of carrying out the experiments should be randomized. But after the responses are obtained, the results are tabulated in "standard order" to complete the analysis of the experiment.

8.2.2 Calculation of a 2^n Factorial. The actual analysis of variance can be carried out from the responses by the general method of analyzing factorial experiments, or by applying the positive and negative signs from a table similar to Table 8.1 for each of the effects. The mean effect will be the total effect divided by 2^{n-1}, since the total effect is the sum of 2^{n-1} responses at one level less the sum of 2^{n-1} responses at the other level. The sum of squares (sum of squares of deviation from the mean) for each effect will be the total effect squared divided by 2^n. (The sum of squares of the deviations of two numbers from their mean is the square of the difference divided by two. The sum of squares of deviation of the sum of 2^{n-1} numbers from the mean of this sum and the sum of another set of 2^{n-1} numbers is the square of their difference divided by $2 \cdot [2^{n-1}]$, or 2^n.) Since each effect is the difference between two numbers (the sum of 2^{n-1} responses at one level and the sum of 2^{n-1} responses at another level), its sum of squares has one degree of freedom. Therefore the mean square for a 2^n factorial, which is the sum of squares divided by the degrees of freedom, is equal to the sum of squares.

When the responses are arranged in the standard order described in the preceding section, there is a relatively simple method for calculating the total effects and the sum of squares. The method involves only the addition and subtraction of pairs of entries in a table of the responses, and finally a squaring of the totals. This method is outlined in the following steps and is illustrated in example 8.1.

Step 1. For a 2^n factorial, prepare a table with $n + 4$ columns. In the first column write the response symbols in standard order: (1), a, b, ab, c, ac, bc, abc, d, ad, bd, abd, cd, acd, bcd, $abcd$, e, ae, etc.

Step 2. In the second column write the actual numerical values of the responses corresponding to the symbols in the first column. In the first row of the second column write the response for the experiment at the low level of all factors. In the second row write the response for the experiment at the upper level of the first factor and the lower levels of the other factors, and so forth. The total of the second column will be the total of all the responses, Σx.

Step 3. In the upper half of the third column write the sum of the responses in the second column taken in pairs. The first row of column three has a value equal to $(1) + a$, the second row has a value equal to $b + ab$, the third row a value equal to $c + ac$, etc. These sums will occupy half of column three. In the bottom half of the column enter the difference between the pairs in column two, taking the difference as the even number responses minus the odd: $2 - 1, 4 - 3, 6 - 5$, etc. The first number in the bottom half of column three will be $a - (1)$, the second number will be $ab - b$, the third $ac - c$, etc.

Step 4. In the fourth column write the sums and differences of the pairs of numbers in the third column. Column four will have the same relation to column three as column three had to column two.

Step 5. The summing and differencing of pairs is continued until it has been done n times for a 2^n factorial. The first row of the last sum and difference column will be the total of all responses, Σx, the same as the total of the second column—giving a slight check to the arithmetic. The other values in the last sum and difference column are the total effects. These are the total effects corresponding to the symbols for the responses in the first column. The value in the second row of the last sum and difference column is the A effect. The value in the third row is the B effect. The value in the fourth row is the AB interaction effect, etc.

Step 6. The mean effects are the total effects divided by 2^{n-1}. A positive sign of the mean effect (and the total effect) indicates that the responses of the experiments represented by the presence of the factor symbol are greater than the responses of experiments represented by the absence of the symbol. Negative mean effects indicate the opposite: responses for experiments represented by the presence of the factor symbol are less than the responses of experiments represented by the absence of the symbol.

Step 7. In the last column the sum of squares for each effect is calculated. The sum of squares, and hence the mean square, since each sum of squares is calculated with one degree of freedom, is the total effect squared divided by 2^n. The analysis of variance testing for significance is carried out by F test in the same manner as discussed in the previous chapter. Without replication, there is no direct estimate of the error variance. The higher-order interaction sums of squares are usually pooled for an error variance estimate which is used for the significance testing of the lower-order interactions and main effects.

The calculation of a 2^4 factorial is illustrated in example 8.1.

example 8.1. In example 7.15 data in triplicate from some electric furnace experiments were analyzed for a completely balanced factorial design example. The mean of the replicates is used in this example for

THE 2^n FACTORIAL

the analysis of a 2^4 factorial designed experiment. (The actual replicated data could be used in a calculation similar to that illustrated in this example, as will be discussed in the next section, but to simplify this example, the means of the replicates are used.)

The design of the experiment involved obtaining data on power consumption for electric furnace melts in high and low roof furnaces; with 700-lb and 1,000-lb scrap charges; with tubular scrap and with plate scrap; and with high power input and with low power input at the start of the melt. In the following calculation, the symbols are:

r = low roof   no symbol = high roof
c = 1,000-lb charge   no symbol = 700-lb charge
s = plate scrap   no symbol = tubular scrap
p = low power input   no symbol = high power input

The units of the response values are kwhr/ton of product. The calculation is detailed in Table 8.2.

TABLE 8.2 Analysis of Variance Calculation for Electric Furnace Data a 2^4 Factorial

Step 1	Step 2	Step 3	Step 4	Step 5			Step 6	Step 7
Experiments	Responses	First sum and difference	Second sum and difference	Third sum and difference	nth sum and difference = total effect		Mean effect	Sum of squares†
(1)	715	1,408	2,910	6,626	13,123			
c	693	1,502	3,716	6,497	−331		−41.38	6,847.56
s	775	1,801	2,986	−18	479		59.88	14,340.06
cs	727	1,915	3,511	−313	−155		−19.38	1,501.56
p	862	1,455	−70	208	1,331		166.38	10,722.56
cp	939	1,531	52	271	137		17.12	1,173.06
sp	970	1,658	−164	−128	139		17.38	1,207.56
csp	945	1,853	−149	−27	−235		−29.38	3,451.56
r	785	−22	94	806	−129		−16.12	1,040.06
cr	670	−48	114	525	−295		−36.88	5,439.06
sr	790	77	76	122	63		7.88	248.06
csr	741	−25	195	15	101		12.62	637.56
pr	843	−115	−26	20	−281		−35.12	4,925.06
cpr	815	−49	−102	119	−107		−13.38	715.56
spr	987	−28	66	−76	99		12.38	612.56
$cspr$	866	−121	−93	−159	−83		−10.38	430.56
	13,123							

† (Step 5)$^2/2^n$.

Discussion. The "sum of squares" (in the eighth column of Table 8.2) for each effect can be tested for significance by the F test. For an error variance estimate against which to test the sum of squares, the highest-order inaction sum of squares is used—unless there is an independent estimate of the error variance from some other source. Inasmuch as each sum of squares in the 2^n factorial computation is calculated with one degree of freedom, a more sensitive F test is obtained if all the sums of squares smaller than or of the same magnitude as the highest order interaction sum of squares are pooled, i.e., added and divided by the total degrees of freedom, one for each sum of squares included in the pooling. In Table 8.2, the highest-order interaction is the $CSPR$ effect, which has a sum of squares of 430.56. The SPR, CPR, CSR and SR sums of squares are all not significantly different from the $CSPR$ value of 430.56, so these five sums of squares are pooled (added) to give an error variance estimate of 528.86 with five degrees of freedom.

Table 8.3 shows the F test results for the effects not pooled to obtain the error variance estimate. The scrap factor and the power factor are significant at less than the 0.01 level, and the charge factor, the charge-roof interaction and the power-roof interaction are significant between the 0.01 and 0.05 level. The other factors are not significant.

TABLE 8.3   Variance Ratio Tests for Results from Table 8.2

Effect	Mean square	Degrees of freedom	F ratio	Significance level
C	6,847.56	1	12.95	0.05
S	14,340.06	1	27.12	0.01
P	110,722.56	1	209.36	0.01
R	1,040.06	1	1.97	—
CS	1,501.56	1	2.84	—
CP	1,173.06	1	2.22	—
CR	5,439.06	1	10.28	0.05
SP	1,207.56	1	2.28	—
PR	4,935.06	1	9.33	0.05
CSP	3,451.56	1	6.52	—
"Error"	528.86	5		

$F_{0.01,5,1} = 16.26$

$F_{0.05,5,1} = 6.61$

8.2.3   Further Discussion of example 8.1. The sign of the total effects and the mean effect is to be interpreted in terms of which factor level was designated "upper" and which "lower." A positive sign of the factor effect indicates that the responses designated by the presence of the factor symbol were higher than the responses designated by the absence of the symbol. In example 8.1, the C factor had a negative effect and the S factor had a positive effect. The s symbol was used for plate scrap, and the s symbol was omitted for the runs with tubular scrap. The results indicate that the plate scrap had higher power consumption than the tubular scrap. The c symbol was used for the 1,000-lb charges, and the c symbol was omitted for the runs with 700-lb charges. The negative C effect indicates that the 700-lb charges had a higher power consumption than the 1,000-lb charges.

In example 8.1 the responses were assumed to be single values obtained from individual experiments. Actually the values used in the example were the means of three determinations. The effect of this replication on the calculation is discussed in the next paragraph. Continuing the assumption that the responses used in example 8.1 were single values, the mean effects are the mean differences between eight values at one level of the factor and eight values at the other level. The estimated error variance obtained from the pooled sums of squares of the higher-order interactions is the error variance of a single observation. The confidence range of the mean effects can be obtained from an appropriate t factor and the error variance divided by 2^{n-2}. In example 8.1 the mean C effect is -41.38. The error variance estimate is 528.86. The confidence range of the C effect is $-41.38 \pm t_{\alpha,\nu}(s_0[x]/\sqrt{2^{n-2}}) = -41.38 \pm 2.571 \times \sqrt{528.86/4} = -41.38 \pm 29.56$ at the 0.05 level. t at the 0.05 level, and 5 degrees of freedom—the number of degrees of freedom associated with the error variance estimate—is 2.571. If t is taken at the 0.01 level, the value at 5 degrees of freedom is 4.032, and the confidence interval is $\pm 4.032 \times 11.50$ or 46.36, which would overlap zero and confirm the F test reported in example 8.1.

When experiments are replicated the analysis of variance procedures described in the preceding paragraphs are only slightly affected. The replications provide a direct estimate of the error variance. This is the pooled sum of squares of deviations of each set of replicates from the mean for the set divided by the total degrees of freedom for all sets. And this is identical with the formula from Table 7.39:

$$\frac{\Sigma x^2 - \sum \frac{(\text{replicate total})^2}{\text{no. of replicates}}}{\Sigma(\text{no. of replicates} - 1)}$$

The F tests for the factor sums of squares are made with the measured error variance rather than the estimated error variance. The degrees of freedom associated with the error variance calculated from actual replicates is the denominator in the above equation.

The actual analysis of variance calculation can be made with the means of the replicates or with the totals, if the number of replicates is the same for each experiment. If the number of replicates is different for different experiments, the analysis of variance is most simply calculated using the means of each set of data. When the number of replicates is the same for each set, the only difference in the calculation is in computing the mean effects. When the totals of the responses at each experimental condition are used to calculate the analysis of variance, the mean effects are the total effects divided by $k \times 2^{n-1}$ where k is the number of replicates. When the means of each set of replicates are used, the mean effects are equal to the total effects divided by 2^{n-1}. The sums of squares calculated from the totals of the replicates will be equal to k times the sums of squares calculated from the means of the replicated responses. This can be seen by comparing the results obtained in example 8.1 using the means of the three results at each set of conditions with the results given in example 7.15 where the individual values were used.

8.3 Confounding

If the number of experiments involved in a particular design is too large for all of the experiments to be carried out in a single group or block, it may be necessary to carry out the program in several stages. Half of the experiments may be run one day and the other half the next. Or the experiments might be divided among four analysts, each of them doing one quarter of the total. We will call these groups of experiments "blocks." In the first case, one block would be done on one day and another block on the second day. In the second case, each analyst would do a block of experiments. If inadvertently all of the experiments at the upper level of one factor were done in one block and all the experiments at the lower level were done in another block, it would not be possible to differentiate between the factor effect and any block effect that might be present.

Examination of Table 8.1 will disclose that there is an equal number of plus signs and minus signs for each effect heading. Each effect is calculated from the difference between the sum of responses of half the experiments and the sum of the responses of the other half. If a design is carried out in two block, and one block consists of all the experiments under one effect heading having plus signs and the other block consists of all the experiments having minus signs, any difference between blocks will be measured by the same difference that is used to calculate the particular effect involved. These two measurements will be "confounded."

For example, referring to Table 8.1, if the eight experiments require two days to run, any difference between days can be confounded with the highest order interaction by doing all the experiments with plus signs under the ABC heading on one day and all those with minus signs on the next day. Experiments (1), ab, ac, and bc would be run on one day and experiments a, b, c, abc would be run on the next day. Any difference between days would be measured by the same difference that would be used to measure the ABC effect. It will also be true that all the other effects will be free of any difference between days, because in each case the other effects will be determined by differences between experiments that are divided equally between the two day-blocks. This statement can be illustrated by further reference to Table 8.1.

If any "day" effect was confounded with the ABC interaction in the blocking arrangement just described, this day effect would not be present in any of the other measured effects. If there were a difference between days, and the second day added a factor, say D, to the results, this factor would cancel in the calculation of all the effects except that one with which days was confounded. This fact is demonstrated in Eqs. (8.2) which follow and which illustrate the calculation of some of the effects. The same procedure can be applied to all the effect calculations with similar results. The experiments done on the second day are (1), ab, ac, and bc, and we are assuming that each of the responses would have an increment D added due to some effect of the second day's operation. The responses would therefore be $(1) + D$, $ab + D$, $ac + D$, and $bc + D$. In the calculation of each of the effects other than ABC, the D increments will cancel as shown in Eqs. (8.2) below:

$$A = (a + [ab + D] + [ac + D] + abc)$$
$$- ([(1) + D] + b + c + [bc + D])$$
$$B = (b + [ab + D] + [bc + D] + abc)$$
$$- ([(1) + D] + a + c + [ac + D]) \quad (8.2)$$
$$AB = ([(1) + D] + [ab + D] + c + abc)$$
$$- (a + b + [ac + D] + [bc + D])$$
etc.

In each case the D factor is added to two responses in each group and is therefore eliminated in the calculation of the difference between the groups.

In experimental designs which are run in two blocks, it usually will be most advantageous to confound the blocks with the highest-order interaction and leave all the lower-order interactions and main effects free of any block effect. The interaction or other effect confounded with the block effect is called the *defining contrast*. Thus, if an experiment is run in two blocks and blocks are confounded with the $ABCD$ interaction, the defining contrast is as shown in Eq. (8.3).

$$\text{Blocks} = ABCD \quad (8.3)$$

You are not limited to only two blocks. However, if more than two blocks of experiments are run, more than one effect will be confounded with the block effect, and the defining contrast will have more than one effect term. With r block, $r - 1$ degrees of freedom are utilized in calculating the differences among blocks. Since each effect in a 2^n factorial is calculated with one degree of freedom, r blocks will be confounded with $r - 1$ factor effects, and the defining contrast will involve $r - 1$ sets of factor effect symbols. For the 2^4 factorial illustrated in Table 8.2, if carried out in four blocks, the defining contrast might be

$$\text{Blocks} = RCSP,\ SP,\ RC \tag{8.4}$$

All the effects confounded with the block effect cannot be independently chosen. If there are to be r blocks, where $r = 2^m$, then m of the confounded effects can be independently chosen, and the other $r - m - 1$ effects that will be confounded with blocks will be set by the m chosen effects. The total defining contrast is *generated* from the chosen (generating) effects in the following manner. If we are to have 2^m blocks, m effects are selected to be confounded with the block effect, and these m effects are multiplied under the rule that squared terms equal unity to form the other effects that will also be confounded with blocks. In Eq. (8.4), where four blocks are to be used, $m = 2$, and two effects may be chosen, say $RCSP$ and SP. These two combinations of symbols are multiplied to give RCS^2P^2, which under our rule is equal to RC, which is the third effect confounded with the blocks.

The effects initially chosen are called the *generating contrast*, and the total set generated is the defining contrast.

For example, if a 2^7 factorial, consisting of factors A, B, C, D, E, F, and G is to be carried out in $8 = 2^3$ blocks, three effects may be chosen as the generating contrast. If $ABFG$, ACF, and $BCEF$ are chosen, the total defining contrast is made up as follows:

$$\begin{aligned}
(ABFG)(ACF) &= (A^2BCF^2G) = BCG \\
(ABFG)(BCEF) &= (AB^2EF^2G) = AEG \\
(ACF)(BCEF) &= (ABC^2EF^2) = ABE \\
(ABFG)(ACF)(BCEF) &= (A^2B^2C^2EF^2FG) = EFG
\end{aligned} \tag{8.5}$$

or

$\text{Blocks} = \underline{ABFG},\ \underline{ACF},\ \underline{BCEF},\ BCG,\ AEG,\ ABE,\ EFG$

where the underlined effects are the generating contrast.

Care must be taken in selecting the generating contrasts so that the resulting defining contrast does not include effects that it would be better to keep free of the block effect. If a 2^4 factorial, consisting of factors

A, B, C, and D, is to be carried out in four blocks, two effects may be chosen as the generating contrast. If the generating contrast is selected as the $ABCD$ and ABD effects, the total defining contrast will be $ABCD$, ABD, and $(ABCD)(ABD) = (A^2B^2CD^2) = C$, and the main effect C will be confounded with the block effect. In this case, if the highest order interaction was selected as one of the generating contrast effects, a two-factor interaction, or first-order interaction, effect would have to be selected to avoid confounding one of the main effects with the block effect.

In carrying out a blocked factorial experiment, the particular experiments to be performed in each block can be determined by the signs from a table similar to Table 8.1. For an experiment in two blocks, the experiments in each block are those carrying the same sign under the effect used for the defining contrast. For a design carried out in four or more blocks, each block will consist of those experiments having the combination of signs formed from the signs in the columns of the generating contrast effects. In a four-block experiment, with two effects in the generating contrast, the combinations of signs will be $++$, $+-$, $-+$, and $--$. In an eight-block experiment with three effects in the generating contrast, the combinations of signs will be $+++$, $++-$, $+-+$, $-++$, $+--$, $-+-$, $--+$, and $---$.

For example, if RC and SP are chosen for the generating contrast for a four block design of the experiment of Table 8.4, the four blocks, corresponding to $++$, $+-$, $-+$, and $--$ combination of signs, would be: $((1), sp, cr, cspr)$, (s, p, csr, cpr), (c, csp, r, spr), and (cs, cp, sr, rp).

Table 8.4 illustrates the determination of the combination of signs for the selection of blocks. The experiments are listed in standard order and the signs for the main effects are tabulated. The signs for the generating contrast effects are obtained by multiplying the corresponding signs for the main effects. This is the same procedure that was used to make up the table of signs described in Sec. 8.2. It is not necessary to make up the entire table of signs in order to establish the blocks of experiments, but if the table of signs is available, it can be used directly by selecting the signs tabulated under the generating contrast effect symbols.

The arrangement of experiments within a block, and the arrangement of blocks within a program should be randomized once the particular design and blocking are established. The calculation of the factor and block effects is carried out in the manner described in Sec. 8.2.2. When the calculation of the analysis of variance is completed, the calculated effects are identified by the symbols in the calculation table. Those not included in the defining contrast are free of block effects and may be so reported. If the defining contrast factors have effects that are significant, these can be pooled and reported as a block effect. For example, if the experiment of Table 8.2 has been run in four block, with blocks confounded with SP, RC, and $RCSP$, then these three effects would be pooled, with three degrees of freedom, and reported as the block effect. The other effects would be free of any block effect and would be as originally reported.

TABLE 8.4 Generation of Four Blocks for a 2^4 Factorial

	Main effects				Generating contrast		Blocks
	C	S	P	R	RC	SP	
(1)	−	−	−	−	+	+	1
c	+	−	−	−	−	+	3
s	−	+	−	−	+	−	2
cs	+	+	−	−	−	−	4
p	−	−	+	−	+	−	2
cp	+	−	+	−	−	−	4
sp	−	+	+	−	+	+	1
csp	+	+	+	−	−	+	3
r	−	−	−	+	−	+	3
cr	+	−	−	+	+	+	1
sr	−	+	−	+	−	−	4
csr	+	+	−	+	+	−	2
pr	−	−	+	+	−	−	4
cpr	+	−	+	+	+	−	2
spr	−	+	+	+	−	+	3
$cspr$	+	+	+	+	+	+	1

Block 1 = ++, Block 2 = +−, Block 3 = −+, Block 4 = −−

8.4 Fractional Replicates of 2^n Factorials

A factorial design involving n factors at two levels each requires 2^n experiments for the complete design. Sometimes it is possible to obtain a major portion of the required information by doing a fraction of the total design. Since the same information may be obtained by doing any one of equal fractions of the experiment, the fractions are termed "replicates." If a subset of 2^{n-p} experiments of a 2^n is carried out, this is a $1/2^p$ fraction of the total design, where n and p are integers, $n > p$. The National Bureau of Standards (Applied Mathematics Series—48) has published the designs of fractional replicates for n from 5 to 16 and for p from 1 to 8. The purpose of this section is to explain the setting up and calculation of a fractional factorial. Particular designs for most occasions can be taken from the reference publication or from other sources of tabulated designs. Some of the simpler designs will be obvious from the discussion which follows.

If a $1/2^p$ fraction of a 2^n factorial is carried out, there will be only 2^{n-p} experiments performed and hence only $2^{n-p} - 1$ degrees of freedom available for estimating the effects. Since a 2^n factorial has $2^n - 1$ factor effects, including all main effects and interactions, it is not possible to estimate all the effects from the fraction of the total experiment. In fact, in a $1/2^p$ fraction, each calculated total effect will be the algebraic sum of 2^p effects. If you carried out a $1/4$, $(1/2^2)$, fraction of a 2^6, each calculated total effect would be the sum of 2^2 or 4 effects. The effects included in each group are called *aliases*. The fractional designs are usually set up so that the main effects and lower-order interactions are aliased with higher order interactions. The designs given in the Bureau of Standards' publication referred to have the main effects free from being aliased with first-order interactions. The determination of the aliases is probably best illustrated by an example.

8.4.1 Half Replicates. A 2^3 full factorial can be used to find effects and interactions of three factors. If it is desired to use a 2^3 as a half replicate $(1/2^1)$, of a 2^4, i.e., a 2^{4-1} factorial involving four factors, each total effect will measure a pair of aliases. The aliases and the experiments to be run can be obtained by setting up the table of signs for a three factor experiment. Table 8.5 illustrates the determination of both the aliases and the experimental design.

The table of signs is set up for three factors, say A, B, and C, in the manner previously described. The fourth factor, D, will be aliased with the highest order interaction, ABC, so that D is set equal to ABC as indicated in the eighth column of Table 8.5. The ABC interaction effect

TABLE 8.5 Table of Signs for a ½ Replicate of 2^4 Factorial

Experiments	A	B	C	AB	AC	BC	$D = ABC$	Effects
(1)	−	−	−	+	+	+	−	Total and $ABCD$
(d) a	+	−	−	−	−	+	+	A and BCD
(d) b	−	+	−	−	+	−	+	B and ACD
ab	+	+	−	+	−	−	−	AB and CD
(d) c	−	−	+	+	−	−	+	C and ABD
ac	+	−	+	−	+	−	−	AC and BD
bc	−	+	+	−	−	+	−	BC and AD
(d) abc	+	+	+	+	+	+	+	ABC and D

is measured by the difference between the sum of the responses indicated by plus signs under the ABC symbol and the sum of the responses with the minus signs. Since the D effect is to be aliased with the ABC interaction, the D effect will be measured by this same difference. Therefore the experiments with the plus signs are done at the upper level of the D

factor and those with the minus sign are done at the lower level. This is indicated by the *d* in parentheses in the experiment column at the left of the table. The experiments to be done are (1), *ad, bd, ab, cd, ac, bc,* and *abcd*. If the other half of the 2^4 were done, *D* would be aliased with $-ABC$, and the experiments with the minus sign in the *ABC* column would be done at the upper level of *D*. In that case the experiments would be *d, a, b, abd, c, acd, bcd, abc*.

The alias of one of the factors is selected by equating this factor effect to the effect with which it is to be aliased. In the example, *D* is aliased with the *ABC* interaction: $D = ABC$. Multiplying both sides of this equality by *D*, following the rule that squared terms equal unity, results in the relation $D^2 = ABCD$ or

$$I = ABCD \tag{8.6}$$

which is the defining contrast for the ½ fractional factorial.

The aliases of all the other effects are determined by multiplying the defining contrast by the effect symbols.

$$\begin{aligned} A \text{ times } (I = ABCD) \text{ gives } A &= BCD \\ B \text{ times } (I = ABCD) \text{ gives } B &= ACD \\ &\cdot \\ &\cdot \\ BC \text{ times } (I = ABCD) \text{ gives } BC &= AD \\ &\cdot \\ &\cdot \\ ABC \text{ times } (I = ABCD) \text{ gives } ABC &= D \end{aligned} \tag{8.7}$$

The aliases for all of the effects are those listed in the right column of Table 8.5. If *D* had originally been equated to $-ABC$ for the other half replicate, the defining contrast would have been $I = -ABCD$ and each of the alias pairs would have carried one negative sign. The negative sign does not affect the calculation of the total effects. After the effects have been calculated, if the total effect is positive and one term in the alias carried a negative sign, it would indicate that the responses with negative signs in the table of signs were larger than the responses with positive signs. In the example, if *D* were equated to $-ABC$ and the calculation of the results gave a positive result for the *ABC* effect, it would mean that the experiments done at the lower level of *D* gave larger responses than the experiments at the higher level of *D*.

The calculation of the effects is performed in the usual way. With the responses listed in standard order, the sums and differences are calculated in the manner described in the preceding chapter. In a half replicate, a 2^{n-1} factorial, involving *n* factors, only $n - 1$ of the factors are

THE 2^n FACTORIAL   251

taken into account when arranging the responses in standard order. One of the response symbols is ignored in arranging the order of the responses for calculation and identification of the effects. If n is an even number and the added factor is aliased with the positive value of the highest-order interaction, the experiment (1) at the low level of all factors will be included in the design and the standard order for calculation purposes may be set using any $n - 1$ of the main factor symbols; i.e., any one factor symbol may be ignored. If n is odd, the same situation holds if the added factor is aliased with the negative value of the highest order interaction. If n is even and the added factor is aliased with the negative value of the highest-order interaction, or if n is odd and the added factor is aliased with the positive value of the highest-order interaction, the experiment (1) will *not* be included in the fraction to be run. In this case, in order to arrange the responses in standard order for computation purposes, when one of the factor symbols is ignored, a null term will result. This term can be considered (1) for calculation purposes. This situation is illustrated in example 8.2 which follows.

When the experimental responses are arranged in standard order for computation purposes, the end result is a series of values which we called the *total effects*. The factors and combination (interaction) of factors associated with each total effect are those corresponding to the response symbols used in the standard order—with one of the symbols disregarded. The pairs of aliases associated with each total effect are determined by multiplying the defining contrast by the factor symbol terms from the analysis of variance calculation.

example 8.2. An experimental program is to be run to determine the effect of temperature, pressure, flow rate, catalyst concentration, and catalyst size. Each of these variables is to be used at two levels. If the complete factorial were to be carried out, $2^5 = 32$ experiments would be required. We will do a half replicate, 16 experiments, and alias one of the main effects with the third order interaction of the other four. The factors will be temperature T, pressure P, flow rate F, catalyst concentration C, and catalyst size S. We will let $S = TPFC$. (The interested student can determine that the experimental program will be the same whichever main factor we alias with the third-order interaction of the other four.)

In Table 8.6 the signs for calculating the $TPFC$ interaction effect are determined. Since the S factor is to be aliased with this interaction effect, these signs will also establish which experiments are to be run at the upper level of the S variable and which at the lower level. The T, P, F, and C factor symbols are arranged in standard order and the signs for the $TPFC$ effect are determined. These are shown on the left side of the table. The experiments which are to be run at the upper level of the S factor are those with a $+$ sign in the $TPFC$ column. The actual experiments for the half replicate with $S = TPFC$ are shown, and the actual experimental results are also given in the right-hand column. The signs in the $TPFC$ column are the algebraic product of the signs in the individual factor columns as explained in the discussion of the table of signs.

TABLE 8.6

	T	P	F	C	$S = TPFC$	Experiments to be run	Experimental results
(1)	−	−	−	−	+	s	1
t	+	−	−	−	−	t	10
p	−	+	−	−	−	p	16
tp	+	+	−	−	+	tps	24
f	−	−	+	−	−	f	7
tf	+	−	+	−	+	tfs	16
pf	−	+	+	−	+	pfs	21
tpf	+	+	+	−	−	tpf	33
c	−	−	−	+	−	c	1
tc	+	−	−	+	+	tcs	9
pc	−	+	−	+	+	pcs	16
tpc	+	+	−	+	−	tpc	24
fc	−	−	+	+	+	tcs	8
tfc	+	−	+	+	−	tfc	17
pfc	−	+	+	+	−	pfc	23
$tpfc$	+	+	+	+	+	$tpfcs$	31

The calculation of the effects is illustrated in Table 8.7. The procedure by sums and differences is identical with that described in Sec. 8.2.2. In arranging the responses in standard order for computation purposes, the symbol for the factor aliased with the highest order interaction is ignored, and the remaining response symbols are arranged in standard order. Any other factor symbol could have been ignored in arranging the responses for calculation with identical results. One of the problems at the end of the chapter suggests this as an example. In deciding on the half replicate, the S effect was equated to the $TPFC$ interaction. The defining contrast is therefore $I = STPFC$. Multiplying this equation by each of the other factor effects determines its alias. All of the aliases are shown in the right column of Table 8.7.

Discussion. The last sum and difference column in Table 8.7 gives the total effects for each factor and its alias. The mean effects will be the total effects divided by 2^{n-p-1}, or $2^{5-1-1} = 8$ in this example. The mean temperature effect is $71/8 = 8.88$. The mean pressure effect is $^{119}\!/_8 = 14.88$, etc. The flow rate effect is the only other factor that appears to be significant. All of the others may be pooled for an error variance estimate. The variance estimates would be the total effects squared divided by 2^{n-p} or 16, and divided by the degrees of freedom. In this example, the pooled error sum of squares is 7.75 with 12 degrees of freedom (pooling the 12 total effects equal to 5 or less),

TABLE 8.7 Calculation of Effects of a Half Replicate of a 2^5 Factorial with $S = TPFC$

Experiments	Results	First sum and difference	Second sum and difference	Third sum and difference	Last sum and difference	Effects and aliases
$(1)(s)$	1	11	51	128	257	total
t	10	40	77	129	71	$T = FCSP$
p	16	23	50	38	119	$P = FCST$
$tp(s)$	24	54	79	33	1	$TP = FCS$
f	7	10	17	60	55	$F = CSPT$
$tf(s)$	16	40	21	59	5	$TF = CSP$
$pf(s)$	21	25	16	2	1	$PF = CST$
tpf	33	54	17	-1	3	$TPF = CS$
c	1	9	29	26	1	$C = FSPT$
$tc(s)$	9	8	31	29	-5	$TC = FSP$
$pc(s)$	16	9	30	4	-1	$PC = FST$
tpc	24	12	29	1	-3	$TPC = FS$
$fc(s)$	8	8	-1	2	3	$FC = SPT$
tfc	17	8	3	-1	-3	$TFC = SP$
pfc	23	9	0	4	-3	$PFC = ST$
$tpfc(s)$	31	8	-1	-1	-5	$TPFC = S$

to give an error variance estimate of 0.65. The mean squares of the temperature, pressure, and flow rate factors are highly significant when tested against this error variance.

8.4.2 Lower-order Fractionals than Half. 2^{n-p}, $p > 1$. With a half replicate, the effects are measured in pairs, called aliases. The pair of aliases is determined by multiplying the defining contrast by one set of factor effect symbols. This relation is shown in Eqs. (8.6) and (8.7). The defining contrast can be selected by choosing any one of the pair of aliases. With a one-quarter replicate, the aliases occur in groups of four; i.e., four effects are measured by each total from the analysis of variance calculation. The defining contrast is made up of the identity element and three sets of effect symbols, two of which may be independently chosen, and the other is generated from the multiplication of the first two, following the multiplication rule that squared terms equal unity.

In general, in a 2^{n-p} fractional replicate of a 2^n factorial, the aliases consist of 2^p sets of factor symbols—each total effect measures 2^p items—and the defining contrast consists of the identity element and $2^p - 1$ sets of factor symbols. p effects for aliasing may be independently chosen, and the others are generated by multiplication. The p selected sets are called the generating contrast, and the total group generated is the defining contrast.

As with block confounding, when a quarter or lower-order fraction is desired, care must be exercised in selecting the generating contrast so that the main effects are aliased with high-order interactions and that none of the main effects are aliased with each other. If a one-quarter $(1/2^2)$ replicate of a 2^6 (a 2^{6-2}) is desired, the plan can be set up by setting up the table of signs for a 2^4, for four factors, and equating the fifth and sixth factors with two of the interactions from the four-factor design. If the six factors are designated A, B, C, D, E, and F; A, B, C, and D can be used to set up the four-factor table of signs, and E set equal to $ABCD$ and F equal to ABC so that the generating members of the defining contrast are $ABCDE$ and $ABCF$. These will generate the third member of the defining contrast: $(ABCDE)(ABCF) = (A^2B^2C^2DEF) = DEF$, or

$$I = ABCDE = ABCF = DEF \tag{8.8}$$

With this defining contrast, the aliases for the other factors can be determined by multiplying the defining contrast by the factor symbols.

$$\begin{aligned}
A &= BCDE = BCF = ADEF \\
B &= ACDE = ACF = ADEF \\
C &= ABDE = ABF = CDEF \\
D &= ABCE = ABCDF = EF \\
E &= ABCD = ABCEF = DF \\
F &= ABCDEF = ABC = DE
\end{aligned} \tag{8.9}$$

It can be seen from Eqs. (8.9) that the main effects D, E, and F are aliased with first-order interactions. It is possible to avoid having any of the main effects aliased with first-order interactions by selecting a different set of generating contrasts. If E is equated to ABC, and F is equated to ACD so that the members of the generating contrast are $ABCE$ and $ACDF$, the defining contrast will be

$$I = ABCE = ACDF = BDEF \tag{8.10}$$

The $BDEF$ term is the product of the multiplication of the $ABCE$ and the $ACDF$ terms. It can be seen that the defining contrasts of all the main

THE 2^n FACTORIAL

effects will contain no effect of lower-order interaction than two, and no main effect will be aliased with first-order interactions. When the defining contrast contains terms of four or more symbols, the main effects will be aliased with second- or higher-order interactions. When the defining contrast contains terms with only three symbols, some of the main effects will be aliased with first-order interactions. If the defining contrast contains any terms with only two symbols, at least two of the first-order effects will be aliased with each other.

The actual experimental design, i.e., the experiments to be run, can be obtained from the table of signs used to establish the generating contrasts as just described. With a one-quarter replicate of a 2^6, where the factors are designated A, B, C, D, E, and F, if the generating contrasts are selected by letting $E = ABC$ and $F = ACD$ to give the defining contrast of Eq. (8.10), the actual experimental design is determined by writing the design for the factors A, B, C, and D and adding the factors E and F

TABLE 8.8 Experimental Design for One-quarter Replicate of a 2^6 Factorial
defining contrast $I = ABCE = ACDF = BDEF$

	A	B	C	D	$E = ABC$	$F = ACD$	Experiments to be run	Aliases			
(1)	−	−	−	−	−	−	(1)	total,	$ABCE$	$ACDF$	$BDEF$
a	+	−	−	−	+	+	$a(ef)$	A	BCE	CDF	$ABDEF$
b	−	+	−	−	+	−	$b(e)$	B	ACE	$ABCDF$	DEF
ab	+	+	−	−	−	+	$ab(f)$	AB	CE	$BCDF$	$ADEF$
c	−	−	+	−	+	+	$c(ef)$	C	ABE	ADF	$BCDEF$
ac	+	−	+	−	−	−	ac	AC	BE	DF	$ABCDEF$
bc	−	+	+	−	−	+	$bc(f)$	BC	AE	$ABDF$	$ACDEF$
abc	+	+	+	−	+	−	$abc(e)$	ABC	E	BDF	$CDEF$
d	−	−	−	+	−	+	$d(f)$	D	$ABCDE$	ACF	BEF
ad	+	−	−	+	+	−	$ad(e)$	AD	$BCDE$	CF	$ABEF$
bd	−	+	−	+	+	+	$bd(ef)$	BD	$ACDE$	$ABCF$	EF
abd	+	+	−	+	−	−	abd	ABD	CDE	BCF	AEF
cd	−	−	+	+	+	−	$cd(e)$	CD	$ABDE$	AF	$BCEF$
acd	+	−	+	+	−	+	$acd(f)$	ACD	BDE	F	$ABCEF$
bcd	−	+	+	+	−	−	bcd	BCD	ADE	ABF	CEF
$abcd$	+	+	+	+	+	+	$abcd(ef)$	$ABCD$	DE	BF	$ACEF$

corresponding to the signs for the interactions ABC and ACD, respectively. This is illustrated in Table 8.8. The experimental condition corresponding to the upper level of E is employed when the sign for the ABC factor is positive, and the experimental condition corresponding to the upper level of F is employed when the sign for the ACD factor is positive. The lower levels of E and F are used when the respective signs are negative.

In calculating the effects from the experimental results, the same sum and difference procedure previously described is followed. To arrange the responses in standard order for computation purposes, the experiment symbols used to set up the table of signs are employed and the additional symbols that were used to set the generating contrasts are ignored. (Actually, in a ¼ replicate any two factors in most cases can be ignored and the remaining symbols of the experimental design can be arranged in standard order. The only restriction is that two symbols representing one experiment cannot both be ignored as this would leave two (1) terms to account for. In Table 8.8, the pairs of symbols that could not be ignored at the same time are b,e or a,c or d,f.)

In Table 8.8 the eighth column gives the actual experiments to be run for the particular one-quarter replicate. To carry out the calculation from the responses, the results could be arranged in the order shown in the table, ignoring the e and f symbols. If the a and b symbols were ignored in their place, the standard order of the responses would be (1), $e(b)$, $f(ab)$, $ef(a)$, $c(a)$, $ce(ab)$, $cf(b)$, cef, $d(ab)$, $de(a)$, df, $def(b)$, $dc(b)$, dce, $dcf(a)$, $dcef(ab)$. The interested reader can demonstrate for himself by supplying some hypothetical data that ignoring different pairs of symbols does not alter the results.

The design of any 2^{n-p} fractional factorial can readily be determined by setting up the table of signs as illustrated in Table 8.8. However, there are tables of fractional factorial designs available so that the actual procedure of forming the defining contrast is not necessary. The designs available usually give the defining contrast and the experiment to be run. The most extensive set of designs is that published by the U.S. National Bureau of Standards, and referred to in Sec. 8.4. The designs given in the Bureau of Standards' publication give the defining contrast, the experiment to be run, and also arrange the experiment so that block confounding can be employed if desired. The designs given show the experiment to be run with the symbols in rows and columns and with information as to which effects are confounded with the rows, columns, and blocks of experiments. The experiment fraction listed in the Bureau of Standards' tables is that fraction containing the experiment designated (1), i.e., all factors at their lower levels. If for some reason this fraction cannot be run, another fraction can be obtained by multiplying the symbols in the fraction given by a symbol or combination of symbols not included in the fraction given. In carrying out the multiplication, the rule that squared terms equal unity is followed.

The arrangement of the experiments in the Bureau of Standards' publication is that columns, rows, and blocks are confounded with some

THE 2^n FACTORIAL

indicated effect. To put the results in "standard order" for calculation purposes, in a 2^{n-p} fraction, p response symbols are ignored and the remaining symbols are arranged as previously described. The only restriction on the selection of the symbols to be ignored is, as previously mentioned, that when all p symbols are disregarded at one time, no more than one (1) symbol designated experiment is obtained. Once the standard order is obtained, the calculation by sums and differences is the same as before, and the factor effects and their aliases are those identified by the response symbols used in the calculation. All these operations are illustrated in Example 8.3 following.

example 8.3. The Bureau of Standards' publication lists a $\frac{1}{8}$ replicate of a seven-factor design—a 2^{7-3}—as follows:
Factors: A, B, C, D, E, F, G.
Defining contrast: $I = ABEG = ACFG = BCEF = ABCD = CDEG$
$= BDFG = ADEF$.
Blocks confounded with EG, FG, EF
Rows confounded with AE, ABF, BEF
Plan:

	Blocks		
1	2	3	4
(1)	abeg	bdfg	adef
abcdefg	cdf	ace	bcg
abcd	cdeg	acfg	bcef
efg	abf	bde	adg

Problem 1. We find we cannot run the *abcdefg* experiment because of the physical nature of our equipment. We cannot run all the factors at their high levels at the same time. We therefore want to run an alternate fraction, one containing the *abcdef* experiment.

Problem 2. After we run the alternate fraction, we need to arrange the results in standard order for purposes of calculation.

Solution:
The first column of Table 8.9 shows the fraction proposed in the plan. In the second column is the fraction obtained by multiplying the original fraction by the *abcdef* symbol combination. In reporting the results of the multiplication, squared terms were dropped. The second item in the second column is $ga^2b^2c^2d^2e^2f^2$, but all of the squared terms are equal to unity. The experiment listed in the second column is the one that will be run. These experiments in the same order as the proposed plan will have the same block and row confounding. In column three the symbols *def* are isolated from each term where they appear.

TABLE 8.9

Original fraction	Original fraction multiplied by abcdef†	Revised fraction with 3 factors isolated	Revised fraction in standard order
(1)	abcdef	abc(def)	(1)(ef)
abcdefg	g	g	g
abcd	ef	(1)(ef)	c(de)
efg	abcdg	abcg(d)	cg(df)
abeg	cdfg	cg(df)	b(df)
cdf	abe	ab(e)	bg(de)
cdeg	abfg	abg(f)	bc
abf	cde	c(de)	bcg(ef)
bdfg	aceg	acg(e)	a(d)
ace	bdf	b(df)	ag(def)
acfg	bdeg	bg(de)	ac(f)
bde	acf	ac(f)	acg(e)
adef	bc	bc	ab(e)
bcg	adefg	ag(def)	abg(f)
bcef	ad	a(d)	abc(def)
adg	bcefg	bcg(ef)	abcg(d)

† Squared terms have been dropped.

Inasmuch as this is a 1/8 fraction, a $1/2^3$ fraction, three symbols need to be ignored when the data are arranged in standard order. The fourth column lists the experiments in standard order with respect to all factors except d, e, and f.

The experiments to be run are those listed in column two. These responses, when obtained, are arranged in the order of column four for calculation of the effects. The effects calculated will have the aliases listed in Table 8.10.

TABLE 8.10 Aliases of 2^{7-3} Fractional Factorial of Table 8.9

I	= ABEG	= ACFG	= BCEF	= ABCD	= CDEG	= BDFG	= ADEF
G	= ABE	= ACF	= BCEFG	= ABCDG	= CDE	= BDF	= ADEFG
C	= ABCEG	= AFG	= BEF	= ABD	= DEG	= BCDFG	= ACDEF
CG	= ABCE	= AF	= BEFG	= ABDG	= DE	= BCDF	= ACDEFG
B	= AEG	= ABCFG	= CEF	= ACD	= BCDEG	= DFG	= ABDEF
BG	= AE	= ABCF	= CEFG	= ACDG	= BCDE	= DF	= ABDEFG
BC	= ACEG	= ABFG	= EF	= AD	= BDEG	= CDFG	= ABCDEF
BCG	= ACE	= ABF	= EFG	= ADG	= BDE	= CDF	= ABCDEFG
A	= BEG	= CFG	= ABCEF	= BCD	= ACDEG	= ABDEG	= DEF
AG	= BE	= CF	= ABCEFG	= BCDG	= ACDE	= ABDF	= DEFG
AC	= BCEG	= FG	= ABEF	= BD	= ADEG	= ABCDFG	= CDEF
ACG	= BCE	= F	= ABCEFG	= BDG	= ADE	= ABCDF	= CDEFG
AB	= EG	= BCFG	= ACEF	= CD	= ABCDEG	= ADFG	= BDEF
ABG	= E	= BCF	= ACEFG	= CDG	= ABCDE	= ADF	= BDEFG
ABC	= CEG	= BFG	= AEF	= D	= ABDEG	= ACDFG	= BCDEF
ABCG	= CE	= BF	= AEFG	= DG	= ABDE	= ACDF	= BCDEFG

THE 2^n FACTORIAL

8.5 Problems

1. With the following response symbols and values, set the responses in standard order and calculate the highest-order interaction effect. $t = 10$, $p = 16$, $f = 7$, $tp = 24$, $tf = 16$, $pf = 21$, $tpf = 33$, $(1) = 1$. Ans.: $TPF = 4$, total effect.

2. Use the data of example 8.1, but arrange the responses in standard order, starting as follows: $(1), r, p, rp, s, \ldots$, and calculate the total effects.

3. If the experiment of example 8.1 had been done in two blocks, with the block effect confounded with the $SCPR$ interaction effect, and if the first block contains the r and s experiments, in which block is the c experiment? In which block is the rcs experiment?

4. Assume that the data of example 8.1 were actually from a 2^{6-2} fractional factorial in which additional factors A and B were included. If the defining contrast was $I = SPAR = SCBR = ABCP$, what is the total A effect? What is the total B effect? Ans.: 99, 101.

5. On the same assumption as Problem 4, which experiments would have been run at the upper level of A?

6. In example 8.2, using the responses given in Table 8.7, arrange the values in standard order ignoring the f factor response, and calculate the total effects.

7. In a one-quarter replicate of a 2^5 factorial involving factors A, B, C, D, and E, D is aliased with BC and E is aliased with AC. The responses arranged in standard order (ignoring the symbols a and b) are as follows: $(1) = 7.2$, $c = 8.4$, $d = 2.0$, $cd = 3.0$, $e = 6.7$, $ce = 9.2$, $de = 3.4$, $cde = 3.7$. What are the total A and B effects? What effects, if any, are significant?

chapter nine
CORRELATION—REGRESSION

Engineers are all familiar with the business of plotting data and drawing a line through the plot to see if there is any correlation. When there does not appear to be a correlation for the data plotted on ordinary coordinate paper, the procedure for plotting on semilog or on log-log paper is well established. We shall assume that we are this far along and shall discuss methods for drawing the best line through the data, establishing whether the correlation is better than might be expected by chance, setting confidence limits to the best line, and making confidence statements about predictions drawn from the correlation.

The plotting of experimental results to see if there is any orderly relation between variables is usually referred to by engineers as "correlating the data." Correlation refers to the degree of association between one variable and another or between one variable and several others. We shall have occasion later to refer to a *correlation coefficient* as a measure of the amount of relation between variables. A correlation coefficient of 1.0 indicates a perfect association between the variables; a correlation coefficient of 0.0 indicates a completely random relation. Nothing is said about the dependence or independence of the variables, and nothing is said about the nature of the relation between the variables. You might correlate tide with time; or you might correlate octane number with the aniline point; or the octane number with the aniline point, the bromine number, and the initial boiling point of a gasoline. None of these variables can be said to be independent in the sense that they may be varied at the will of the experimenter (except perhaps the initial boiling point), while the correlations might be reasonably good. The correlation is a measure of the predictability of the variation in one variable from the variation in another or other variables.

Regression, on the other hand, deals with the nature of the relation between variables. With two variables that can be correlated linearly, the *regression coefficient* is the slope of the line used to correlate the variables. If a higher-order correlation is used, there are regression coefficients for each of the ordered terms relating the variables. If more than two variables are involved in the correlation, there are regression coefficients for each order of each of the variables used to predict the variation in some other variable. In the calculation of the regression coefficients for the correlations discussed in this book the assumption is made that one of the variables is imprecisely known and the other or others are precisely known. It is possible to calculate a regression when both variables are subject to error.[1] However, the more common engineering practice is to find the relation of one variable in terms of others assumed to be known or measured without error.

9.1 Linear Correlation of Two Variables

9.1.1 The Least-squares Line.

It has been shown in Chap. 4 that one of the attributes of the mean of a group of measurements is that it is the value about which the sum of squares of deviations of the individual measurements is a minimum. Likewise, with data that can be correlated by a straight line, there is one straight line from which the sum of squares of deviations of one of the variables is a minimum. This is the *least-squares line*.

If the pairs of values of the variables associated with each data point are designated x_i and y_i, with y assigned to the variable which is imprecisely known, or to the dependent variable, a straight line through the data is expressed as

$$\hat{y} = a + bx \tag{9.1}$$

where $\hat{y}$ = estimated value of y for an observed value of x
a = intercept, giving estimated value of y at $x = 0$
b = slope of line, identical with regression coefficient

[1] A. Wald, *Ann. Math. Statis.*, **11**:284 (1940).

The values of a and b corresponding to the line with the minimum-squared deviation of y from $\hat{y}$ have been well established. These are:

$$a = \bar{y} - b\bar{x} \tag{9.2}$$

$$b = \frac{\Sigma(x - \bar{x})(y - \bar{y})}{\Sigma(x - \bar{x})^2} \tag{9.3}$$

Following the nomenclature established in Chap. 4, Eq. (9.3) may be written

$$b = \frac{\Sigma' xy}{\Sigma' x^2} \tag{9.4}$$

or

$$b = \frac{\Sigma xy - \dfrac{\Sigma x \Sigma y}{N}}{\Sigma x^2 - \dfrac{(\Sigma x)^2}{N}} \tag{9.5}$$

where N is the number of data points.

Equations (9.2) and (9.3) can be derived by setting a variable equal to the sum of squares of deviations of y from $\hat{y}$, and finding the minimum expression for this variable by equating its first derivative to zero.

$$Q = \Sigma(y - \hat{y})^2 = \Sigma(y - a - bx)^2$$

$$\frac{\partial Q}{\partial a} = -2\Sigma(y - a - bx) = 0 = \Sigma y - Na - b\Sigma x$$

$$a = \bar{y} - b\bar{x} \tag{9.2}$$

$$\frac{\partial Q}{\partial b} = -2\Sigma x(y - a - bx) = 0 = \Sigma xy - a\Sigma x - b\Sigma x^2$$

(substituting $\bar{y} - b\bar{x}$ for a)

$$0 = \Sigma xy - \bar{y}\Sigma x + b\bar{x}\Sigma x - b\Sigma x^2$$

$$b = \frac{\Sigma xy - \bar{y}\Sigma x}{\Sigma x^2 - \bar{x}\Sigma x}$$

or

$$b = \frac{\Sigma' xy}{\Sigma' x^2} \tag{9.4}$$

From Eq. (9.5) it can be seen that all that is needed to calculate the least-squares line is Σxy, Σx^2, Σx, and Σy. These quantities can all be obtained in one, or at most two, sets of operations on a desk calculator.

This is not the whole story, as we shall see, and with a little more calculation it is possible to obtain a great deal of information from the least-squares line. However, if all that is desired is to draw a straight line through the data without any statistical analysis of the result, it does not seem unreasonable to emphasize that only a small amount of work is involved in drawing the best line compared to drawing a line "by eye." Certainly order of magnitude less time is required to calculate the least-squares line than was expended in getting the data originally.

When a plot on rectangular coordinate paper is evidently nonlinear, it is sometimes possible to obtain a linear relation on semilog or on log-log paper, indicating a relation of

$$\hat{Y} = A(10)^{BX} \tag{9.6}$$

or

$$\hat{Y} = AX^B \tag{9.7}$$

Either of these equations may be readily transposed into the standard form by "taking" logarithms. Thus

$$\log \hat{Y} = \log A + BX \tag{9.8}$$

or

$$\log \hat{Y} = \log A + B \log X \tag{9.9}$$

and hence

$$\hat{y} = a + bx \tag{9.1}$$

The least-squares line corresponding to a transformation of Eq. (9.8) or (9.9) minimizes the squares of the deviations of $\log Y$ from $\log \hat{Y}$, and not the deviations of Y from $\hat{Y}$. This modification is usually not of importance to the engineer; nevertheless, it is a fact that should be kept in mind.

If a plot of the data indicates that a linear relation is obviously not valid, or if some irregular curve from point to point is what is of interest, then the least-squares line is not involved. Methods of obtaining the constants for some simple curvilinear equations are outlined later in this chapter.

example 9.1. The data in Table 9.1 give the weight per cent gaseous hydrocarbons formed in the noncatalytic hydrogenolysis of coal at 400°C, together with the weight per cent benzene insolubles remaining after hydrogenolysis for some experiments run by the U.S. Bureau of Mines.[1] These data are plotted in Fig. 9.1.

[1] M. G. Pelipetz et al., *Ind. Eng. Chem.*, **47**:2101 (1955).

TABLE 9.1

Wt. per cent gaseous hydrocarbons, y	Wt. per cent benzene insolubles, x
3.98	89.16
4.94	84.16
6.47	74.79
3.75	86.47
4.99	82.87
7.09	57.77
4.79	79.07
10.00	46.40
6.16	59.51
9.32	23.09

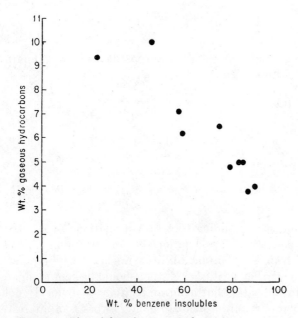

Fig. 9.1 Plot of data from example 9.1.

Inspection of Fig. 9.1 indicates that a linear relation exists between the two variables. The best straight line through these data is calculated from Eq. (9.5) and (9.2) as follows:

$$N = 10$$

$$\Sigma x = \Sigma(89.16 + 84.61 + \cdots + 23.09) = 683.74$$

$$\Sigma y = \Sigma(3.98 + 4.94 + \cdots + 9.32) = 61.49$$

$$\Sigma x^2 = \Sigma[(89.16)^2 + (84.61)^2 + \cdots + (23.09)^2] = 50{,}863.3856$$

$$\Sigma xy = \Sigma[(89.16)(3.98) + (84.61)(4.94) + \cdots + (23.09)(9.32)]$$
$$= 3{,}828.6203$$
$$\Sigma' x^2 = \Sigma x^2 - \frac{(\Sigma x)^2}{N} = 50{,}863.3856 - \frac{(683.74)^2}{10} = 4{,}113.34684$$
$$\Sigma' xy = \Sigma xy - \frac{\Sigma x \Sigma y}{N} = 3{,}828.6203 - \frac{(683.74)(61.49)}{10}$$
$$= -375.69696$$
$$\bar{x} = \frac{\Sigma x}{N} = \frac{683.74}{10} = 68.374$$
$$\bar{y} = \frac{\Sigma y}{N} = \frac{61.49}{10} = 6.149$$
$$b = \frac{\Sigma' xy}{\Sigma' x^2} = \frac{-375.69696}{4{,}113.34684} = -0.091336$$
$$a = \bar{y} - b\bar{x} = 6.149 - (-0.091336)(68.374) = 12.380$$
$$y = a + bx = 12.380 - 0.091x \tag{9.10}$$

Equation (9.10) gives the least-squares line. This equation is plotted in Fig. 9.2, together with the original data. When the a value,

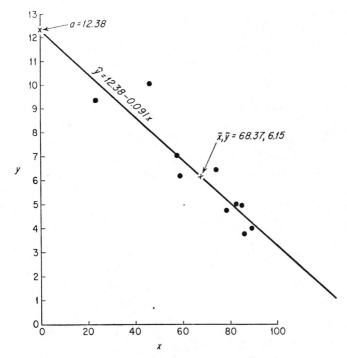

Fig. 9.2 Least-squares line drawn from example 9.1.

or intercept, is within the range of the plot, the simplest way is to draw the line through a and $\bar{x}$, $\bar{y}$. If point a is not on the plot, then the line is most easily drawn through $\bar{x}$, $\bar{y}$ with slope b.

The details of example 9.1 may make the calculation of the least-squares line appear somewhat formidable. Actually, with an electric desk calculating machine, the complete solution can be obtained in a few minutes. In using a desk calculator, Σx and Σx^2 can be obtained in one series of operations, and likewise Σy and Σxy can be obtained in another. If the data have only one or two significant figures, all four of these summations can be obtained in one operation. After calculating $\bar{x}$ and $\bar{y}$, $\Sigma' x^2$ and $\Sigma' xy$ can be obtained in two more operations by noting that

$$\Sigma' x^2 = \Sigma x^2 - \frac{(\Sigma x)^2}{N} = \Sigma x^2 - \bar{x} \Sigma x$$

and

$$\Sigma' xy = \Sigma xy - \frac{\Sigma x \Sigma y}{N} = \Sigma xy - \bar{x} \Sigma y$$

9.1.2 Significance of a Linear Correlation. There are times (all too many) when it is not obvious from a plot of the data whether or not it is reasonable to draw a straight line through the points. In cases of this kind, one is inclined to use the statistician's term "scatter diagram" to describe the plot. More important than the descriptive nomenclature, the statistician has a criterion for establishing whether a correlation line is significant.

In discussing the t test in Chap. 4, we discussed the procedure of setting up a hypothesis and then calculating from the experimental data a statistic for which the probability distribution is known. From the value of the calculated statistic, we could accept or reject the initial hypothesis with a measurable chance of error, depending on whether the magnitude of the statistic was more or less than we could expect by chance. The corresponding hypothesis for a correlation is that there is no relation between the variables and the statistic involved is the correlation coefficient r, which is related to t and F in a way that will be explained.

If the sum of squares of deviation of y from $\bar{y}$ is expressed as $\Sigma' y^2$, and if the sum of squares of deviation of y from the correlation line, $\hat{y}$, is expressed as $\Sigma' \hat{y}^2$, $\Sigma' \hat{y}^2 = \Sigma (y - \hat{y})^2$; then if a straight line correlation is employed it can be shown, by substituting for $\hat{y}$ its equivalent $a + bx$, that

$$\Sigma' \hat{y}^2 = \Sigma' y^2 - b^2 \Sigma' x^2 = \Sigma' y^2 - b \Sigma' xy \tag{9.11}$$

It can be seen from Eq. (9.11) that $b^2\Sigma'x^2$ (or $b\Sigma'xy$) represents the portion of the sum of squares of deviation of the original data that has been removed by the linear correlation, $\Sigma'\hat{y}^2$ being the remainder still unaccounted for and usually attributed to error. The sum of squares of deviation removed by the correlation may be designated $\Sigma'c^2$ and expressed in the alternative forms

$$\Sigma'c^2 = b^2\Sigma'x^2 = b\Sigma'xy \tag{9.12}$$

The ratio of the sum of squares removed by the correlation to the sum of squares of the original data is a measure of the goodness of the correlation. The square root of this ratio is the correlation coefficient r

$$r^2 = \frac{\Sigma'c^2}{\Sigma'y^2} = 1 - \frac{\Sigma'\hat{y}^2}{\Sigma'y^2} = \frac{b^2\Sigma'x^2}{\Sigma'y^2} = \frac{\Sigma'y^2 - \Sigma'\hat{y}^2}{\Sigma'y^2} \tag{9.13}$$

When there is perfect correlation between x and y, there is no residual deviation of y from $\hat{y}$ and r^2 equals 1.0. When there is no correlation and none of the sum of squares of deviation is removed by the linear relationship, r^2 equals 0.0. The correlation coefficient ranges from ± 1.0 to 0.0 depending on the goodness of the fit of the line. Negative values of the correlation coefficient indicate a negative slope. This relation can be seen if r is expressed in an alternative form:

$$r = b\sqrt{\frac{\Sigma'x^2}{\Sigma'y^2}} = \frac{\Sigma'xy}{(\Sigma'x^2\Sigma'y^2)^{1/2}} \tag{9.14}$$

If the hypothesis is made that the correlation coefficient is equal to zero, Fisher[1] has shown that r is related to t (of Chap. 6). It is therefore possible to tabulate the values of r corresponding to various probability levels and degrees of freedom based on the hypothesis that there is no correlation between the two variables involved. This is done in Table 9.2 for probability values from 0.001 to 0.1 and for degrees of freedom from 1 to 50.

Values of r in Table 9.2 are the maximum values of the correlation coefficient that can be expected by chance for the amount of data involved if there were no correlation. The probability level indicates the chance of getting a value of r as large as the tabulated value when there is no correlation. The 0.10 level means there is only a 10 per cent chance of getting a value of r as large as those in the 0.10 column when no correlation exists. The degrees of freedom have the same meaning as in the other statistical tests and represent the number of independent categories that exist. In

[1] R. A. Fisher, "Statistical Methods for Research Workers," Oliver & Boyd Ltd., Edinburgh and London, 1948.

calculating the least-squares line, two restrictions are involved: the slope and the intercept, or the slope and the mean. On either basis there are $N - 2$ degrees of freedom for the least-squares line, where N is the number of data points.

TABLE 9.2 Correlation Coefficient r†

D.F.	Probability of a larger value of r				
	0.1	0.05	0.02	0.01	0.001
1	0.988	0.997	1.0	1.0	1.0
2	0.900	0.950	0.980	0.990	1.0
3	0.805	0.878	0.934	0.959	0.991
4	0.729	0.811	0.882	0.917	0.974
5	0.669	0.754	0.833	0.874	0.951
6	0.622	0.707	0.789	0.834	0.925
7	0.582	0.666	0.750	0.780	0.898
8	0.549	0.632	0.716	0.765	0.872
9	0.521	0.602	0.685	0.735	0.847
10	0.497	0.576	0.658	0.708	0.823
12	0.458	0.532	0.612	0.661	0.780
14	0.426	0.497	0.574	0.623	0.742
16	0.400	0.468	0.542	0.590	0.708
18	0.378	0.444	0.516	0.561	0.679
20	0.360	0.423	0.492	0.537	0.652
25	0.323	0.381	0.445	0.487	0.597
30	0.296	0.349	0.409	0.449	0.554
35	0.275	0.325	0.381	0.418	0.519
40	0.257	0.304	0.358	0.393	0.490
45	0.243	0.288	0.338	0.372	0.465
50	0.231	0.273	0.322	0.354	0.443

† Abridged from Table VI of Statistical Tables for Biological, Agricultural and Medical Research by R. A. Fisher and Frank Yates, Oliver & Boyd, Ltd., Edinburgh and London, 1953, reprinted with permission of publishers.

The use of Table 9.2 may be illustrated as follows. The tabulated value of r for 25 degrees of freedom at the 0.05 probability level is 0.381. This means that there is only a 0.05 chance of getting an r value of 0.381 or larger when there is no correlation between the variables. If in calculating the correlation coefficient from 27 observations a value greater than 0.381 is obtained, then the hypothesis that there is no correlation may be rejected with only 0.05 chance of being wrong.

example 9.2. Figure 9.3 shows a plot of hydrogen consumption against per cent conversion for an experimental pilot-plant run.[1] The data are given in Table 9-3. Because of the scatter of the points at

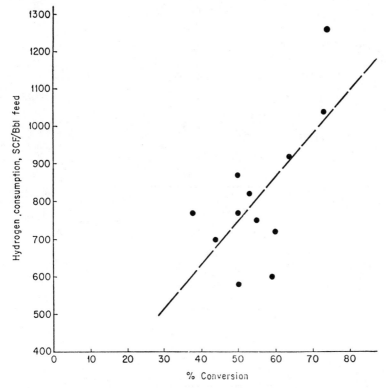

Fig. 9.3 Plot of data from example 9.2.

the lower conversions, there might be some doubt as to whether a correlation between the variables is significant. The calculation of the correlation coefficient is as follows:

Calculation

$$N = 12$$
$$\Sigma x = \Sigma(38 + 44 + \cdots + 74) = 670$$
$$\Sigma y = \Sigma(770 + 700 + \cdots + 1{,}260) = 9{,}800$$
$$\Sigma x^2 = \Sigma[(38)^2 + (44)^2 + \cdots + (74)^2] = 38{,}696$$
$$\Sigma y^2 = \Sigma[(770)^2 + (700)^2 + \cdots + (1{,}260)^2] = 8{,}398{,}000$$
$$\Sigma xy = \Sigma[(38)(770) + (44)(700) + \cdots + (74)(1{,}260)] = 562{,}410$$

[1] Unpublished data, Hydrocarbon Research, Inc.

$$\Sigma'x^2 = \Sigma x^2 - \frac{(\Sigma x)^2}{N} = 38{,}696 - \frac{(670)^2}{12} = 1{,}287.890$$

$$\Sigma'y^2 = \Sigma y^2 - \frac{(\Sigma y)^2}{N} = 8{,}398{,}000 - \frac{(9{,}800)^2}{12} = 394{,}663.400$$

$$\Sigma'xy = \Sigma xy - \frac{\Sigma x \Sigma y}{N} = 562{,}410 - \frac{(670)(9{,}800)}{12} = 15{,}246.600$$

$$r = \frac{\Sigma'xy}{(\Sigma'x^2 \Sigma'y^2)^{1/2}}$$

$$= \frac{15{,}246.600}{[(1{,}287.890)(394{,}663.400)]^{1/2}} = 0.676$$

Data

TABLE 9.3

Hydrogen consumption, y	Per cent conversion, x
770	38
700	44
580	50
770	50
870	50
820	53
750	55
600	59
720	60
920	64
1,040	73
1,260	74

Discussion. The calculated value of r, 0.676, is compared with the values given in Table 9.2 for $N - 2 = 10$ degrees of freedom. Since it exceeds the tabulated value of 0.658 at 0.02 probability level but is less than the 0.708 value at the 0.01 level, we can say with 98 per cent certainty but not with 99 per cent certainty that there is a correlation.

If the actual correlation line is desired, then $\bar{x}$, $\bar{y}$, and b are calculated as in Example 9.1, and the line is drawn accordingly.

$$\bar{x} = {}^{670}\!/_{12} = 55.83$$

$$\bar{y} = {}^{9800}\!/_{12} = 816.67$$

$$b = \frac{15{,}246.600}{1{,}287.890} = 11.838$$

The equation of the least-squares line is then

$\hat{y} = [816.67 - (11.838)(55.83)] + 11.838x$

$\hat{y} = 155.7 + 11.84x$

This is the line drawn in Fig. 9.3.

In calculating the correlation coefficient and slope for the least-squares line, some time may be saved by coding the data to simplify the arithmetic. For example, all the y values in example 9.2 could have been reduced by a factor of 10 without altering the results. A constant may be subtracted from all values (not necessarily the same constant for both variables) without affecting the results. However, coding is not of much value when an electric desk calculator is employed, and it does offer one more opportunity for an arithmetic error. If the original data are coded for computation purposes, then the final least-squares equation must be uncoded to give the correct answer although the slope and the correlation coefficient are not affected.

If the original data were of the order of

$y = 8{,}500,\ 8{,}700,\ 9{,}000$, etc.

$x = 0.075,\ 0.095,\ 0.105$, etc.

and these were coded by dividing y by 100 and multiplying x by 200 to give

$y' = 85,\ 87,\ 90$, etc.

$x' = 15,\ 19,\ 21$, etc.

that is,

$y' = \dfrac{y}{100}$ and $x' = 200x$

then the final least-squares equation

$\hat{y}' = \bar{y}' - b\bar{x}' + bx'$

must be uncoded to give

$\hat{y} = 100\bar{y}' - b\dfrac{\bar{x}'}{200} + b\dfrac{x'}{200}$

9.1.3 Variance Test of Correlation.

The sum of squares of deviation from the mean divided by the degrees of freedom has been defined in

Chap. 4 as the best estimate of the variance. Likewise, the sum of squares of deviation from the least-squares line divided by the degrees of freedom of the correlation gives a measure of the variance of the estimated $\hat{y}$ values. It is possible to test the estimated variance removed by the linear correlation against the estimated variance remaining after correlation in a manner similar to that explained in detail in Chap. 7 on Analysis of Variance.

The total sum of squares of deviation of the dependent variable from its mean is $\Sigma(y - \bar{y})^2$, and the total degrees of freedom are $N - 1$. Since r^2 is the fraction of the sum of squares of deviation removed by the correlation line, $(1 - r^2)\Sigma(y - \bar{y})^2$ is the sum of squares of deviation from the least-squares line, equal to $\Sigma(y - \hat{y})^2$, with $N - 2$ degrees of freedom. $r^2\Sigma(y - \bar{y})^2$ is the sum of squares of deviation attributable to the correlation, with the remaining single degree of freedom. If we designate the estimate of variance removed by the correlation as $s^2(c)$, and the amount still remaining as $s^2(\hat{y})$, that is, the *variance of estimate*, then the analysis of variance of the least-squares line may be written:

TABLE 9.4

Source of variation	"Sum of squares"	Equivalent formulas	Degrees of freedom	Variance estimate
Least-squares line (variation accounted for by the correlation)	$r^2\Sigma(y - \bar{y})^2$	$b^2\Sigma'x^2$ $b\Sigma'xy$ $(\Sigma'xy)^2/\Sigma'x^2$ $\Sigma'c^2$	1	$\sigma^2(c)$
Residual variation from line	$(1 - r^2)\Sigma(y - \bar{y})^2$	$\Sigma(y - \hat{y})^2$ $\Sigma y^2 - a\Sigma y - b\Sigma xy$	$N - 2$	$\sigma^2(\hat{y})$
Total	$\Sigma(y - \bar{y})^2$		$N - 1$	

The ratio of $s^2(c)/s^2(\hat{y})$ may be tested by the F ratio test for 1 and $N - 2$ degrees of freedom to see whether the variance removed by the correlation line is significant when compared to the residual variance of estimate. This test is equivalent to testing the significance of r, the correlation coefficient, since $s^2(c)/s^2(\hat{y}) = r^2(N - 2)/(1 - r^2)$ and the only variable involved is r.

Since F at 1 degree of freedom is equal to t^2, the F ratio test for the correlation, $r^2(N - 2)/(1 - r^2)$, can be expressed in terms of t; that is, $t = r\sqrt{N - 2}/\sqrt{1 - r^2}$. This t test, the F test, and the tabulated significant values for the correlation coefficient in Table 9.2 will all give identical results.

example 9.3. Using the data from example 9.2, the analysis of variance of the linear correlation is calculated as follows:

$$N = 12$$
$$r^2 = (0.676)^2 = 0.4570$$
$$\Sigma(y - \bar{y})^2 = \Sigma'y^2 = 394,663$$

TABLE 9.5

Source of variation	Sum of squares	D.F.	Variance estimate
Least-squares line..	$(0.457)(394,663) = 180,361$	1	180,361
Residual.........	$(0.543)(394,663) = 214,302$	10	21,430
Total..........	394,663	11	

$$F = \frac{180,361}{21,430} = 8.41$$

$$F_{0.05,1,10} = 4.96 \quad \text{(Table 7.8)}$$
$$F_{0.01,1,10} = 10.04 \quad \text{(Table 7.9)}$$

The correlation is significant at the 0.05 level but not at the 0.01 level, the same as was deduced from the r value.

Using the t relation,

$$t = \frac{0.676\sqrt{10}}{\sqrt{0.5430}} = \frac{2.1377}{0.7369} = 2.9009$$

$$t_{0.02,10} = 2.764 \quad \text{(Table 6.1)}$$
$$t_{0.01,10} = 3.169 \quad \text{(Table 6.1)}$$

This is the same result obtained by the r and F tests; i.e., the correlation is significant at the 0.02 level but not at the 0.01 level.

9.1.4 Confidence Limits of Slope and Least-squares Line. The variance of estimate, $s^2(\hat{y})$, may be used to set confidence limits on the least-squares line in a way similar to that in which the variance of a set of measurements is used to set confidence limits on the mean of several measurements or on a single measurement. The confidence limits have been described in Chap. 6 as $\pm ts$, where t is selected at the proper degrees of freedom and the desired probability level, and s is the estimated standard deviation, the square root of the variance, of the function involved.

The following formulas give the estimated variances of the various quantities involved in a linear correlation. Some discussion of these variances is included, but for rigorous derivations more theoretical texts are recommended.[1]

$$\text{Variance of } \hat{\bar{y}}, \; s^2(\hat{\bar{y}}) = \frac{s^2(\hat{y})}{N} \tag{9.15}$$

$$\text{Variance of slope, } s^2(b) = \frac{s^2(\hat{y})}{\Sigma' x^2} \tag{9.16}$$

Variance of any estimated value, $\hat{y}_i$,

$$s^2(\hat{y}_i) = s^2(\hat{y}) \left[\frac{1}{N} + \frac{(\bar{x} - x_i)^2}{\Sigma' x^2} \right] \tag{9.17}$$

Variance of any predicted value, $\hat{Y}_i$,

$$s^2(\hat{Y}_i) = s^2(\hat{y}) \left[1 + \frac{1}{N} + \frac{(\bar{x} - x_i)^2}{\Sigma' x^2} \right] \tag{9.18}$$

The variance of estimate, $s^2(\hat{y})$, is a measure of the average deviation of the data points from their estimated value according to the correlation line. Actually it is the sum of the squares of deviation divided by the degrees of freedom available for calculating the correlation.

$$s^2(\hat{y}) = \frac{\Sigma(y - \hat{y})^2}{N - 2} \tag{9.19}$$

In the discussion of the variance of several measurements, it was pointed out (Chap. 6) that the variance of the mean was equal to the variance of the individual measurements divided by the number of measurements included in the mean:

$$s^2(\bar{x}) = \frac{s^2(x)}{N} \tag{6.7}$$

In the same way, the variance of the estimated mean $\hat{\bar{y}}$ value is equal to $s^2(\hat{y})/N$, as indicated in Eq. (9.15). The standard deviation of the estimated mean is equal to the square root of this expression, $s(\hat{y})/\sqrt{N}$. Any desired confidence range can be set for $\hat{\bar{y}}$ by multiplying the standard deviation by the proper t value selected from Table 6.1. We have indicated that the least-squares line can be drawn through the intercept and

[1] O. W. Davies (ed.), "Statistical Methods in Research and Production," Oliver & Boyd Ltd., Edinburgh and London, 1954.

$\bar{x}$, $\bar{y}$, or through $\bar{x}$, $\bar{y}$, with slope b. The confidence range of $\hat{\bar{y}}$ sets a band on the $\bar{y}$ value at $\bar{x}$ through which the line may be drawn.

There is also a confidence range to the slope. This may be calculated from the standard deviation of the slope, obtained from the square root of the expression given in Eq. (9.16) and the proper t value. The confidence band of the slope is a fan-shaped area converging on $\bar{x}$, $\bar{y}$, with slopes of $b \pm ts(b)$. The confidence area of the correlation can be closely approximated by drawing smooth curves asymptotic to the confidence band of the slope and through the extremities of the confidence range of $\hat{\bar{y}}$. This procedure is illustrated in example 9.4.

It can be seen from Eq. (9.16) that as $\Sigma' x^2$ increases the variance of the slope will decrease. Therefore to obtain the tightest confidence range of the slope with a fixed amount of data, the points should be grouped evenly at the two extremes of the x variable range to make $\Sigma' x^2$ as large as possible. When this is done, however, it is not possible to test for linearity of the correlation. The data are best distributed evenly throughout the range of the independent variable unless the linearity of the correlation has already been established.

The variance of an estimated value of the dependent variable at any value of the independent variable is given by Eq. (9.17). This can be expressed as $s^2(\hat{y}_i) = s^2(\hat{\bar{y}}) + [s^2(b)](\bar{x} - x_i)^2$, where x_i is the value of the x variable at the point x_i, y_i; that is, the variance of any value estimated from the least-squares line is equal to the variance of the mean of the dependent variable, plus the product of the variance of the slope and the square of the deviation of the independent variable from its mean. The further we are from the mean, the greater is the variance of estimated value. This expression can be used to calculate the exact confidence range of the correlation by obtaining the standard deviation of the estimated value of y_i, $s(\hat{y}_i) = \sqrt{s^2(\hat{\bar{y}}) + [s^2(b)](\bar{x} - x_i)^2}$, at a number of values of x_i, and multiplying by the proper t value to obtain the confidence range. A smooth curve may then be drawn through the extremes of the various confidence ranges.

The confidence range of any single predicted value of the dependent variable will be wider than the confidence range of a value estimated from the correlation, wider by a measure of the confidence of the correlation itself. In other words, to the variance of an estimated value of y_i as given by Eq. (9.17), we add the variance of estimate, $s^2(\hat{y})$, defined in Eq. (9.18), to get the variance of a value predicted from the correlation. In example 9.4, which follows, a correlation of bursting pressure against rupture disk thickness is calculated. From this correlation it is possible to estimate what the bursting pressure is for a disk of any thickness taken from the items used for the correlation, and the confidence range of the prediction would be established from the relationship shown in Eq. (9.17).

If it is desired to predict the bursting pressure of some future disk, the confidence range of this prediction would include a term to account for the variance of the correlation itself, $s^2(\hat{y})$, as shown in Eq. (9.18).

example 9.4. Table 9.6[1] gives the bursting pressures and rupture disk thicknesses from some experimental determinations made with

TABLE 9.6

Disk thickness, in.	Bursting pressure, psi, y	Disk thickness, coded, x
0.001	1	1
0.002	5	2
0.003	15	3
0.0045	21	4.5
0.005	22	5
0.008	47	8
0.010	57	10

aluminum foil. The disk thicknesses are coded by multiplying by 1,000 to simplify the calculations.

Calculation

$$N = 7$$
$$\Sigma x = 33.5; \bar{x} = 4.786$$
$$\Sigma y = 168.0; \bar{y} = 24.0$$
$$\Sigma x^2 = 223.25; \Sigma' x^2 = 62.919$$
$$\Sigma y^2 = 6,634; \Sigma' y^2 = 2,602$$
$$\Sigma xy = 1,206.5; \Sigma' xy = 402.5$$
$$b = \frac{\Sigma' xy}{\Sigma' x^2} = 6.397$$
$$a = \bar{y} - b\bar{x} = -6.616$$
$$\hat{y} = -6.616 + 6.397x$$
$$r = \frac{\Sigma' xy}{(\Sigma' x^2 \Sigma' y^2)^{1/2}} = 0.9948; r^2 = 0.9896$$
$$s^2(\hat{y}) = \frac{(1 - r^2)\Sigma' y^2}{N - 2} = \frac{\Sigma' y^2 - b\Sigma' xy}{N - 2} = 5.4323$$

[1] P. B. Stewart and R. T. Fox, Jr., *Chem. Eng. Prog.* **52**:115-M (1956).

$$s^2(b) = \frac{s^2(\hat{y})}{\Sigma' x^2} = 0.0863; \ s(b) = 0.2938$$

$$s^2(\bar{y}) = \frac{s^2(\hat{y})}{N} = 0.7760; \ s(\bar{y}) = 0.8809$$

$t_{0.05, N-2} = 2.571$ (Table 6.1)

95% confidence range of $\bar{y} = 24.0 \pm (2.571)(0.8809)$
$= 21.74$ to 26.26

95% confidence range of $b = 6.397 \pm (2.571)(0.2938)$
$= 5.642$ to 7.152

Discussion. The r value of 0.9948 for 5 degrees of freedom, $7 - 2$, is larger than the 0.01 value of 0.874 in Table 9.1, hence there is less than one chance in a hundred of being in error in concluding that there is correlation between these variables.

The r^2 value of 0.9896 indicates that about 99 per cent of the vari-

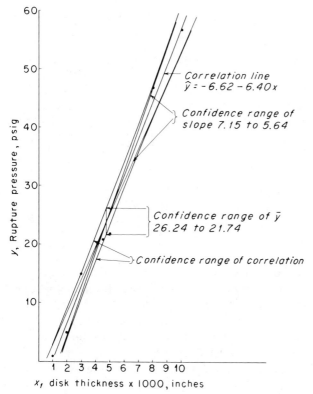

Fig. 9.4 Plot of data from example 9.4.

ability in bursting pressure is removed by the linear correlation with disk thickness.

The confidence ranges of $\bar{y}$ and b are used to draw the confidence range of the correlation as illustrated in Fig. 9.4. Lines with maximum and minimum slope are drawn through $\bar{x}$, $\bar{y}$, and the 95 per cent confidence range of $\hat{\bar{y}}$ is drawn at $\bar{x}$. Curves through the extremes of $\hat{\bar{y}}$ are drawn asymptotic to the lines with minimum and maximum slope. The area between the two curved lines is the confidence range of the correlation. The 95 per cent confidence range on the estimated $\hat{y}$ value at an x value of 6.0, for example, is obtained from the curve, or from the variance of Eq. (9.17), as follows:

$$\hat{y}_{x=6} = -6.616 + (6.397)(6.0) = 31.766$$

$$s(\hat{y}_{x=6}) = \left\{5.4323\left[\frac{1}{7} + \frac{(6.0 - 4.786)^2}{62.919}\right]\right\}^{1/2}$$

$$= [0.7760 + (0.0863)(6.0 - 4.786)^2]^{1/2}$$

$$= 0.9504$$

95% confidence range of $\hat{y}_{x=6} = 31.766 \pm (2.571)(0.9504) = 29.323$ to 34.209

If the predicted rupture pressure for any disk of a particular thickness is to be estimated from the correlation of example 9.4, the confidence range of this estimation is obtained from the variance of Eq. (9.18). For example, if the rupture pressure for a disk of 0.0075 in. is desired, $x = 7.5$, $\hat{Y} = -6.616 + (6.397)(7.5) = 41.362$ psi. The 99 per cent confidence range of this prediction is:

$$41.362 \pm t_{0.01,5}s(\hat{Y})$$

$$41.362 \pm 4.032\left\{5.4323\left[1 + \frac{1}{7} + \frac{(7.5 - 4.4323)^2}{62.919}\right]\right\}^{1/2}$$

41.362 ± 10.548

30.814 to 51.903 psi

9.1.5 Comparison of Two Slopes.

As explained in the preceding section, the standard deviation of the slope, $s(b)$, can be employed to calculate the confidence range of the slope in the same manner that the standard deviation of the mean is used to set confidence ranges to that quantity. The standard deviation of the slope may also be made use of to test whether the slope is significantly different from zero, whether it is different from some other preassigned value, and also whether two slopes are significantly different from each other. Each of these tests is made by comparison with the t distribution in a manner similar to that described in Chap. 6.

CORRELATION—REGRESSION

The t test to establish whether the slope is significantly different from zero is

$$t = \frac{b - 0}{s(b)} \qquad (9.20a)$$

This is identical with the expression for t in terms of the correlation coefficient, $t = (r\sqrt{N-2})/\sqrt{1-r^2}$, discussed in Sec. 9.1.3. The identity can be established by noting that $b = \Sigma'xy/\Sigma'x^2$, $s(b) = s(\hat{y})/\sqrt{\Sigma'x^2}$, and $s(\hat{y}) = [(1-r^2)\Sigma'y^2/(N-2)]^{1/2}$.

The t test to establish whether the slope is significantly different from some prescribed slope, β, is

$$t = \frac{b - \beta}{s(b)} \qquad (9.20b)$$

This formulation is similar to Eq. (6.1) used to test whether a mean value is significantly different from some preestablished value.

For example, the correlation of bursting pressure with disk thickness in example 9.4 gave a slope of 6.397 and a standard deviation of slope of 0.2938. There were 5 degrees of freedom associated with $s(b)$. If it had been desired to test whether the slope was significantly different from 7.0, we could have used the t test of Eq. (9.20b):

$$t = \frac{7.0 - 6.397}{0.2933} = 2.056$$

$$t_{0.05, 5} = 2.571$$

$$t_{0.10, 5} = 2.015$$

The calculated slope is significantly different from 7.0 at the 0.10 level but not at the 0.05 level. Put another way, we would have a 90 per cent chance of not being wrong if we said that the slope was different from 7.0, but not a 95 per cent chance. It can be seen that the 95 per cent confidence range of the slope calculated in example 9.4 included the 7.0 value.

The t test to establish whether there is significant difference between two slopes requires an expression for the pooled estimate of the standard deviation of the slope, similar to the pooled estimate of the standard deviation of means in Eq. (6.26). The two variances of estimate from the two correlations are pooled by weighting according to their respective degrees of freedom to give a pooled variance of estimate:

$$\bar{s}^2(\hat{y}) = \frac{(N_1 - 2)s^2(\hat{y}_1) + (N_2 - 2)s^2(\hat{y}_2)}{(N_1 - 2) + (N_2 - 2)} \qquad (9.21)$$

If we designate the pooled variance of estimate as $\bar{s}^2(\hat{y})$, the t test for the difference between two slopes is

$$t = \frac{b_1 - b_2}{\bar{s}(\hat{y}) \left(\frac{1}{\Sigma' x_1^2} + \frac{1}{\Sigma' x_2^2} \right)^{1/2}} \quad (9.22)$$

with $N_1 + N_2 - 4$ degrees of freedom.

After having established that there is no significant difference between two slopes, the slopes may be pooled

$$\bar{b} = \frac{\Sigma' x_1 y_1 + \Sigma' x_2 y_2}{\Sigma' x_1^2 + \Sigma' x_2^2} \quad (9.23)$$

example 9.5. The data[1] in Table 9.7 give the evaporation rate of two organic solids in terms of the final per cent saturation of solid in the

Data

TABLE 9.7

ψ	$1/\psi$	ϕ	$\log 1/\psi$	$\log(100\,\phi)$
Acetamide				
3.865×10^{-3}	258	0.0284	2.411620	0.453318
9.54×10^{-3}	105	0.0535	2.021189	0.728354
1.542×10^{-2}	64.8	0.0733	1.811575	0.865104
3.182×10^{-2}	31.4	0.1475	1.496930	1.168792
1.81×10^{-3}	552	0.0160	2.741939	0.204120
1.29×10^{-3}	775	0.0150	2.889302	0.176051
3.00×10^{-2}	33.3	0.1348	1.522444	1.129690
Camphene				
2.55×10^{-3}	444	0.0485	2.647383	0.685742
5.86×10^{-3}	171	0.0822	2.232996	0.914872
9.12×10^{-3}	110	0.1089	2.041393	1.037028
2.70×10^{-2}	37.0	0.280	1.568202	1.447158
2.245×10^{-3}	445	0.0485	2.648360	0.685742
1.073×10^{-3}	930	0.0308	2.968483	0.488551
7.51×10^{-4}	1330	0.0265	3.123852	0.423246
1.835×10^{-2}	54.5	0.226	1.736397	1.354108
6.41×10^{-3}	156	0.0935	2.193125	0.970812

air, ϕ, and a modified form of the Graetz number ψ, which is a function of the physical dimensions of the system. The theoretical relationship

[1] A. C. Plewes et al., *Chem. Eng. Progr.*, **50**:77 (1954).

is $\phi = 1.45\psi^{2/3}$. The following example gives the procedure for testing whether the data are satisfactorily fitted by the 2/3 exponent and whether individual exponents fitted to each set of data for the individual compounds are significantly different from each other. If they are not significantly different from each other, they will be pooled for a better estimate of the exponent.

An equation of the form $\phi = A\psi^B$ will be fitted; the constants A and B will be calculated from the least-squaring operations. In order to do this, the equation is transformed to a linear form in terms of logarithms, log ϕ = log A + B log ψ. To simplify the arithmetic, i.e., eliminate the necessity of employing negative characteristics of the logarithms, the solution will be calculated in terms of 100 ϕ and $1/\psi$, and the desired equation obtained by decoding the solution.

Calculation. From the data of Table 9.7, the following terms are calculated, where x = log $(1/\psi)$ and y = log (100 ϕ).

TABLE 9.8

	Acetamide	Camphene
N =	7	9
Σx =	14.894998	21.160191
Σx^2 =	33.607851	52.030476
Σy =	4.725429	8.007259
Σy^2 =	4.199335	8.141072
Σxy =	8.670411	17.313624
$\Sigma' x^2 = \Sigma x^2 - (\Sigma x)^2/N$ =	1.913425	2.280074
$\Sigma' y^2 = \Sigma y^2 - (\Sigma x)^2/N$ =	1.009382	1.017054
$\Sigma' xy = \Sigma xy - (\Sigma x \Sigma y)/N$ =	−1.384626	−1.512499
$b = \Sigma' xy / \Sigma' x^2$ =	−0.7236	−0.6634
$r^2 = b\Sigma' xy / \Sigma' y^2$ =	0.9927	0.9865
r =	0.9963***	0.9932***
$s^2(\hat{y}) = (1 - r^2)\Sigma' y^2 / (N - 2)$ =	0.001483	0.001961
$s^2(b) = s^2(\hat{y}) / \Sigma' x^2$ =	0.000775	0.000860
$s(b)$ =	0.0278	0.0293
$t_{0.05}$ (from Table 6.1) =	2.571	2.363
$ts(b)$ =	0.0716	0.0693
95% range of b =	−0.652 to −0.795	−0.594 to −0.733

Discussion. Both of the correlations are highly significant, as indicated by the values of the correlation coefficient r, compared with Table 9.2. Both of the calculated slopes, which, from the form of the linear equation employed, are identical with the exponents, include in the 95 per cent range the theoretical value of 0.667. Although the 95 per cent confidence ranges of the two slopes overlap, this fact does not establish that there is no significant difference between them. The difference between the slopes must be tested by the t test, as given by

Eqs. (9.21) and (9.22). For the results in Table 9.8, this test is as follows:

Pooled variance of estimate, $\bar{s}^2(\hat{y})$, from Eq. (9.21), is

$$\bar{s}^2(\hat{y}) = \frac{(5)(0.001483) + (7)(0.001961)}{12} = 0.001762$$

The t test, from Eq. (9.22), is

$$t = \frac{0.7236 - 0.6634}{[(0.001762)(1/1.9134 + 1/2.2801)]^{1/2}}$$

$$= \frac{0.0602}{(0.001693)^{1/2}} = 1.46$$

This corresponds to a t value at 12 degrees of freedom of between 0.1 and 0.2 probability level. Stated another way, you could expect a difference between slopes of the observed magnitude to occur 1 to 2 times in 10 from data with the variance of estimate encountered in this example when there was no difference between the slopes.

Since there is no significant difference between the slopes, they may be pooled for a better over-all estimate of the slope. (In this example it might be more desirable to use the theoretical slope of 2/3, but that is a matter of interpretation of the problem, and we shall proceed to pool the slopes to illustrate the method.) The equation for pooling slopes is

$$\bar{b} = \frac{\Sigma' x_1{}^2 b_1 + \Sigma' x_2{}^2 b_2}{\Sigma' x_1{}^2 + \Sigma' x_2{}^2} \qquad (9.23)$$

The pooled slope for the results in Table 8.8 is

$$\bar{b} = \frac{(-0.7236)(1.9134) + (-0.6634)(2.2801)}{1.9134 + 2.2801} = -0.6909$$

The variance of the pooled slope is

$s^2(\bar{b}) = 0.001693$ (from the denominator in the t test)

$s(\bar{b}) = 0.0411$

$t_{0.05,13} = 2.160$

The 95 per cent confidence range of the pooled slope is $\bar{b} = ts(\bar{b})$ = -0.602 to -0.780, slightly better than the combined 95 per cent confidence ranges of the two separate slopes, and still including the theoretical value of 0.667.

The least-squaring has been done for an equation in the form $y = \bar{y}$

$+ b(x - \bar{x})$, where $y = \log(100\phi)$ and $x = 1/\psi$. To put the result in the form $\phi = A\psi^B$, it is only necessary to put the calculated constants in the linear logarithmic equation and then to transform the result to the exponential form. This procedure is illustrated for the acetamide data. The interested student can follow a parallel procedure for the camphene data. In the solution that follows, the pooled slope is used for b in the linear equation. It is obvious that using the individual values of the slope, or the theoretical value of 0.667, will give slightly different results, but the statistical analysis has shown that considering only these data, the pooled slope is the best estimate to use.

$$y = \bar{y} + b(x - \bar{x})$$

$$\log(100\phi) = 0.6751 - 0.6909 \left(\log \frac{1}{\psi} - 2.1279\right)$$

$$\log(100\phi) = 2.1453 - 0.6909 \log \frac{1}{\psi}$$

$$100\phi = 139.7 \left(\frac{1}{\psi}\right)^{-0.691}$$

$$\phi = 1.40(\psi)^{+0.691}$$

9.1.6 Comparison of Several Slopes. In a manner similar to the expansion of the t test for the comparison of two means to the F test for the comparison of several means, the test for the comparison of two slopes can be expanded to test more than two. When two or more sets of data are at hand and a plot of these data indicates that separate straight lines might be drawn through each set of data, there are several questions that might be asked:

1. Do the individual straight lines make a significantly better correlation than one line?
2. Does a better correlation result if each of the individual lines is drawn with its own slope, or should all of the lines be drawn with the same slope?
3. Is there significant displacement in the Y direction between the lines?

It is possible sometimes to obtain obvious answers to some of these questions by examining the plot. At other times, the answers are not so obvious. Statistics, however, offers a quantitative method for an objective analysis of the data.

The statistical method is to determine the sum of squares of deviations from the best straight line through all the data, the best straight line through the individual sets with a pooled estimate of the slope, and the best straight line through the individual sets, each with its own slope. The comparison of each of the sums of squares at its degrees of freedom with the minimum sum of squares from each line separately provides a variance ratio which can be tested by the F test. If the decrease in the sum of squares of deviation when using several lines with individual slopes compared with using a pooled slope for all lines is not significantly larger than might be expected by chance for the number of degrees of freedom available, then for the data involved there is no basis for using different slopes for the different sets of data. A detailed discussion of the procedure is given in Snedecor.[1] We shall merely outline the method and illustrate it with example 9.6.

If there are several sets of data, it is necessary to calculate the sums of squares and sums of cross products required for the computation of the best straight line through each set. It is also necessary to make the same calculation for the total data and for the means of each set of data. Table

TABLE 9.9 Sums of Squares for Comparison of Several Linear Correlations

Total.........	$\Sigma'x^2$	$\Sigma'y^2$	$\Sigma'xy$	$\Sigma'c^2$	$\Sigma'\hat{y}^2$
Means........	$\Sigma'\bar{x}^2$	$\Sigma'\bar{y}^2$	$\Sigma'\bar{x}\bar{y}$	$\Sigma'\bar{c}^2$	$\Sigma'\hat{y}_m^2$
Difference....	$\Sigma'x_w^2$	$\Sigma'y_w^2$	$\Sigma'x_wy_w$	$\Sigma'c_w^2$	$\Sigma'\hat{y}_w^2$
Set:					
1	$\Sigma'x_1^2$	$\Sigma'y_1^2$	$\Sigma'x_1y_1$	$\Sigma'c_1^2$	$\Sigma'\hat{y}_1^2$
2	$\Sigma'x_2^2$	$\Sigma'y_2^2$	$\Sigma'x_2y_2$	$\Sigma'c_2^2$	$\Sigma'\hat{y}_2^2$
3	$\Sigma'x_3^2$	$\Sigma'y_3^2$	$\Sigma'x_3y_3$	$\Sigma'c_3^2$	$\Sigma'\hat{y}_3^2$
4	$\Sigma'x_4^2$	$\Sigma'y_4^2$	$\Sigma'x_4y_4$	$\Sigma'c_4^2$	$\Sigma'\hat{y}_4^2$
Sum.........	$\Sigma\Sigma'x^2$	$\Sigma\Sigma'y^2$	$\Sigma\Sigma'xy$	$\Sigma\Sigma'c^2$	$\Sigma\Sigma'\hat{y}^2$

$$\Sigma'c^2 = \frac{(\Sigma'xy)^2}{\Sigma'x^2} = b\Sigma'xy; \quad \Sigma'c_w^2 = \frac{(\Sigma\Sigma'xy)^2}{\Sigma\Sigma'x^2}$$

$$\Sigma'\hat{y}^2 = \Sigma'y^2 - \Sigma'c^2; \quad \Sigma'\hat{y}_w^2 = \Sigma'y_w^2 - \Sigma'c_w^2$$

9.9 gives symbolically the terms for four sets of data. Two new terms in this table are the sum of squares for means and the sum of squares for differences.

The sums of squares for means, $\Sigma'\bar{x}^2$, $\Sigma'\hat{y}_m^2$, etc., are the respective values for the best straight line drawn through the means of each set of data weighted according to the number of data points which make up each set. If each set of data has the same number of data points, the sum of squares for means may be calculated for k sets of data by noting that

$$\Sigma'\bar{x}^2 = k\Sigma(\bar{x} - \bar{\bar{x}})^2 \tag{9.24}$$

[1] George W. Snedecor, "Statistical Methods," 5th ed., Iowa State College Press, Ames, Iowa, 1956.

where $\bar{x}$ is the over-all mean. If the number of data points varies from set to set, the sum of squares of means is calculated in a manner similar to that used in the analysis of variance for getting the sums of squares for row and column means. For example,

$$\Sigma'\bar{x}^2 = \frac{(\Sigma x_1)^2}{n_1} + \frac{(\Sigma x_2)^2}{n_2} + \cdots + \frac{(\Sigma x_k)^2}{n_k} - \frac{(\Sigma x)^2}{N} \qquad (9.25)$$

$$\Sigma'\bar{y}^2 = \frac{(\Sigma y_1)^2}{n_1} + \frac{(\Sigma y_2)^2}{n_2} + \cdots + \frac{(\Sigma y_k)^2}{n_k} - \frac{(\Sigma y)^2}{N} \qquad (9.26)$$

$$\Sigma'\bar{x}\bar{y} = \frac{\Sigma x_1 \Sigma y_1}{n_1} + \frac{\Sigma x_2 \Sigma y_2}{n_2} + \cdots + \frac{(\Sigma x_k \Sigma y_k)}{n_k} - \frac{\Sigma x \Sigma y}{N} \qquad (9.27)$$

where $N = \Sigma n$.

The difference between the means sum of squares, and the total sum of squares is equal to the sum of the individual sums of squares. This gives a check to the calculation.

$\Sigma'\hat{y}^2$ is the sum of squares of deviation from the best straight line through all the data. $\Sigma\Sigma'\hat{y}^2$ is the sum of the sums of squares of deviation from the best straight line through each set of data separately. This quantity is the minimum deviation that can be obtained by straight-line correlation of the data and is used as an error estimate for testing the significance of the other deviations. $\Sigma'\hat{y}_w^2$ is the sum of squares of deviation from separate lines drawn through each set of data, but all drawn with the same slope. $\Sigma'c_w^2$ is calculated from $(\Sigma'x_w y_w)^2/\Sigma'x_w^2$, and $\Sigma'\hat{y}_w^2$ is calculated from $\Sigma'y_w^2 - (\Sigma'x_w y_w)^2/\Sigma'x_w^2$, which is identical with $\Sigma\Sigma'y^2 - (\Sigma\Sigma'xy)^2/\Sigma\Sigma'x^2$. The pooled slope has a value equal to $\Sigma\Sigma'xy/\Sigma\Sigma'x^2$. $\Sigma'\hat{y}_m^2$ is the sum of squares of deviation of the best straight line through the means of the sets of data and is a measure of any trend from one set of data to the next.

The difference between $\Sigma\Sigma'\hat{y}^2$ and $\Sigma'\hat{y}_w^2$ represents the improvement in the correlation obtained by using individual slopes over using a pooled slope. This difference can be tested for significance by the usual F test. If the difference between the deviations using the pooled slope and using individual slopes is not significant, the lines may be drawn parallel, with the pooled slope equal to $\Sigma\Sigma'xy/\Sigma\Sigma'x^2$ and $\Sigma'\hat{y}_w^2$ used to estimate the error.

Table 9.10 shows the analysis-of-variance form for both of these cases. The significance of $\Sigma'\hat{y}_m^2$ when compared to the error sum of squares indicates whether a linear trend exists between the means of the sets of data. The significance of the balance sum of squares with 1 degree of freedom is a measure of the difference between the slope of the means and the slope of the data within sets and is therefore an indication of the displacement in the y direction of the correlations. The exact interpretation of this type of analysis depends entirely on the data under examination.

If two different responses are measured at several levels of some factor and an analysis of variance is carried out for each, it may be found that in both cases the "between group" effect is significant when compared with the error variance. However, the two responses may be correlated and may not necessarily represent two independently significant responses. This relation may be tested by an analysis of covariance.

TABLE 9.10 Analysis of Variance of Several Linear Correlations

k = sets of data

n_i = data points per set

N = total data points

	Sum of squares	
With individual slopes		D.F.
Source of variance	Symbol	
Means correlation.........	$\Sigma' \hat{y}_m^2$	$k - 2$
Difference between means slope and pooled slope......	(Balance = $\Sigma' \hat{y}^2 - \Sigma' \hat{y}_m^2 - \Sigma' \hat{y}_w^2$)	1
Between slopes............	$\Sigma' \hat{y}_w^2 - \Sigma\Sigma' \hat{y}^2$	$k - 1$
Error....................	$\Sigma\Sigma' \hat{y}^2$	$\Sigma(n_i - 2)$
Total....................	$\Sigma' \hat{y}^2$	$N - 2$
With pooled slope		
Means....................	$\Sigma' \hat{y}_m^2$	$k - 2$
Difference................	Balance	1
Error....................	$\Sigma' \hat{y}_w^2$	$N - k - 1$
Total....................	$\Sigma' \hat{y}^2$	$N - 2$

With the two responses designated x and y, the "difference sums of squares" and the "means sums of squares" of Table 9.9 are identical with the "error" and the "between groups" sums of squares of an analysis of variance calculation. The means residual $\Sigma' \hat{y}_m^2$ may be tested against

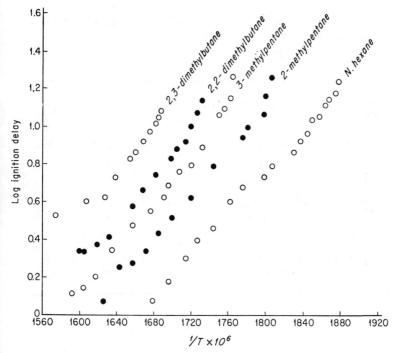

Fig. 9.5   Plot of data from example 9.6.

the difference residual $\Sigma'\hat{y}_w{}^2$ corrected for the proper degrees of freedom as indicated in Table 9.10. This is equivalent to adjusting the y response for the variation in x, assuming a constant linear relation at all levels of the factor being studied. Lack of significance of $\Sigma'\hat{y}_m{}^2$ indicates covariance of y and x.

example 9.6. Figure 9.5 is a plot of some ignition-delay data against reciprocal temperature for several hexanes from some work done by the Naval Research Laboratory.[1] The slope of these plots is used to derive the apparent activation energies required for ignition. This example illustrates the method of determining whether separate slopes are warranted, considering the amount of scatter of the individual sets. (It must be emphasized that statistical conclusions are drawn from the data only and there may be other reasons—previous experience, theoretical basis, experimental information not evaluated—which may either substantiate a doubtful correlation or override an apparently reliable one.)

[1] J. Enoch Johnson et al., *Ind. Eng. Chem.*, **46**:1512 (1954).

We shall designate the log ignition delay as y and $1/T \times 10^6$ as x. The original data may then be summarized as follows:

TABLE 9.11

Compound	Hexane	2-methyl-pentane	3-methyl-pentane	2,2-dimethyl-butane	2,3-dimethyl-butane
Set no....	1	2	3	4	5
$N =$	18	13	15	13	11
$\Sigma x =$	32,415	22,417	25,375	21,770	18,168
$\Sigma x^2 =$	58,448,059	38,705,021	42,974,661	36,483,448	30,020,272
$\Sigma y =$	13.749	8.752	10.181	9.319	9.216
$\Sigma y^2 =$	12.694567	7.702862	8.866479	7.656971	8.095858
$\Sigma xy =$	25,160.948	15,389.736	17,528.814	15,766.486	15,289.627

The sums of squares calculated from these data are:

TABLE 9.12

	$\Sigma' x^2$	$\Sigma' y^2$	$\Sigma' xy$	$\Sigma' c^2$	$\Sigma' \hat{y}^2$	Slope b
Total........	419,749.235	7.542743	1,228.998705	3.598505	3.944235	0.002928
Means †	207,147.581	0.231863	-5.081554	0.000127	0.231736	-0.000025
Difference	212,601.654	7.310880	1,234.080259	7.163416	0.147464 ‡	0.005804 ¶
Set:						
1	74,057.305	2.192627	401.301305	2.174652	0.017975	0.005419
2	49,504.872	1.810744	297.916673	1.792863	0.017881	0.006018
3	48,610.875	1.956298	305.964125	1.925738	0.030560	0.006294
4	27,079.450	0.976682	160.748580	0.954204	0.022478	0.005936
5	13,349.152	0.374527	69.149576	0.347904	0.026623	0.005105
Sum	212,601.654	7.310878	1,234.080259	7.195361	0.115517	

† The means sums of squares are calculated as follows (for the first set):

$$\Sigma' \bar{x}^2 = \frac{(32,415)^2}{18} + \frac{(22,417)^2}{13} + \cdots + \frac{(18,168)^2}{11} - \frac{(120,145)^2}{70} = 207{,}147.581$$

‡ The sum of squares of deviation from the pooled slope through each set of data, $\Sigma' \hat{y}_w^2 = [7.310880 - (1{,}234.080259)^2/(212{,}601.654)] = 0.147{,}464.$

¶ The pooled slope b_p is calculated from $\Sigma\Sigma' xy / \Sigma\Sigma' x^2 = 1{,}234.080259/212{,}601.654 = 0.005804.$

The equivalence of the sum of the sums of squares of the sets and the difference between the sums of squares of totals and means represents a check on the arithmetic.

The analysis of variance, corresponding to Table 9.10, is as follows:

TABLE 9.13

	Sum of squares	D.F.	Mean square
Means............	0.231736	3	0.077245
Difference.........	3.565035	1	3.565035
Between slopes....	0.147464 − 0.115517 =		
	0.031947	4	0.007987
Error.............	0.115517	60	0.001925
Total...........	3.944235	68	

The ratio of the "between slopes" mean square and the "error" mean square of $7{,}987/1{,}925 = 4.15$ is tested against F at 4 and 60 degrees of freedom (Tables 7.8 and 7.9). The F values are 2.52 at the 0.05 level and 3.65 at the 0.01 level, indicating that with respect to these data the amount of deviation removed by using individual least-squares lines for each set over that removed by using a pooled slope for all the sets is highly significant. There is less than a 1 per cent chance of not being in error if the lines are considered to have a common slope, 0.005804.

The highly significant value of the "difference" mean square indicates what is obvious from the plot of the data, that there is a displacement in the y direction between the separate lines.

The usefulness of the "means" mean square depends on physical interpretation of the experiment. The negative slope of the log ignition delay against reciprocal temperature when going from normal hexane to the more complex isomers is evidently significant. Statistical analysis can only point it out; some other discipline must explain it.

9.1.7 Linear Regression with Replication. When more than one determination of the dependent variable is made at several or all levels of the independent variable, then the check of the replications can be used as an estimate of the error. Up to now, for a linear correlation, we have been using the sum of squares of the deviation from the line as an estimate of the error. If the error as estimated from the deviations from the line is significantly larger than the error estimated from the replicates, the interpretation is that the straight-line correlation does not account for all the variability in the dependent variable. This statement can be illustrated by an analysis-of-variance table. If $\Sigma\Sigma' y_i^2$ designates the total sum of squares of deviation of each set of replicates from the set mean with $\Sigma(k_i - 1)$ degrees of freedom, where k_i is the number of replicates of the ith set of data, and $\Sigma'c^2$ designates the sum of squares of deviation removed by the linear correlation, with 1 degree of freedom, then the total sum of squares of deviation may be broken down as follows:

TABLE 9.14

	Sum of squares	D.F.
Correlation line..........	$\Sigma'c^2 = (\Sigma'xy)^2/\Sigma'x^2$	1
Residual from correlation line.................	$\Sigma'\hat{y}^2 = \Sigma'y^2 - \Sigma\Sigma'y_i^2 - \Sigma'c^2$	$N - [\Sigma(k_i - 1) + 2]$
Error.................	$\Sigma\Sigma'y_i^2$	$\Sigma(k_i - 1)$
Total................	$\Sigma'y^2$	$N - 1$

If the residual mean square, i.e., the residual sum of squares divided by its degrees of freedom, is significantly larger than the error mean square when tested by the F test, then there still remains some variation of the data which is not accounted for by either the correlation or the error of the measurements.

The "best straight line" drawn through replicated data is identical with the "best straight line" through the means of the sets of replicated datum points weighted by the number of replicates in each set. This equivalence is evident from the equations used in example 9.7 following.

The variance, and hence the precision or error, estimated from replicates can always be used as a test for the significance of the variance calculated from some correlation, or in the investigation of some suspected factor. The replicates need not be from the same experiment, although some provision should be made to ensure that the replicate variance is applicable. For example, if the variance of an analysis has been well established by past experience, this variance can be used to test the residue of a correlation of the analytical result with some independent variable, without the necessity of running replicates in the experiment.

example 9.7. The data in Table 9.15 are from the study of interfacial areas in liquid-liquid agitation.[1] The independent variable is the stirring speed in revolutions per minute and the dependent variable is light transmission in per cent for a 40 per cent CCl_4 solution in water. The data represent a number of replicate determinations at each stirring speed.

[1] Theodore Verneulen et al., *Chem. Eng. Progr.*, **51**:85-F (1955).

CORRELATION—REGRESSION   291

Calculation. The problem will be solved both in terms of x and of $\log x$ to see if any improvement in the correlation results from the use of logarithms. The original x' values will be coded by dividing by 100 to simplify the arithmetic. Note that in the calculations that follow $\Sigma'x^2$ is not simply the sum of squares of deviation of the four x

TABLE 9.15
Stirring speed,

x'	Per cent light transmission, y
400	6.5, 7.6, 7.4, 7.3, 7.4, 6.95
250	9.0, 9.0, 9.75, 9.3, 9.1
150	10.3, 12.5. 13.1, 11.8, 11.9
100	15.0, 15.2, 18.6, 18.1, 16.4, 16.3

values, but is a weighted sum of squares of deviation with each x value weighted according to the number of replicates that has been run at that level of x. This, of course, gives the same result that would be obtained if each x value were taken separately and the fact of replication ignored. On the other hand, if all the replicates have been run the same number of times, the weighting term can be factored from the expression for the sum of squares of deviation.

For example, if k_i is the number of replicates in the ith set of runs and there are n sets, the total number of data points $N = \Sigma k_i$, and $\Sigma'x^2 = \Sigma k_i x_i^2 - (\Sigma k_i x_i)^2/N$. If all the k's are equal, $N = nk$ and $\Sigma'x^2 = k[\Sigma x_i^2 - (\Sigma x_i)^2/n]$.

The data and equations for the solution of the stirring problem are:

TABLE 9.16

Data set	1	2	3	4	
x	4.0	2.5	1.5	1.0	
$\log x$	0.60206	0.39794	0.17609	0.00000	
y	6.5	9.0	10.3	15.0	
	7.6	9.0	12.5	15.2	
	7.4	9.3	13.1	18.6	
	7.3	9.1	11.8	18.1	
	6.95	9.75	11.9	16.4	
	7.4			16.3	
Σy_i	43.15	46.15	59.60	99.60	248.50
k_i	6	5	5	6	
$\bar{y}_i$	7.1916	9.23	11.92	16.60	

$$\Sigma'x^2 = (6)(4)^2 + (5)(2.5)^2 + \cdots + (6)(1.0)^2 -$$
$$\frac{[(6)(4) + (5)(2.5) + \cdots + (6)(1.0)]^2}{22} = 144.50 - 113.625$$
$$= 30.865$$

$$\Sigma'(\log x)^2 = (6)(0.60206)^2 + \cdots + (6)(0.00000)^2 -$$
$$\frac{[(6)(0.60206) + \cdots + (6)(0.0)]^2}{22} = 3.121677 - 1.910137$$
$$= 1.21154$$

$$\Sigma'y^2 = (6.5)^2 + (7.6)^2 + \cdots + (16.3)^2 - \frac{(248.50)^2}{22} = 310.0625$$

$$\Sigma\Sigma'y_i{}^2 = \Sigma(\Sigma y_i{}^2 - \bar{y}_i\Sigma_i) = \Sigma y^2 - \Sigma(\bar{y}_i\Sigma y_i) = (6.5)^2 + (7.6)^2$$
$$+ \cdots + (16.4)^2 + (16.3)^2 - [(7.1916)(43.15) + (9.23)(46.15)$$
$$+ \cdots + (16.60)(99.60)] = 3{,}116.5450 - 3{,}100.0050 = 16.5400$$

$$\Sigma'xy = \Sigma xy - \Sigma(k_ix_i)\frac{\Sigma y}{N} = \Sigma(x_i\Sigma y_i) - \Sigma(k_ix_i)\frac{\Sigma y}{N}$$
$$= (4)(43.15) + (2.5)(46.15) + \cdots + (1.0)(99.60)$$
$$- \frac{[(6)(4) + (5)(2.5) + \cdots + (6)(1)](248.50)}{22}$$
$$= 476.975 - 564.772 = -87.797$$

$$\Sigma'y \log x = \Sigma[(\log x_i)\Sigma y_i] - \Sigma(k_i \log x_i)\frac{\Sigma y}{N}$$
$$= (0.60206)(43.15) + (0.39794)(46.15) + (0.17609)(59.60)$$
$$- \frac{[(6)(0.60206) + (5)(0.39794) + (5)(0.17609)](248.50)}{22}$$
$$= 54.8388 - 73.2290 = -18.384$$

$$\Sigma'c^2 \text{ (for } y{:}x \text{ line)} = \frac{(\Sigma'xy)^2}{\Sigma'x^2} = \frac{(-87.797)^2}{30.865} = 249.743$$

$$\Sigma'c^2 \text{ (for } y{:}\log x \text{ line)} = \frac{(\Sigma'y \log x)^2}{\Sigma'(\log x)^2} = \frac{(-18.384)^2}{1.21154} = 278.960$$

$$b_{yx} = \frac{-87.797}{30.865} = -2.842$$

$$b_{y \log x} = \frac{-18.384}{1.21154} = -15.13$$

The analysis-of-variance table may now be set up as Table 9.17 shows on page 293.

Discussion. Comparison of the mean squares of both the $y{:}x$ line and the $y{:}\log x$ line with the error mean square indicates that both of these correlations are highly significant, the log correlation being somewhat better. The residual from the $y{:}x$ line compared to the error mean square indicates that a significant deviation is still unaccounted for. The residual from the $y{:}\log x$ line is just significant

TABLE 9.17

	Sum of squares	D.F.	Mean square
$y:x$ line, $\Sigma'c^2$	249.74	1	249.74
$y:\log x$ line, $\Sigma'c^2$	278.96	1	278.96
Residual from $y:x$ line, $\Sigma'\hat{y}_x^2$	43.78	2	21.89
Residual from $y:\log x$ line, $\Sigma'\hat{y}_{\log x}^2$	14.56	2	7.28
Error, $\Sigma\Sigma'y_i^2$	16.54	18	0.92
Total, $\Sigma'y^2$	310.06	21	

at the 0.01 level; the mean square ratio is 7.91 compared to an F at the 0.01 level for 2 and 18 degrees of freedom of 6.01. Whether it would be worthwhile to try to account for more of the deviations by a higher-order equation depends entirely on the investigation under study.

9.1.8 Straight Line through the Origin. There are many cases of correlations when from the nature of the data the correlation line can be expected to go through the origin. Functions plotted against time; effect of impurities on yield; analytical determinations to find added ingredients; these might all be examples of correlations through the origin. Even though the nature of the data might prescribe a line through the origin, the actual data might indicate otherwise.

It has been pointed out that the least-squares line is the linear regression that gives the minimum sum of squared deviation of the actual data. The sum of squares of deviation from the line through the origin therefore cannot be less than that from the least-squares line. If the difference between the deviations from these two lines is significant when compared to the error as measured by the minimum residual (or by replicates if they are available), then the data may be said to indicate that the line does not go through the origin. What interpretation is made of this statement depends entirely on the analysis involved: maybe it indicates poor measurement or bias in the x measurements; maybe it indicates a bias in the analytical procedure which gives a fixed displacement to the y variable; maybe it indicates the presence of some unknown factor which has not been considered.

If the line goes through the origin, the y intercept is zero, and the equation of the line is

$$\hat{y} = b_0 x \tag{9.28}$$

where the subscript 0 refers to the line through the origin.

TABLE 9.18

	Least-squares line	Line through origin
Equation of line	$\hat{y} = a + bx$	$\hat{y} = b_0 x$
Degrees of Freedom	$N - 2$	$N - 1$
Intercept	$\bar{y} - b\bar{x}$	0
Slope b and b_0	$\dfrac{\Sigma' xy}{\Sigma' x^2} =$ $\dfrac{\Sigma(x - \bar{x})(y - \bar{y})}{\Sigma(x - \bar{x})^2}$	$\dfrac{\Sigma xy}{\Sigma x^2}$
Residual sum of squares of deviation	$\Sigma' y^2 - \dfrac{(\Sigma' xy)^2}{\Sigma' x^2} =$ $\Sigma' y^2 - b^2 \Sigma' x^2 =$ $\Sigma' y^2 - b \Sigma' xy =$ $\Sigma' \hat{y}^2$	$\Sigma y^2 - \dfrac{(\Sigma xy)^2}{\Sigma x^2} =$ $\Sigma y^2 - b_0^2 \Sigma x^2 =$ $\Sigma y^2 - b_0 \Sigma xy =$ $\Sigma \hat{y}_0^2$
Variance of estimate $s^2(\hat{y})$	$\dfrac{\Sigma' \hat{y}^2}{N - 2}$	$\dfrac{\Sigma \hat{y}_0^2}{N - 1}$
Variance of slope $s^2(b)$	$\dfrac{s^2(\hat{y})}{\Sigma' x^2}$	$\dfrac{s^2(\hat{y}_0)}{\Sigma x^2}$
Correlation coefficient r	$\left(\dfrac{b \Sigma' xy}{\Sigma' y^2} \right)^{1/2}$	$\left[\dfrac{(b_0 \Sigma xy - \bar{y} \Sigma y)}{(\Sigma y^2 - \bar{y} \Sigma y)} \right]^{1/2} =$ $\left[\dfrac{\Sigma' y^2 - (\Sigma y^2 - b_0 \Sigma xy)}{\Sigma' y^2} \right]^{1/2}$

The equations for the slope and the other regression terms are given in Table 9.18 for the line through the origin, and in parallel the equations for the least-squares line are repeated to show the similarity between the formulas. Note that the formulas for the line through the origin are without reference to the mean and involve only sums of squares and sums of cross products. [For this reason the prime symbol (′) is omitted from the summation sign.]

It will be noted from the foregoing tabulation that the degrees of freedom for the line through the origin are only one less than the number of data points, while they are two less for the least-squares line. The additional degree of freedom lost to the least-squares line is a consequence of employing the means for the calculation. If the loss of this additional degree of freedom results in a significant decrease in deviation of the data from the line, this can be established by testing the decrease in deviation against the residual variance of estimate from the least-squares line by the F test at 1 and $N - 2$ degrees of freedom.

The following analysis of variance table indicates this type of test.

TABLE 9.19

	Sums of squares	D.F.	Mean square
Residual from line through origin.......	$\Sigma \hat{y}_0^2$	$N - 1$	
Decrease in residual....	$\Sigma \hat{y}_0^2 - \Sigma' \hat{y}^2$	1	$\Sigma \hat{y}_0^2 - \Sigma' \hat{y}^2$
Residual from least-squares line.........	$\Sigma' \hat{y}^2$	$N - 2$	$\dfrac{\Sigma' \hat{y}^2}{N - 2} = s^2(\hat{y})$

F test at 1, $N - 2$ degrees of freedom, $(\Sigma \hat{y}_0^2 - \Sigma' \hat{y}^2)/s^2(\hat{y})$.

example 9.8. The data in Table 9.20 are from some U.S. Bureau of Mines synthetic fuels research work[1] and represent oxygenate analyses

TABLE 9.20

Oxygenate determination, y	Carbon determination, x
42.7	23.1
58.2	30.3
42.5	22.4
22.0	12.0
26.6	13.2
30.1	17.0
15.4	10.8
16.8	8.6
46.8	26.9
40.1	23.2
43.3	24.6
47.2	27.0
57.3	32.4
47.1	25.2
59.2	34.0
41.6	20.9

and corresponding carbon analyses of the water-soluble product. It is reasonable to expect that when the carbon determination is zero, the oxygenates will also be zero, and in the original presentation of the

[1] M. D. Schlesinger et al., *Ind. Eng. Chem.*, **46**:1322 (1954).

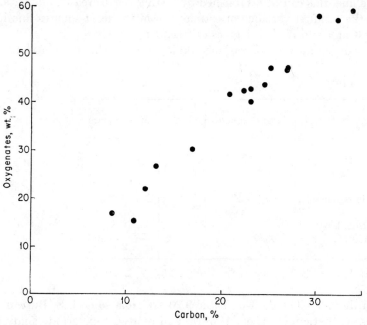

Fig. 9.6 Plot of data from example 9.8.

data the correlation was drawn in this manner. A plot of the data is given in Fig. 9.6.

Calculation

$$N = 16$$

$$\Sigma x^2 = 8{,}624.92$$

$$\Sigma xy = 15{,}593.95$$

$$b_0 = \frac{15{,}593.95}{8{,}624.92} = 1.8080$$

$$\Sigma y^2 = 28{,}271.63$$

$\Sigma \hat{y}_0^2$ = (residual sum of squares of deviation from the line through the origin) = $28{,}271.63 - (1.8080)(15{,}593.95) = 77.61$

$$\Sigma' y^2 = 898.51$$

$$\Sigma' xy = 1{,}598.09$$

$$b = \text{(least-squares slope)} = \frac{1{,}598.09}{898.51} = 1.7786$$

$$\Sigma' y^2 = 2{,}919.06$$

$\Sigma' \hat{y}^2$ = (residual sum of squares of deviation from the least-squares line) = $2{,}919.06 - (1.7786)(1{,}598.09) = 76.70$

Discussion. The difference between the residual sum of squares for the two lines, which is the improvement gained by the least-squares line, as that is the line with the minimum sum of squares of deviation, is $\Sigma\hat{y}_0^2 - \Sigma'\hat{y}^2 = 77.61 - 76.70 = 0.91$. This is obviously not significant when compared to the variance of estimate $s^2(\hat{y})$ obtained from the least-squares line of $76.60/14 = 5.48$. Actually the variance of estimate obtained from the line through the origin is slightly less, although not significantly so, because of the extra degree of freedom, that is, $s^2(\hat{y}_0) = 77.61/15 = 5.17$.

The evidence is strong from these data that the correlation line goes through the origin.

9.2 Linear Regression with More Than Two Variables

In presenting the method for multiple regression, we shall first present the general procedure for any number of variables. Section 9.2.5 will present some simplifications to the general method when applied to regressions with three variables. However, the general method described in Sec. 9.2.1 for any number of variables can, of course, be applied to three variables with the same result. We shall refer to the regressions in terms of one dependent and several independent variables. The regression formulas are not restricted to variables having a dependent-independent relationship, but merely describe in mathematical terms the nature of the relation between the variables. However, in evaluating the degree of the relation, all the error or imprecision is assumed to be in the measurements of one of the variables, and the other variables are assumed to be precisely known. The imprecision is associated with the variable referred to as "dependent," and the "independent" variables are the precise variables.

9.2.1 Regression with More Than Two Variables. The engineer does not need to be taught that correlations often occur which are functions of several variables. Conversions as functions of temperature, pressure, and catalyst activity; fluid-flow and heat-transfer correlations in terms of various sets of dimensionless numbers made up of several variables; general diffusional processes expressed in terms of film resistance, driving potential, and reaction rate are all common examples of multiple correlations. In general, a dependent variable y may be expressed in terms of several independent variables:

$$\hat{y} = a + bx + cz + dw + \cdots \qquad (9.29)$$

An alternative form, in which the independent variables are not necessarily to the first power, can sometimes be modified to give the multiple linear equation. For example, if

$$\hat{y} = ax^b z^c w^d \cdots \qquad (9.30)$$

then

$$\log \hat{y} = \log a + b \log x + c \log z + d \log w + \cdots \qquad (9.31)$$

and, therefore,

$$\hat{Y} = A + bX + cZ + dW + \cdots \qquad (9.32)$$

This section will deal exclusively with multiple linear regressions of the form of Eq. (9.29) or (9.32). Equations of higher order, such as

$$\hat{y} = a + bx + cx^2 + dz + ez^3 + \cdots \qquad (9.33)$$

will be discussed briefly in Sec. 9.3 dealing with curvilinear regressions.

If the degree of fit of a multiple linear regression is measured by the sum of the squares of deviation of the observed from the predicted values, $\Sigma(y_i - \hat{y}_i)^2$, then in a manner directly analogous to that used for a simple linear regression, an expression for the constants of Eq. (9.29) can be obtained that will give the minimum value for $\Sigma(y_i - \hat{y}_i)^2$.

We shall rewrite Eq. (9.29) to simplify the symbols:

$$\hat{y} = a + b_1.x_1 + b_2.x_2 + b_3.x_3 + \cdots \qquad (9.34)$$

where $b_1.$ means $b_{1.23\cdots k}$ and indicates the regression coefficient for y upon x_1 when all the other independent variables, $x_2, x_3, \ldots x_k$ are held constant. $b_1.$ is referred to as the partial regression coefficient. The dot symbol differentiates $b_1.$ from the simple b_1 (without the dot), which is the regression coefficient of y upon x_1 when the other variables are disregarded.

If the sum of squares of deviation between the observed values of the dependent variable y, and those predicted by an equation of the form of Eq. (9.34), is differentiated with respect to each of the constants in the equation, and in each case the derivative is equated to zero, the value of the constant will be obtained that gives the minimum sum of squares of deviation. This was the method used to obtain the constants for the least-squares line for two variables in Eqs. (9.2) and (9.3). For the multivariable case, the equations obtained are those listed as Eqs. (9.35).

$$\begin{aligned}
Na + b_1.\Sigma x_1 + b_2.\Sigma x_2 + b_3.\Sigma x_3 + \cdots &= \Sigma y \\
a\Sigma x_1 + b_1.\Sigma x_1{}^2 + b_2.\Sigma x_1 x_2 + b_3\Sigma.x_1 x_3 + \cdots &= \Sigma x_1 y \\
a\Sigma x_2 + b_1.\Sigma x_2 x_1 + b_2.\Sigma x_2{}^2 + b_3.\Sigma x_2 x_3 + \cdots &= \Sigma x_2 y \\
a\Sigma x_3 + b_1.\Sigma x_3 x_1 + b_2.\Sigma x_3 x_2 + b_3.\Sigma x_3{}^2 + \cdots &= \Sigma x_3 y \\
\cdots &
\end{aligned} \qquad (9.35)$$

where N is the number of sets of data points.

Equations (9.35) can be solved by any method for the solution of simultaneous equations for the evaluation of the regression coefficients $b_1., b_2.$, etc., and for the intercept a.

However, if the first of the equations listed in the (9.35) series is solved for a,

$$a = \bar{y} - b_1.\bar{x}_1 - b_2.\bar{x}_2 - b_3.\bar{x}_3 - \cdots$$

and this value substituted for a in the other equations, a second set of equations results which are more amenable to statistical calculations. These equations, listed as Eqs. (9.36) are in terms of the deviations of each variable from its mean, $\Sigma'x = \Sigma(x - \bar{x})^2$, and $\Sigma'xy = \Sigma(x - \bar{x})(y - \bar{y})$, etc.

$$b_1.\Sigma'x_1{}^2 + b_2.\Sigma'x_1x_2 + b_3.\Sigma'x_1x_3 + \cdots = \Sigma'x_1y$$
$$b_1.\Sigma'x_2x_1 + b_2.\Sigma'x_2{}^2 + b_3.\Sigma'x_2x_3 + \cdots = \Sigma'x_2y \quad (9.36)$$
$$b_1.\Sigma'x_3x_1 + b_2.\Sigma'x_3x_2 + b_3.\Sigma'x_3{}^2 + \cdots = \Sigma'x_3y$$
$$\cdots$$

The expressions for the sum of squares of deviation removed by the regression, the residual sum of squares, and the variance of estimate for the multiple linear regression corresponding to the same terms for the simple least-squares line are as follows:

Sum of squares of deviation removed by the regression:

$$\Sigma'c^2 = b_1.\Sigma'x_1y + b_2.\Sigma'x_2y + b_3.\Sigma'x_3y + \cdots \quad (9.37)$$

Residual sum of squares of deviation:

$$\Sigma'\hat{y}^2 = \Sigma'y^2 - \Sigma'c^2 \quad (9.38)$$

The variance of estimate:

$$s^2(\hat{y}) = \frac{\Sigma'\hat{y}^2}{N - k - 1} \quad (9.39)$$

where k is the number of independent variables.

The significance of the multiple regression can be tested by comparing with the corresponding F value the ratio of the sum of squares removed by the regression to the residual sum of squares each divided by its respective degrees of freedom; i.e.,

$$F = \frac{\Sigma'c^2/k}{\Sigma'\hat{y}^2/(N - k - 1)} \quad \text{at } k \text{ and } N - k - 1 \text{ degrees} \quad (9.40)$$

9.2.2 Gaussian Multipliers. If instead of solving directly for the regression coefficients, as in Eqs. (9.35) or (9.36), the solution is made by means of statistics known as the Gaussian multipliers, several advantages are gained. Sometimes after a regression has been calculated, the engineer wishes to know how some other dependent variable may be related.

For example, after finding a correlation between yield and temperature, pressure, and perhaps some other independent variables, he might be interested to know whether reaction rate or some other dependent variable may be correlated against the same independent variables used for the first correlation; or in rubber compounding, the chemist might be interested in finding whether both tensile strength and abrasion resistance can be correlated against the same independent variables of various component ingredients and curing time. The use of the Gaussian multipliers permits the solution to these types of problems without solving separate sets of simultaneous equations for each dependent variable.

The Gaussian multipliers are used for calculating the variance, and hence the confidence range, of the regression coefficients directly from the variance of estimate. They are also used to set confidence ranges on any value estimated from the regression and for comparing two regression coefficients for significant difference.

In some instances the effect of one factor in a multiple regression will turn out to be nonsignificant, and the engineer will want to delete that variable from the regression. It is much simpler to delete a variable after the regression has been calculated than it is to include another variable under similar circumstances. The Gaussian multipliers are involved in the simple equations for deleting a variable from a regression.

The Gaussian multipliers are identical with the elements of the inverse of the matrix of the "sums of squares" in Eqs. (9.36). The inverse matrix may be calculated in a number of ways. One method of calculation is described in the following paragraphs. Any other method available to the reader may be used with identical results.

The following equations illustrate the calculation of the Gaussian multipliers for three independent variables, x_1, x_2, x_3. The extension to a greater number of variables will be obvious from the form:

$$c_{11}\Sigma' x_1{}^2 + c_{12}\Sigma' x_1 x_2 + c_{13}\Sigma' x_1 x_3 = 1$$
$$c_{11}\Sigma' x_2 x_1 + c_{12}\Sigma' x_2{}^2 + c_{13}\Sigma' x_2 x_3 = 0 \qquad (9.41)$$
$$c_{11}\Sigma' x_3 x_1 + c_{12}\Sigma' x_3 x_2 + c_{13}\Sigma' x_3{}^2 = 0$$

$$c_{21}\Sigma' x_1{}^2 + c_{22}\Sigma' x_1 x_2 + c_{23}\Sigma' x_1 x_3 = 0$$
$$c_{21}\Sigma' x_2 x_1 + c_{22}\Sigma' x_2{}^2 + c_{23}\Sigma' x_2 x_3 = 1 \qquad (9.42)$$
$$c_{21}\Sigma' x_3 x_1 + c_{22}\Sigma' x_3 x_2 + c_{23}\Sigma' x_3{}^2 = 0$$

$$c_{31}\Sigma' x_1{}^2 + c_{32}\Sigma' x_1 x_2 + c_{33}\Sigma' x_1 x_3 = 0$$
$$c_{31}\Sigma' x_2 x_1 + c_{32}\Sigma' x_2{}^2 + c_{33}\Sigma' x_2 x_3 = 0 \qquad (9.43)$$
$$c_{31}\Sigma' x_3 x_1 + c_{32}\Sigma' x_3 x_2 + c_{33}\Sigma' x_3{}^2 = 1$$

CORRELATION—REGRESSION

where the c terms are the Gaussian multipliers in question. Because of the symmetry of the three sets of equations, c_{12} must equal c_{21}, and c_{13} must equal c_{31}, etc. The three sets (9.41), (9.42), and (9.43) are combined into one matrix for simultaneous solution. The combined form of the equations is:

	$j =$	1	2	3
$c_{1j}\Sigma'x_1^2 + c_{1j}\Sigma'x_1x_2 + c_{1j}\Sigma'x_1x_3$	=	1	0	0
$c_{2j}\Sigma'x_2x_1 + c_{2j}\Sigma'x_2^2 + c_{2j}\Sigma'x_2x_3$	=	0	1	0
$c_{3j}\Sigma'x_3x_1 + c_{3j}\Sigma'x_3x_2 + c_{3j}\Sigma'c_3^2$	=	0	0	1

(9.44)

Equations (9.44) do not involve the dependent variable and are functions only of the independent variables. The partial regression coefficients are obtained from the solutions of Eqs. (9.44) by the following relations:

$$b_1. = c_{11}\Sigma'x_1y + c_{12}\Sigma'x_2y + c_{13}\Sigma'x_3y$$
$$b_2. = c_{21}\Sigma'x_1y + c_{22}\Sigma'x_2y + c_{23}\Sigma'x_3y \quad (9.45)$$
$$b_3. = c_{31}\Sigma'x_1y + c_{32}\Sigma'x_2y + c_{33}\Sigma'x_3y$$

It can be seen that Eqs. (9.45) can be altered by using any dependent variable to obtain a new set of regression coefficients without having to recalculate the multipliers.

The results obtained from Eqs. (9.44) and (9.45) are identical with those obtained from the solution of Eq. (9.35) or (9.36). Equations (9.37), (9.38), and (9.39) for the sum of squares removed by the regression, the residual sum of squares, and the variance of estimate also still apply. In addition, the multipliers permit the simple calculation of the variance of the partial regression coefficients or of single predicted values.

The variance of the regression coefficients is

$$s^2(b_1.) = c_{11}s^2(\hat{y})$$
$$s^2(b_2.) = c_{22}s^2(\hat{y}) \quad (9.46)$$
$$\cdots$$

The variance of an average predicted value, $\hat{y}_i$, is

$$s^2(\hat{y}_i) = s^2(\hat{y})\left[\frac{1}{N} + c_{11}(x_1 - \bar{x}_1)^2 + c_{22}(x_2 - \bar{x}_2)^2 + \cdots \right.$$
$$\left. + 2c_{12}(x_1 - \bar{x}_1)(x_2 - \bar{x}_2) + 2c_{13}(x_1 - \bar{x}_1)(x_3 - \bar{x}_3) + \cdots \right] \quad (9.47)$$

TABLE 9.21 Systematic Solution of Equations for Gaussian Multipliers

Solution Steps	Symbolic Equations		$j = $		
			1	2	3
Initial equations	$c_{i1}\Sigma'x_1^2 + c_{i2}\Sigma'x_1x_2 + c_{i3}\Sigma'x_1x_3 =$		1	0	0
	$c_{i1}\Sigma'x_1x_2 + c_{i2}\Sigma'x_2^2 + c_{i3}\Sigma'x_2x_3 =$		0	1	0
	$c_{i1}\Sigma'x_1x_3 + c_{i2}\Sigma'x_2x_3 + c_{i3}\Sigma'x_3^2 =$		0	0	1
Solution					
1 †	$a_{11}c_{i1} + a_{12}c_{i2} + a_{13}c_{i3} =$		1	0	0
2	$a_{21}c_{i1} + a_{22}c_{i2} + a_{23}c_{i3} =$		0	1	0
3	$a_{31}c_{i1} + a_{32}c_{i2} + a_{33}c_{i3} =$		0	0	1
4. 1 divided by a_{11}	$1\ c_{i1} + a_{42}c_{i2} + a_{43}c_{i3} =$		a_{44}	0	0
5. 2 divided by a_{21}	$1\ c_{i1} + a_{52}c_{i2} + a_{53}c_{i3} =$		0	a_{55}	0
6. 3 divided by a_{31}	$1\ c_{i1} + a_{62}c_{i2} + a_{63}c_{i3} =$		0	0	a_{66}
7. 4 − 5	$0 + a_{72}c_{i2} + a_{73}c_{i3} =$		a_{74}	a_{75}	0
8. 5 − 6	$0 + a_{82}c_{i2} + a_{83}c_{i3} =$		0	a_{85}	a_{86}
9. 7 divided by a_{72}	$0 + 1\ c_{i2} + a_{93}c_{i3} =$		a_{94}	a_{95}	0
10. 8 divided by a_{82}	$0 + 1\ c_{i2} + a_{103}c_{i3} =$		0	a_{105}	a_{106}
11. 9 − 10	$0 + 0 + a_{113}c_{i3} =$		a_{114}	a_{115}	a_{116}
12. 11 divided by a_{113}	$0\qquad 0\qquad 1\ c_{i3} =$		a_{124}	a_{125}	a_{126}
	$c_{i3} =$		a_{124}		
	$c_{i3} =$			a_{125}	
	$c_{i3} =$				a_{126}
13. Substitute c_{i3} in 10 (or 9)	$c_{i2} + a_{103}c_{124} =$		0		
	$c_{i2} + a_{103}a_{125} =$			a_{105}	
	$c_{i2} + a_{103}a_{126} =$				a_{106}
14. Transposing 13	$c_{i2} =$		a_{144}		
	$c_{i2} =$			a_{145}	
	$c_{i2} =$				a_{146}
15. Substitute c_{i2} and c_{i3} in 6 (or 5 or 4)	$c_{i1} + a_{62}a_{144} + a_{63}a_{124} =$		0		
	$c_{i1} + a_{62}a_{145} + a_{63}a_{125} =$			0	
	$c_{i1} + a_{62}a_{146} + a_{63}a_{126} =$				a_{66}
16. Transposing 15	$c_{i1} =$		a_{164}		
	$c_{i1} =$			a_{165}	
	$c_{i1} =$				a_{166}

General Solution, $c_{ij} =$

$j =$	1	2	3
$i =$			
1	a_{164}	a_{144}	a_{124}
2	a_{165}	a_{145}	a_{125}
3	a_{166}	a_{146}	a_{126}

$c_{12} = c_{21}, c_{13} = c_{31}, c_{23} = c_{32}; c_{11}, c_{22}, c_{33}$ must be $\geqq 0$.

† Where $a_{11} = \Sigma'x_1^2$, $a_{12} = \Sigma'x_1x_2$, etc.

The difference between any two regression coefficients may be tested by the t test, similar to the method of Eq. (9.22), by the following relation:

$$t = \frac{b_1 - b_2}{s(\hat{y})\sqrt{c_{11} - 2c_{12} + c_{22}}} \tag{9.48}$$

The solution to Eqs. (9.44) may be obtained in any systematic method used for the solving of simultaneous equations. Table 9.21 illustrates one method symbolically. A numerical example of this method is included in example 9.9, which follows. This method is recommended for engineers who may not be solving problems of this type very frequently. The method provides a running check on the arithmetic by including an additional column on the right which is a total of all the numbers in the corresponding row. When the operation performed on the row is performed on this total, the resulting answer must equal the sum of the individual numbers in the succeeding row. Also the substitution of successive solutions in the parallel equations provides a second check on the results. Finally, the symmetry of the matrix, in that $c_{12} = c_{21}$, $c_{23} = c_{32}$, etc., gives an additional check on the answers.

example 9.9. The equation for heat transfer in liquids has been proposed in the form

$$\text{Nu} = A(\text{Re})^{b_1}(\text{Pr})^{b_2}\left(\frac{\mu_a}{\mu_w}\right)^{b_3} \tag{9.49}$$

The value of the exponents has been calculated by multiple correlation using the method of Eqs. (9.35).[1] This example uses the same data but employs the method of Gaussian multipliers. The results are identical.

TABLE 9.22

$\bar{y} = 1.262113$	$\Sigma'y^2 = 2.171305$	$\Sigma'x_1y = 3.795016$
$\bar{x}_1 = 2.210485$	$\Sigma'x_1^2 = 34.639265$	$\Sigma'x_2y = -2.317256$
$\bar{x}_2 = 3.004987$	$\Sigma'x_2^2 = 19.153509$	$\Sigma'x_3y = 10.509764$
$\bar{x}_3 = -0.650749$	$\Sigma'x_3^2 = 88.139543$	$\Sigma'x_1x_2 = -22.750174$
$N = 67$	$\Sigma'x_1x_3 = -4.615744$	$\Sigma'x_2x_3 = -2.922539$

Equation (9.49) is put in linear form by operating with logarithms on both sides:

$$\log \text{Nu} = \log A + b_1 \log \text{Re} + b_2 \log \text{Pr} + b_3 \log \frac{\mu_a}{\mu_w} \tag{9.50}$$

---

[1] Octave Levenspiel et al., *Ind. Eng. Chem.*, **48**:324 (1956).

304 APPLIED STATISTICS FOR ENGINEERS

TABLE 9.23 Calculation of Gaussian Multipliers for Example 9.9

	c_{i1}	c_{i2}	c_{i3}	c_{1j}	c_{2j}	c_{3j}	Checks
1. Initial equations	34.639265	−22.750174	−4.615744	1	0	0	8.273347
2.	−22.750174	19.153509	−2.922539	0	1	0	−5.519204
3.	−4.615744	−2.922539	88.139543	0	0	1	81.601261
4. 1 ÷ 34.639265	1	−0.656774	−0.133252	0.028869	0	0	0.238843
5. 2 ÷ −22.750174	1	−0.841906	0.128462	0	−0.043956	0	0.242600
6. 3 ÷ −4.615744	1	0.633167	−19.095414	0	0	−0.216650	−17.678897
7. 4 − 5	0	0.185132	−0.261714	0.028869	0.043956	0	−0.003757
8. 5 − 6	0	−1.475073	19.223876	0	−0.043956	0.216650	17.924197
9. 7 ÷ 0.185132	0	1	−1.413662	0.155937	0.237431	0	−0.020294
10. 8 ÷ −1.475073	0	1	−13.032491	0	0.029799	−0.146874	−12.149566
11. 9 − 10	0	0	11.618829	0.155937	0.207632	0.146874	12.129272
12. 11 ÷ 11.618829	0	0	1	0.013421	0.017870	0.012641	1.043932
		c_{12}	c_{13}	0.013421			
		c_{22}	c_{23}				
		c_{32}	c_{33}				
13. Substitute c_{i3} in 10			−0.174909	0	0.017870	0.012641	
			−0.233891		0.029799	−0.146874	
			−0.164744				
14. Transposing 13				0.174909			
15. Substitute c_{i2}, c_{i3} in 6	1	0.110747	−0.256280	0	0.262690	0.017870	
	1	0.166327	−0.341235		0	−0.216650	
	1	0.011315	−0.241385				
16. Transposing 15	c_{11}			0.145533	0.174908	0.013421	
	c_{21}						
	c_{31}						

or
$$\hat{y} = a + b_1.x_1 + b_2.x_2 + b_3.x_3 \tag{9.51}$$

The means and the sums of squares of deviations from the means required for the solution are given in Table 9.22. The calculation of the Gaussian multipliers is given in Table 9.23.[1]
Solution: See Table 9.23.

	c_{i1}	c_{i2}	c_{i3}
$c_{1j} =$	0.145533	0.174908	0.013421
$c_{2j} =$	0.174909	0.262690	0.017870
$c_{3j} =$	0.013421	0.017870	0.012641

The regression coefficients are:

$$b_1. = (0.145533)(3.795016) + (0.174909)(-2.317256)\\ + (0.013421)(10.509764) = 0.28804$$

$$b_2. = (0.174909)(3.795016) + (0.262690)(-2.317256)\\ + (0.017870)(10.509764) = 0.24287$$

$$b_3. = (0.013421)(3.795016) + (0.017870)(-2.317256)\\ + (0.012641)(10.509764) = 0.14238$$

$$a = 1.262113 - (0.28804)(2.210485) - (0.24287)(3.004987)\\ - (0.14238)(-0.650749) = -0.011763$$

The final equation in the logarithmic form is

$$\log \text{Nu} = -0.011763 + 0.28804 \log \text{Re}\\ + 0.24287 \log \text{Pr} + 0.14238 \log \frac{\mu_a}{\mu_w}$$

And in the exponential form

$$\text{Nu} = 0.973(\text{Re})^{0.288}(\text{Pr})^{0.243}\left(\frac{\mu_a}{\mu_w}\right)^{0.142} \tag{9.52}$$

The generally accepted form of this equation is

$$\text{Nu} = 0.402(\text{Re})^{1/3}(\text{Pr})^{1/3}\left(\frac{\mu_a}{\mu_w}\right)^{0.14} \tag{9.53}$$

[1] The original data, consisting of 67 sets of measurements can be found in E. N. Sieder and G. E. Tate, *Ind. Eng. Chem.*, **28**:1429 (1936).

Discussion

Sum of squares of deviation from log mean removed by correlation

$$\Sigma' c^2 = (0.28804)(3.795016) + (0.24287)(-2.317256)$$
$$+ (0.14238)(10.509764) = 2.026705$$

Residual sum of squares of deviation $\Sigma' \hat{y}^2 = 2.171305 - 2.026705$
$$= 0.144600$$

$$\text{Variance of estimate } s^2(\hat{y}) = \frac{0.144600}{63} = 0.002295$$

Variances of the three regression coefficients are:

$$s^2(b_1.) = (0.002295)(0.145533) = 0.000334$$
$$s^2(b_2.) = (0.002295)(0.262690) = 0.000603$$
$$s^2(b_3.) = (0.002295)(0.012641) = 0.000029$$

The 95 per cent confidence range of these regression coefficients, taking t at the 95 per cent level for 63 degrees of freedom from Table 6.1 as 1.99, is

$$b_1. = 0.288 \pm (1.99)(0.0183) = 0.252 \text{ to } 0.324$$
$$b_2. = 0.243 \pm (1.99)(0.0246) = 0.194 \text{ to } 0.292$$
$$b_3. = 0.142 \pm (1.99)(0.0054) = 0.131 \text{ to } 0.153$$

The 95 per cent confidence ranges of $b_1.$ and $b_2.$, the exponents of Re and Pr, respectively, do not include the $\frac{1}{3}$ values usually accepted. If the 99 per cent value of t of 2.64 (from Table 6.1) is used to calculate the confidence ranges, these are then 0.240 to 0.336 for $b_1.$ and 0.179 to 0.307 for $b_2.$. From these confidence ranges we may make the following statement: these data indicate the Re exponent is not $\frac{1}{3}$, with between 0.01 and 0.05 chance of being wrong; the Pr exponent is not $\frac{1}{3}$, with less than 0.01 chance of being wrong.

9.2.3 Deletion of a Variable. Sometimes after a regression has been calculated it is evident that one of the independent variables could well be omitted. If the partial regression coefficient is not significantly different from zero, the variable may be deleted. A review of the experimental procedure may suggest the elimination of some poorly measured variable. A study of the regression may indicate that while a variable makes a significant contribution to the correlation, the contribution is small and the simplification of the equation by its elimination would overbalance the increase in residual variance.

The formulas which follow for deleting a variable may only be applied to one variable at a time. After a variable has been deleted, revised regression coefficients and revised Gaussian multipliers must be calculated before a second variable may be removed. However, the procedure is simple and may be repeated as often as is desired. The formulas that follow are for deleting one variable from a four-variable regression: one dependent and three independent variables. The application to any number of variables can be deduced by following a parallel procedure.

If the variable corresponding to the symbol x_2 is deleted, the reduction in the sum of squares of deviation attributable to the regression resulting from the deletion is equal to $b_{2 \cdot 13}{}^2/c_{22}$. (Why the full symbol for the regression coefficient is used will be apparent soon.) The revised sum of squares due to the regression is therefore $\Sigma' c_r{}^2 = \Sigma' c^2 - b_{2 \cdot 13}{}^2/c_{22}$.

It is possible from the above relationship to calculate the relative decrease in regression caused by the removal of any of the independent variables. The sum of all the b^2/c terms should not be expected to equal the value of $\Sigma' c^2$. The value of $b_{i \cdot}{}^2/c_{ii}$ is the decrease in the sum of squares of regression that would occur if the variable x_i were ignored, whereas the contribution of the $b_i \Sigma' x_i y$ term to $\Sigma' c^2$ in Eq. (9.37) is on the basis of holding the other variables constant. If there is correlation between two independent variables, the elimination of either one will have little effect on the over-all regression, and both b^2/c terms will be small even though the regression may be strongly dependent on one of these variables.

Also, in calculating $\Sigma' c^2$ from the sum of the $b_{i \cdot} \Sigma' x_i y$ terms, it may happen that one of these terms is negative, although the sum cannot be negative. It does not follow that elimination of the variable contributing the negative terms will improve the correlation by an equal amount. The significance of a negative term is that within groups in which the other independent variables are held constant the partial regression coefficient is of one sign, while for the regression disregarding the other independent variables, the simple regression coefficient is of the opposite sign.

example 9.10. Study the following table of imaginary data.

TABLE 9.24

y	x_1	x_2	x_3
8	1	2	5
9	1	2	6
12	4	5	3
13	4	5	4
14	6	7	1
15	6	7	2
18	8	10	0

$y = x_1 + x_2 + x_3$. In the pairs where x_1 and x_2 are constant, y increases as x_3 increases, while over the entire range, y increases as x_3 decreases. The individual partial regression coefficients are each 1.0, and the total sum of squares of deviation removed by the correlation is therefore $\Sigma' x_1 y + \Sigma' x_2 y + \Sigma' x_3 y$. The numerical values of these terms are:

$$\begin{aligned}\Sigma' x_1 y &= 53.57\\ \Sigma' x_2 y &= 58.85\\ \Sigma' x_3 y &= -41.00\\ \text{Total} &= 71.42 = \Sigma' c^2\end{aligned}$$

(The actual arithmetic is left to the interested student.)

The removal of the x_3 variable from the correlation would not increase the correlation by the amount of the negative value of $\Sigma' x_3 y$, but would actually result in a decrease in the sum of squares removed by 1.71, which is the value of the $b_{3\cdot}{}^2/c_{33}$ term.

The partial regression coefficient value of 1.0 for $b_{3\cdot}$ is the value for the rate of change in y with a change in x_3 when x_1 and x_2 are held constant. As is evident from the numbers in Table 9.24, if y were correlated against x_3 without regard for the other variables, a negative value of the simple regression coefficient, or slope, would be expected. In fact, $b_3 = -1.46$, while $b_{3\cdot} = 1.0$.

The formulas for deleting a variable and for calculating a revised regression based on the remaining variables are as follows. These formulas are written for removing x_2 from a regression of y upon x_1, x_2, and x_3. The corresponding formulas can be applied to the removal of any variable from any group. If two variables are to be removed, revised partial regression coefficients and Gaussian multipliers must be calculated after the removal of the first variable before removing the second.

$$\text{Decrease in sum of squares of regression} = \frac{b_{2\cdot 13}{}^2}{c_{22}}$$

$$\text{Revised sum of squares of regression, } \Sigma' c_r{}^2 = \Sigma' c^2 - \frac{b_{2\cdot 13}{}^2}{c_{22}}$$

$$\text{Revised value of } c_{11}, (c_{11})_r = c_{11} - \frac{c_{12}{}^2}{c_{22}}$$

$$\text{Revised value of } c_{33}, (c_{33})_r = c_{33} - \frac{c_{23}{}^2}{c_{22}}$$

$$\text{Revised value of } c_{13}, (c_{13})_r = c_{13} - \frac{c_{12} c_{23}}{c_{22}}$$

Revised value of partial regression coefficients:

$$b_{1\cdot 3} = b_{1\cdot 23} - c_{12}\frac{b_{2\cdot 13}}{c_{22}}$$

$$b_{3\cdot 1} = b_{3\cdot 12} - c_{32}\frac{b_{2\cdot 13}}{c_{22}}$$

Removal of a variable from a regression increases by 1 the number of degrees of freedom. A test for the significance of a variable is the ratio of the difference in the sum of squares of regression when this variable is included, $\Sigma' c_r^2 - \Sigma' c^2$ with 1 degree of freedom, and the residual variance of estimate $s^2(\hat{y}) = \Sigma' \hat{y}^2/(N - k - 1)$ at $(N - k - 1)$ degrees of freedom. If the ratio of these two quantities is larger than the corresponding F value at 1 and $(N - k - 1)$ degrees of freedom, then the variable can be considered to make a significant contribution to the correlation with the indicated probability level of being wrong.

example 9.11. We return to the data of example 9.9 from the heat-transfer correlation to see the effect of the removal of one of the variables. The partial regression coefficients, the Gaussian multipliers, and sums of squares of deviation from this correlation were:

TABLE 9.25

$b_1. = 0.28804$	$c_{11} = 0.14553$
$b_2. = 0.24287$	$c_{22} = 0.26269$
$b_3. = 0.14238$	$c_{33} = 0.01264$
$\Sigma' c^2 = 2.02671$	$c_{21} = 0.17491$
$\Sigma' \hat{y}^2 = 0.14460$	$c_{23} = 0.01787$

The loss in sum of squares of correlation corresponding to the deletion of each variable is:

$$\text{Re}, x_1 = \frac{b_1.^2}{c_{11}} = \frac{(0.28804)^2}{0.14553} = 0.5701 = 26.3\%$$

$$\text{Pr}, x_2 = \frac{b_2.^2}{c_{22}} = \frac{(0.24287)^2}{0.26269} = 0.2245 = 10.3\%$$

$$\frac{\mu_a}{\mu_w}, x_3 = \frac{b_3.^2}{c_{33}} = \frac{(0.14238)^2}{0.01264} = 1.6038 = 73.9\%$$

The percentages are of the total squared log deviations, $\Sigma' y^2$. The four-variable correlation accounted for 93.4 per cent; that is, $\Sigma' c^2/\Sigma' y^2 = 2.0267/2.1713 = 0.934$, of the total squared deviation of log Nu. The above tabulation means that if the x_1 variable Re was omitted, the correlation would be reduced to $93.4 - 26.3 = 67.1$ per cent of the variation in log Nu. If the x_2 variable Pr were omitted and Re were retained, then $93.4 - 10.3 = 83.1$ per cent of the variation in log Nu would be accounted for. The question the engineer would have to answer is whether the simplification of the correlation to two independent variables which account for 83 per cent of the variation is better than the more complicated correlation with three independent variables that accounts for 93 per cent. If the only criterion for retaining the variable is significance of correlation, then the x_2 variable would not be discarded. The following F test shows it to be highly significant.

TABLE 9.26

	Sum of squares	D.F.	Mean square
x_2 variable..................	0.2245	1	0.2245
Residual, $\Sigma'\hat{y}^2$.............	0.1446	63	0.0023
$\Sigma'c_r^2$; x_1, x_3 correlation.....	1.8022	2	0.9011
Total, $\Sigma'y^2$..............	2.1713	66	

$$F = \frac{0.2245}{0.0023} = 97.6***$$

$$F_{0.001,1,60} = 12.0$$

If, for the sake of simplicity, the x_2 variable is deleted, the revised partial regression coefficients are calculated as follows:

$$b_{1\cdot 3} = 0.2880 - \frac{(0.1749)(0.2429)}{0.2627} = 0.1263$$

$$b_{3\cdot 1} = 0.1424 - \frac{(0.0179)(0.2429)}{0.2627} = 0.1279$$

The revised variance of estimate is the new residual divided by the new degrees of freedom:

$$s^2(\hat{y})_r = \frac{0.1446 + 0.2245}{64} = 0.0058$$

The revised c terms are:

$$(c_{11})_r = 0.1455 - \frac{(0.1749)^2}{0.2627} = 0.0291$$

$$(c_{33})_r = 0.0126 - \frac{(0.0179)^2}{0.2627} = 0.0114$$

The variances and confidence ranges of the revised partial regression coefficients are:

$s^2(b_{1 \cdot 3}) = (0.0058)(0.0291) = 0.00017$

$s^2(b_{3 \cdot 1}) = (0.0058)(0.0114) = 0.00006$

$b_{1 \cdot 3} = 0.1263 \pm (1.99)(0.013) = 0.1005$ to 0.1521

$b_{3 \cdot 1} = 0.1279 \pm (1.99)(0.003) = 0.1219$ to 0.1339

The Re exponent without the Pr factor in the correlation at the 95 per cent confidence level is of the order of 0.1 instead of 0.3 of the first correlation, and the exponent of the viscosity terms is now just below the 0.14 value of the original four-variable correlation.

9.2.4 Correlation Coefficients. The statistic for testing the significance of a simple two-variable linear correlation has been defined in Eq. (9.14) and is the correlation coefficient r. r^2 is the fraction of the sum of squares of deviation that is removed by the correlation. The corresponding term for a multiple correlation is designated R^2 and has the same relation as r^2 for the simple correlation.

$$R^2 = \frac{\Sigma' c^2}{\Sigma' y^2} = 1 - \frac{\Sigma' \hat{y}^2}{\Sigma' y^2} \tag{9.54}$$

The F test for a multiple correlation indicated in Eq. (9.40) can be rewritten in terms of R^2, since $\Sigma' c^2 = R^2 \Sigma' y^2$ and $\Sigma' \hat{y}^2 = (1 - R^2)\Sigma' y^2$, to give

$$F = \frac{R^2(N - k - 1)}{(1 - R^2)k} \tag{9.55}$$

From this relation it is possible to tabulate significant values of R for various combinations of k and N from the F table. Such a table is available,[1] but with a multiple correlation other variance tests are usually desired, so that the direct F test in the form of Eq. (9.40) or (9.55) is very readily performed.

The correlation coefficient for a two-variable system as previously given is

$$r_{yx} = \frac{\Sigma' xy}{(\Sigma' x^2 \Sigma' y^2)^{1/2}} \tag{9.14}$$

This same type of relation holds for any two variables in a multiple correlation whether they be the dependent or the independent variables. r_{yx_1}, r_{yx_2}, $r_{x_1x_2}$, etc., represent the correlation coefficients for the variables

---

[1] G. W. Snedecor, "Statistical Methods," 5th ed., Iowa State College Press, Ames, Iowa, 1956.

indicated by the subscripts and are evaluated by the numerical value of the corresponding ratio, $\Sigma' x_1 y/(\Sigma' x_1{}^2 \Sigma' y^2)^{1/2}$, $\Sigma' x_2 y/(\Sigma' x_2{}^2 \Sigma' y^2)^{1/2}$, etc. The significance of any of these r terms can be determined by reference to Table 9.2 at $N-2$ degrees of freedom, in the same way that the two-variable linear correlation coefficient was tested.

The multiple correlation coefficient can be expressed in terms of the simple correlation coefficients in the form

$$R = \left[r_{y_1} b_1 \cdot \left(\frac{\Sigma' x_1{}^2}{\Sigma' y^2} \right)^{1/2} + r_{y_2} b_2 \cdot \left(\frac{\Sigma' x_2{}^2}{\Sigma' y^2} \right)^{1/2} + \cdots \right]^{1/2} \qquad (9.56)$$

The simple correlation coefficient r can be either positive or negative depending on the sign of the slope. R, however, is taken only as positive as the negative value, for a multiple correlation would have no significance. R from Eq. (9.56) can have any value from 0 to 1, but must be greater than any of the simple correlation coefficients, r_{yx_1}, r_{yx_2}, etc. R need not be larger than the correlation coefficient between independent variables, $r_{x_1 x_2}$, $r_{x_2 x_3}$, etc. If there is strong correlation between the independent variables, evidenced by a relatively large value of $r_{x_1 x_2}$, then the multiple correlation coefficient will not be much larger than the simple linear correlation coefficient with either variable separately.

If, for example, three variables have correlation coefficients:

$$r_{yx_1} = 0.708$$
$$r_{yx_2} = 0.714$$
$$r_{x_1 x_2} = 0.836$$

where the 0.836 value for $r_{x_1 x_2}$ indicates a high degree of correlation between the two independent variables, the multiple correlation coefficient R for y upon x_1 and x_2 will only equal 0.738, not much better than for either independent variable taken separately. If, on the other hand, the correlation coefficient for the two independent variables $r_{x_1 x_2}$ was fairly small, say 0.190, then the multiple correlation coefficient would be 0.922, larger than for either simple correlation coefficient taken separately. In other words, where there is good correlation between two independent variables, not much is gained by having a multiple correlation in terms of both of them. When there is little correlation between two independent variables, a multiple correlation containing them both will be improved over a simple correlation on either one.

example 9.12. Refer to the data of example 9.9, the multiple heat-transfer correlation. x_1 was log Re, x_2 was log Pr, and x_3 was log (μ_a/μ_w). The sums of squares and cross products for these three independent variables are:

$\Sigma' x_1^2 = 34.639265 \quad \Sigma' x_1 x_2 = -22.750174$

$\Sigma' x_2^2 = 19.153509 \quad \Sigma' x_1 x_3 = -4.615744$

$\Sigma' x_3^3 = 88.139543 \quad \Sigma' x_2 x_3 = -2.922539$

Calculation of the three correlation coefficients is as follows:

$r_{12} = -22.750174/\sqrt{(34.639265)(19.153509)} = -0.883$

$r_{23} = -2.922539/\sqrt{(88.139543)(19.153509)} = -0.771$

$r_{13} = -4.615744/\sqrt{(88.139543)(34.639265)} = -0.083$

The relatively high value of correlation coefficient between log Re and log Pr, r_{12}, indicates that a correlation on both of these terms as independent variables will not be a large improvement over a correlation on only one of them. This is the same conclusion established quantitatively in example 9.11 wherein the relative contribution of each variable was evaluated.

Corresponding to the partial regression coefficients $b_i.$, and having a relation to them similar to that which r has to b, are the partial correlation coefficients, designated $r_{yi.}$, the dot again signifying that this is the correlation coefficient of y upon the ith variable when all the other variables are held constant. The partial correlation coefficients are used more often in agricultural and biological work than in engineering, but these few comments are included for the sake of completeness.

The partial correlation coefficients can be calculated from the total correlation coefficients by the following relation for three variables:

$$r_{12.3} = \frac{r_{12} - r_{13} r_{23}}{\sqrt{(1 - r_{13}^2)(1 - r_{23}^2)}} \qquad (9.57)$$

Parallel formulas apply for the other partial correlation coefficients. If more than three variables are involved, the partial correlation coefficients may be calculated stepwise by eliminating one variable at a time. For example, if four variables are involved, $r_{12.4}$, $r_{13.4}$, and $r_{23.4}$ are first calculated from formulas similar to Eq. (9.57). The second variable is then eliminated in a similar manner:

$$r_{12.34} = \frac{r_{12.4} - r_{13.4} r_{23.4}}{\sqrt{(1 - r_{13.4}^2)(1 - r_{23.4}^2)}} \qquad (9.58)$$

In cases where the data are not under the control of the experimenter, for example in studying cloud formation, wind velocity, and rainfall, the correlation of one variable on another when the third is held constant can be calculated from the simple correlation coefficients of Eqs. (9.57) and (9.58). Even in cases where the data can be obtained at conditions set by the experimenter, it is usually more advantageous to vary the operating conditions over the range of interest and calculate the partial correlation coefficients at constant values of some of the variables than to hold the operating conditions constant in order to gather the necessary data. In the latter case, only those data at constant conditions would be available for the calculation, while in the former all the data are used to calculate the partial correlation coefficients.

The information conveyed by the partial correlation coefficients depends of course on the type of correlation involved. The significance of a partial correlation coefficient may be tested with Table 9.2 using $N - k - 1$ degrees of freedom, where k is the number of independent variables. However, the significance of the partial correlation coefficient must be examined in relation to the other correlation coefficients. If two independent variables are highly correlated with each other, then the partial correlation coefficient of the dependent variable with either of the independent variables will be small, even though the over-all multiple correlation is large. This situation develops from the fact that if one independent variable is held constant and it is correlated with the other independent variable, both will be constant and the partial correlation

TABLE 9.27

	Case I High correlation between independent variables	Case II Poor correlation between independent variables
r_{yx_1}	0.708	0.708
r_{yx_2}	0.714	0.714
$r_{x_1x_2}$	0.836	0.190
	Partial correlation coefficients	
$r_{yx_1 \cdot 2}$	0.314	0.836
$r_{yx_2 \cdot 1}$	0.289	0.833
	Multiple correlation	
R	0.738	0.922

will be small. On the other hand, if two independent variables are not correlated, then the partial correlation of the dependent variable with either one separately will be relatively high compared to the over-all multiple correlation. Table 9.27 shows this relation for the two sets of simple correlation coefficients given earlier in this section.

9.2.5 Correlation with Two Independent Variables. The most frequent multiple correlation that the engineer will encounter is one with one dependent variable and two independent variables. The general formulas developed in Sec. 9.2.1 for more than two variables apply equally well to a correlation with two independent variables. However, because of the frequency of its use and because of the relative simplicity of the equations, Table 9.28 is presented which gives partial regression coefficients and other factors of the three-variable equation in terms of the sums of squares of the data.

TABLE 9.28

Equation: $\hat{y} = a + b_1.x_1 + b_2.x_2$

Partial regression coefficients: $b_1. = \dfrac{\Sigma' y x_1 \Sigma' x_2^2 - \Sigma' y x_2 \Sigma' x_1 x_2}{\Sigma' x_1^2 \Sigma' x_2^2 - (\Sigma' x_1 x_2)^2}$

$b_2. = \dfrac{\Sigma' y x_2 \Sigma' x_1^2 - \Sigma' y x_1 \Sigma' x_1 x_2}{\Sigma' x_1^2 \Sigma' x_2^2 - (\Sigma' x_1 x_2)^2}$

Intercept: $a = \bar{y} - b_1.\bar{x}_1 - b_2.\bar{x}_2$

Correlation coefficient: $R = \left(\dfrac{b_1.\Sigma' y x_1 + b_2.\Sigma' y x_2}{\Sigma' y^2} \right)^{1/2}$

Sum of squares of regression: $\Sigma' c^2 = b_1.\Sigma' y x_1 + b_2.\Sigma' y x_2$ D.F. = 2

Residual sum of squares: $\Sigma' \hat{y}^2 = \Sigma' y^2 - \Sigma' c^2$ D.F. = $N - 3$

Variance of estimate: $s^2(\hat{y}) = \dfrac{\Sigma' \hat{y}^2}{N - 3}$

Gaussian multipliers: $c_{11} = \dfrac{\Sigma' x_2^2}{\Sigma' x_1^2 \Sigma' x_2^2 - (\Sigma' x_1 x_2)^2}$

$c_{22} = \dfrac{\Sigma' x_1^2}{\Sigma' x_1^2 \Sigma' x_2^2 - (\Sigma' x_1 x_2)^2}$

$c_{12} = \dfrac{\Sigma' x_1 x_2}{\Sigma' x_1^2 \Sigma' x_2^2 - (\Sigma' x_1 x_2)^2}$

Variance of regression coefficients: $s^2(b_1.) = s^2(\hat{y}) c_{11}$

$s^2(b_2.) = s^2(\hat{y}) c_{22}$

Variance of an average estimated value of y:

$$s^2(\hat{y}_i) = s^2(\hat{y}) \left[\dfrac{1}{N} + c_{11}(x_1 - \bar{x}_1)^2 + c_{22}(x_2 - \bar{x}_2)^2 + 2 c_{12}(x_1 - \bar{x}_1)(x_2 - \bar{x}_2) \right]$$

The use of these equations is illustrated in example 9.13, which follows. The analysis-of-variance test for the significance of the regression and the other remarks about multiple regressions all apply to the case of two independent variables solved by means of the equations in Table 9.28. If after the regression has been calculated one of the variables is to be deleted, it can be done either by the equations of Sec. 9.2.3, or perhaps more directly, by calculating the simple regression coefficient b as $\Sigma'xy/\Sigma'x^2$.

example 9.13. The following data give gasoline octane numbers from a correlation against catalyst purity and weight per cent carbon on the catalyst. In the left columns are the original data, and in the right columns are the data coded for ease of calculation. From each of the octane numbers 80 has been subtracted; the purities have all been subtracted from 100, and the carbon levels are repeated in their original form.

TABLE 9.29

Octane number y'	Purity, % x'_1	Carbon, W, % x_2	$y =$ $(y' - 80)$	$x_1 =$ $(100 - x'_1)$	x_2
88.6	99.8	3	8.6	0.2	3
88.4	99.7	10	8.4	0.3	10
87.2	99.6	7	7.2	0.4	7
88.4	99.5	2	8.4	0.5	2
87.2	99.4	5	7.2	0.6	5
86.8	99.3	6	6.8	0.7	6
86.1	99.2	8	6.1	0.8	8
87.3	99.1	3	7.3	0.9	3
86.4	99.0	5	6.4	1.0	5
86.6	98.9	4	6.6	1.1	4
87.1	98.8	2	7.1	1.2	2

Calculation

$N = 11$

$\Sigma'y^2 = 6.32000 \quad\quad \Sigma'x_1x_2 = -3.0000 \quad\quad \bar{y} = 7.1000$

$\Sigma'x_1^2 = 1.1000 \quad\quad \Sigma'x_1y = -1.2400 \quad\quad \bar{x}_1 = 0.7000$

$\Sigma'x_2^2 = 66.0000 \quad\quad \Sigma'x_2y = -13.4000 \quad\quad \bar{x}_2 = 5.0000$

$$b_1. = \frac{(-1.24)(66.0) - (-13.4)(-3.0)}{(1.1)(66.0) - (9.0)} = -1.9188$$

$$b_2. = \frac{(-13.4)(1.1) - (-1.24)(-3.0)}{(1.1)(66.0) - (9.0)} = -0.2903$$

$$a = 7.1 - (-1.9188)(0.7) - (-0.2903)(0.5) = 9.8947$$

$$R = \left[\frac{(1.9188)(1.24) + (0.2903)(13.4)}{6.32}\right]^{1/2} = 0.9959$$

$$\Sigma'c^2 = (1.9188)(1.24) + (0.2903)(13.4) = 6.2694$$

$$\Sigma'\hat{y}^2 = 6.32 - 6.2694 = 0.0506$$

$$s^2(\hat{y}) = \frac{0.0506}{11 - 3} = 0.0063$$

$$c_{11} = \frac{66.0}{(1.1)(66.0) - (9.0)} = 1.0377$$

$$c_{22} = \frac{1.10}{(1.1)(66.0) - (9.0)} = 0.0173$$

$$c_{12} = \frac{-3.0}{(1.1)(66.0) - (9.0)} = -0.0472$$

$$s^2(b_1.) = (0.0063)(1.0377) = 0.0065$$

$$s^2(b_2.) = (0.0063)(0.0173) = 0.0001$$

Discussion. The equation of the multiple correlation line in terms of the coded variables is

$$\hat{y} = 9.8947 - 1.9188x_1 - 0.2903x_2$$

Substituting the original variables, we get

$$\hat{y}' = 80 + 9.9847 - 1.9188(100 - x'_1) - 0.2903x_2$$
$$= -101.99 + 1.92x'_1 - 0.29x_2$$

With a t value of 2.306 from Table 6.1 for 8 degrees of freedom at the 0.05 level, the 95 per cent confidence ranges on the regression coefficients are:

$$b_1. = -1.91 \pm (2.306)(0.08) = -1.91 \pm 0.18$$
$$b_2. = -0.29 \pm (2.306)(0.01) = -0.29 \pm 0.02$$

The correlation is obviously significant with an R value of 0.9959. The F test for this significance is:

TABLE 9.30

Sum of squares	D.F.	Mean square
$\Sigma'c^2 = 6.2694$	2	3.1347
$\Sigma'\hat{y}^2 = 0.0506$	8	0.0063
$\Sigma'y^2 = 6.3200$	10	

$$F = \frac{3.1347}{0.0063} = 497.5^{***}$$

$F_{0.001,2,8} = 18.5$

The 95 per cent confidence range of an average octane number at carbon level 7 per cent, purity 99 per cent, would be:

$$\hat{y}_i = 101.99 + (1.92)(99) - (0.29)(7) = 86.06$$
$$x'_{1i} = 99.0, \; x_{2i} = 7.0$$
$$s^2(\hat{y}_i) = (0.0063)[1/11 + 1.0377(99 - 99.3)^2 + 0.0173(7 - 5)^2$$
$$- 2(0.0472)(99 - 99.3)(7 - 5)] = 0.001954$$
$$s(\hat{y}_i) = 0.0442$$
$$\hat{y}_i = 86.06 \pm (2.306)(0.0442) = 86.16 \text{ to } 85.96$$

9.3 Curvilinear Regression

The regression formulas discussed up to now were all in the form of a linear relation between the dependent variables and the independent variable, i.e., first-order functions of the independent variable or variables. This happy relation, as every engineer knows who has plotted data, does not always give a satisfactory correlation. If an analytic expression relating the variables is desired, it is often possible to improve a correlation by adding second- and third-degree terms of the independent variables. The next three sections of this chapter describe three different methods of solving curvilinear regressions. The first method, described in Sec. 9.3.1, is a general method that is applicable to any equation of the form

$$\hat{y} = \sum_{i=0}^{n} \sum_{j=0}^{m} b_{ij} x_i^j \tag{9.59}$$

The second and third methods are in fact two modifications of a single method developed by Fisher[1] applicable to equations with one independent variable of the form

$$\hat{y} = \sum_{j=0}^{m} b_j x^j \qquad (9.60)$$

These last two methods can be applied only when the independent variable is taken at some regular intervals which can be coded to give a series of positive integers, 1, 2, 3, etc. The simplicity and advantage of these latter two methods, outlined in Secs. 9.3.2 and 9.3.3, will make it evident that the work of subsequent analysis will be greatly lessened if the original data are taken at regular intervals of the independent variable.

The numerical example illustrating each of the methods of solution of curvilinear regression will deal with the same data to show the extent of agreement between the methods.

9.3.1 General Solution to Curvilinear Regressions. The method of multivariable regression of Sec. 9.2.1 can readily be applied to curvilinear regression if the higher-degree functions of the independent variable are considered as separate variables. For instance, Eq. (9.34) can be written in terms of one variable, as

$$\hat{y} = a + b_1 \cdot x + b_2 \cdot x^2 + b_3 \cdot x^3 + \cdots \qquad (9.61)$$

Equation (9.61) can be solved in the same manner as Eq. (9.34), either directly, using Eqs. (9.35), or by use of the Gaussian multipliers. For a solution of this type all the sums of higher-degree x values and cross products, $\Sigma x x^2$, $\Sigma x^2 x^3$, etc., must be obtained. These terms are not necessary where the x values are taken at regular intervals.

The regression equation need not be in terms of one independent variable. The nature of the equation will of course depend upon the data involved. If more than two independent variables are involved and a regression desired at one or more powers of the variables (not necessarily the same degree for all variables), then the general method of solution will have to be employed. The simplified methods are given for only one independent variable.

Example (9.14) gives the general solution for an equation of the second degree with one independent variable, using the formulas of Table 9.28. The extension of this procedure to more involved correlations will follow from the preceding discussion of multiple correlation.

---

[1] R. A. Fisher, "Statistical Methods for Research Workers," Oliver & Boyd Ltd., Edinburgh and London, 1948.

TABLE 9.31

Pressure, psi P	x	x^2	Gas volume–total volume ratio y
200	2	4	0.846
400	4	16	0.573
600	6	36	0.401
800	8	64	0.288
1,000	10	100	0.209
1,200	12	144	0.153
1,400	14	196	0.111
1,600	16	256	0.078

example 9.14. The data in Table 9.31, extracted from a paper by Kay and Stern,[1] give the vapor-phase volume to total volume ratio for different equilibrium pressures at 105°C for nitric acid.

[1] Webster B. Kay and S. Alexander Stern, *Ind. Eng. Chem.*, **47**:1463 (1955).

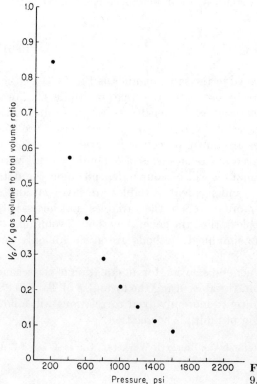

Fig. 9.7 Data from example 9.14, rectangular coordinates.

CORRELATION—REGRESSION

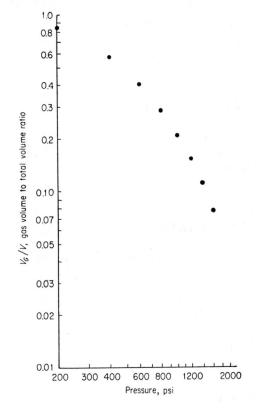

Fig. 9.8 Data from example 9.14, logarithmic coordinates.

The data are plotted in Fig. 9.7 on regular coordinate axes and in Fig. 9.8 on log-log axes. Both of these plots show marked deviation from linearity so that neither a simple linear equation of the form $y = a + bx$ nor an exponential equation of the form $y = ax^b$ can be expected to be completely satisfactory. An equation of the form $y = a + bx + cx^2$ will be fitted.

To simplify the arithmetic of the solution, the original pressure data have been coded by dividing by 100 to give the values tabulated in the column headed x.

Solution

$$N = 8 \qquad \Sigma y = 2.659 \qquad \Sigma x^2 = 816$$
$$\Sigma x = 72 \qquad \bar{y} = 0.3324 \qquad (\bar{x}^2) = 102$$
$$\bar{x} = 9 \qquad \Sigma y^2 = 1.373285 \qquad \Sigma(x^2)^2 = 140{,}352$$
$$\Sigma x^2 = 816 \qquad \Sigma' y^2 = 0.489500 \qquad \Sigma'(x^2)^2 = 57{,}120$$
$$\Sigma' x^2 = 168 \qquad \Sigma x^2 y = 130.076 \qquad \Sigma xx^2 = 10{,}368$$
$$\Sigma xy = 15.422 \qquad \Sigma' x^2 y = -141.142 \qquad \Sigma' xx^2 = 3.024$$
$$\Sigma' xy = -8.509$$

$$b = \frac{(-8.509)(57{,}120) - (-141.142)(3{,}024)}{(168)(57{,}120) - (3{,}024)^2} = -0.131{,}140$$

$$c = \frac{(-141.142)(168) - (-8.509)(3{,}024)}{(168)(57{,}120) - (3{,}024)^2} = 0.004{,}472$$

$$a = 0.3324 - (-0.131140)(9) - (0.004472)(102) = 1.0565$$

Calculated equation, in coded form:

$$\hat{y} = 1.0565 - 0.131140x + 0.004472x^2$$

In terms of the original pressure data P, where $x = P/100$,

$$\hat{y} = 1.0565 - (1.311)(10^{-3})P + (4.472)(10^{-7})P^2$$

The sum of squares removed by the regression, $\Sigma'c^2$, is equal to $(-0.131140)(-8.509) + (0.004472)(-141.142) = 0.484683$.

The fraction of the total sum of squares removed by the correlation, R^2, is equal to $0.484683/0.489500 = 0.9901$, i.e., 99 per cent of the sum of squares of deviation is accounted for by the second-degree regression.

Discussion. If it is desired to compare the second-degree regression with a simple linear first-degree regression, this can readily be done by an analysis-of-variance test. The sum of squares removed by the linear regression, $\Sigma'c_1^2 = (\Sigma'xy)^2/\Sigma'x^2$ (Eq. 9.12) equals

$$(-8.502)^2/168 = 0.430971$$

The difference between this quantity and the sum of squares of the second-degree regression, 0.484683, with 1 degree of freedom is com-

TABLE 9.32

	Sum of squares	D.F.	Mean square
$\Sigma'c^2$ for second-degree equation =	0.484683	2	
$\Sigma'c_1^2$ for linear equation =	0.430971	1	
Difference =	0.053712	1	0.053712
$\Sigma'\hat{y}^2$ residual from second-degree equation =	0.004817	5	0.000963
$\Sigma'y^2$ total....................	0.489500	7	

$$F = \frac{53{,}712}{963} = 55.7^{***}$$

$$F_{0.001,1,5} = 47.0$$

pared with the residual sum of squares of the second-degree regression with 5 degrees of freedom.

The analysis-of-variance table indicates that the additional sum of squares removed by the second-degree regression over a first-degree linear regression is greater than would be expected by chance unless the one-in-a-thousand event has occurred.

9.3.2 Curvilinear Regression by Orthogonal Polynomials.

Orthogonality has a precise meaning in mathematics. The background needed for this precise definition is beyond the scope of this text. For our purposes, if we have a polynomial in the form:

$$y = b_0 + b_1 x + b_2 x^2 + b_3 x^3 + \cdots \tag{9.61}$$

and we can find a function, ξ, $\xi_1(x)$, $\xi_2(x^2)$, etc., such that when we substitute ξ for x in Eq. (9.61) we obtain a polynomial

$$y = B_0 + B_1 \xi_1 + B_2 \xi_2 + B_3 \xi_3 + \cdots \tag{9.62}$$

in which the coefficients B are independent, the ξ functions are orthogonal. The usefulness of this is that with the substitution of the orthogonal polynomials the coefficients of each term of Eq. (9.62) can be calculated independently of the other terms. In other words, in the orthogonal form, the first-degree correlation can be calculated, and then the second-degree correlation, and subsequently higher degrees, each regression coefficient being independent of the others.

For the case where the x values are at regular intervals, as with data taken in time series, or in equal increments of temperature or pressure, the values of the orthogonal polynomials were introduced by Fisher.[1] These values permit the stepwise solution of Eq. (9.61) so that the sum of squares removed by each successive degree of x can be found and the calculation terminated at any level when the regression is deemed sufficiently significant.

A discussion of this method can be found, besides in the original presentation of Fisher, in statistics texts by Goulden,[2] Snedecor,[3] Anderson and Bancroft,[4] and Bennett and Franklin,[5] to name only a few. We shall limit ourselves to a description of the solution method with numerical examples. Two types of solution will be presented. The method of this

[1] R. A. Fisher, "Statistical Methods for Research Workers," Oliver & Boyd Ltd., Edinburgh and London, 1948.
[2] C. H. Goulden, "Methods of Statistical Analysis," 2d ed., John Wiley & Sons, Inc., New York, 1952.
[3] *Op. cit.*
[4] R. L. Anderson and T. A. Bancroft, "Statistical Theory in Research," McGraw-Hill Book Company, New York, 1952.
[5] C. A. Bennett and N. L. Franklin, "Statistical Analysis in Chemistry and the Chemical Industry," John Wiley & Sons, Inc., New York, 1954.

section makes use of tables of the orthogonal polynomials of Fisher and Yates,[1] of which a small part is reproduced in Table 9.33. The original tables of the cited reference contain the terms for the solution of fifth-order equations for data containing up to 75 points.

In the next section the method of calculating the polynomials is presented for cases when the tables are not available. The use of the tables is by far the simpler of the two.

The solution will be obtained in the form of Eq. (9.62). The procedure is as follows:

Step 1. The dependent variable is tabulated in order of increasing value of the independent variable.

$$x = 1, 2, 3, 4, 5, 6, 7, 8, \ldots, n$$

$$y = y_1, y_2, y_3, y_4, y_5, y_6, y_7, y_8, \ldots, y_n$$

(If all that is desired is the degree of the correlation that best fits the data without actually calculating the regression coefficients, then it is not necessary actually to code the x values to give the series of positive integers as long as such coding is possible. On the other hand, if the actual regression equation is desired, it will be necessary actually to solve the equation in terms of the coded x's and subsequently to decode into the form of the original data.)

Step 2. The constants of the equation in the orthogonal form, Eq. (9.62), are obtained from the tabulated h's of Table 9.33, or from the Fisher and Yates tables if there are more than 10 data points, by the following relations:

$$B_0 = \bar{y}$$

$$B_1 = \frac{\sum_{j=1}^{n} h_{1j} y_j}{\sum_{j=1}^{n} h_{1j}^2} \lambda_1$$

$$B_2 = \frac{\sum_{j=1}^{n} h_{2j} y_j}{\sum_{j=1}^{n} h_{2j}^2} \lambda_2 \qquad (9.63)$$

$$B_i = \frac{\sum_{j=1}^{n} h_{ij} y_j}{\sum_{j=1}^{n} h_{ij}^2} \lambda_i$$

[1] R. A. Fisher and Frank Yates, "Statistical Tables," 4th ed., Hafner Publishing Company, Inc., New York, 1953.

CORRELATION—REGRESSION

TABLE 9.33 Orthogonal Polynomial Calculation Table†

h_{ij}

No. of data points, n	i	$j=1$	2	3	4	5	6	7	8	9	10	$\sum_j h_i^2$	λ_i
3	1	−1	0	1								2	1
	2	1	−2	1								6	3
4	1	−3	−1	1	3							20	2
	2	1	−1	−1	1							4	1
	3	−1	3	−3	1							20	10/3
5	1	−2	−1	0	1	2						10	1
	2	2	−1	−2	−1	2						14	1
	3	−1	2	0	−2	1						10	5/6
	4	1	−4	6	−4	1						70	35/12
6	1	−5	−3	−1	1	3	5					70	2
	2	5	−1	−4	−4	−1	5					84	3/2
	3	−5	7	4	−4	−7	5					180	5/3
	4	1	−3	2	2	−3	1					28	7/12
	5	−1	5	−10	10	−5	1					252	21/10
7	1	−3	−2	−1	0	1	2	3				28	1
	2	5	0	−3	−4	−3	0	5				84	1
	3	−1	1	1	0	−1	−1	1				6	1/6
	4	3	−7	1	6	1	−7	3				154	7/12
	5	−1	4	−5	0	5	−4	1				84	7/20
8	1	−7	−5	−3	−1	1	3	5	7			168	2
	2	+7	1	−3	−5	−5	−3	1	7			168	1
	3	−7	5	7	3	−3	−7	−5	7			264	2/3
	4	7	−13	−3	9	9	−3	−13	7			616	7/12
	5	−7	23	−17	−15	15	17	−23	7			2184	7/10
9	1	−4	−3	−2	−1	0	1	2	3	4		60	1
	2	28	7	−8	−17	−20	−17	−8	7	28		2772	3
	3	−14	7	13	9	0	−9	−13	−7	14		990	5/6
	4	14	−21	−11	9	18	9	−11	−21	14		2002	7/12
	5	−4	11	−4	−9	0	9	4	−11	4		468	3/20
10	1	−9	−7	−5	−3	−1	1	3	5	7	9	330	2
	2	6	2	−1	−3	−4	−4	−3	−1	2	6	132	1/2
	3	−42	14	35	31	12	−12	−31	−35	−14	42	8580	5/3
	4	18	−22	−17	3	18	18	3	−17	−22	18	2860	5/12
	5	−6	14	−1	−11	−6	6	11	1	−14	6	780	1/10

i = polynomial index, $B_i\xi_i$; $b_i x^i$
j = variable order index, x_j, y_j
n = number of observations, $j = 1, n$

† Abridged from Table XXIII of "Statistical Tables for Biological, Agricultural and Medical Research," 4th ed., 1953, by R. A. Fisher and Frank Yates, reprinted with permission of Oliver & Boyd, Ltd., Edinburgh and London, publishers.

Step 3. The sums of squares removed by the successive degrees of the linear correlation can be calculated from the B terms, or directly from the tabulated h values. If we designate $\Sigma'c_1{}^2$ as the sum of squares removed by the first degree term, $\Sigma'c_2{}^2$ as the *additional* sum of squares removed by the second degree term, and $\Sigma'c_i{}^2$ as the additional sum of squares removed by the ith term over that removed by all of the terms up to and including the $i - 1^{\text{th}}$ term, these sums of squares are calculated as follows:

$$\Sigma'c_1{}^2 = \frac{B_1 \sum_{j=1}^{n} h_{1j}y_j}{\lambda_1} = \frac{\left(\sum_{j=1}^{n} h_{1j}y_j\right)^2}{\sum_{j=1}^{n} h_{1j}{}^2}$$

$$\Sigma'c_2{}^2 = \frac{B_2 \sum_{j=1}^{n} h_{2j}y_j}{\lambda_2} = \frac{\left(\sum_{j=1}^{n} h_{2j}y_j\right)^2}{\sum_{j=1}^{n} h_{2j}{}^2} \qquad (9.64)$$

$$\Sigma'c_i{}^2 = \frac{B_i \sum_{j=1}^{n} h_{ij}y_j}{\lambda_i} = \frac{\left(\sum_{j=1}^{n} h_{ij}y_j\right)^2}{\sum_{j=1}^{n} h_{ij}{}^2}$$

By means of these relations, and since each successive degree involves the use of 1 degree of freedom, the significance of each step in the regression can be tested, and when the incremental sum of squares is no longer significant compared to the residual at the corresponding degrees of freedom, the further expansion of the equation can stop. This is not a hard and fast rule, as sometimes the higher-degree term may account for a greater portion of the sum of squares than a low degree. Judgment has to be exercised, depending on the size of the residual sum of squares compared with the initial $\Sigma'y^2$, before a decision is made to terminate the regression.

Step 4. If, after the constants of the orthogonal equation have been calculated, it is desired to convert from the form of Eq. (9.62) to (9.61), the following values of ξ in terms of the integer values of x are substituted. In these formulas n is the number of data points. If the original data had been coded to obtain the integer values of x, after substituting the x values of ξ, to obtain an equation of the form (9.61), it must then be decoded to obtain the regression equation in terms of the original data.

CORRELATION—REGRESSION

$$\xi_1 = (x - \bar{x})$$

$$\xi_2 = (x - \bar{x})^2 - \frac{n^2 - 1}{12}$$

$$\xi_3 = (x - \bar{x})^3 - (x - \bar{x})\frac{3n^2 - 7}{20} \quad (9.65)$$

$$\xi_4 = (x - \bar{x})^4 - (x - \bar{x})^2 \frac{3n^2 - 13}{14} + \frac{3(n^2 - 9)(n^2 - 1)}{560}$$

$$\xi_5 = (x - \bar{x})^5 - (x - \bar{x})^3 \frac{5(n^2 - y)}{18} + (x - \bar{x}) \frac{15n^4 - 230n^2 + 407}{1,008}$$

$$\xi_r = \xi_{r-1}\xi_1 - \frac{r^2(n^2 - r^2)}{4(4r^2 - 1)} \xi_{r-2} \quad (9.66)$$

where $\xi_0 = 1$.

example 9.15. The data of example 9.14 will be used, with the original pressure values divided by 200, to give x values in the form of positive integers.

Data

TABLE 9.34

x	y
1	0.846
2	0.573
3	0.401
4	0.288
5	0.209
6	0.153
7	0.111
8	0.078

Solution

$$\Sigma' y^2 = \Sigma y^2 - \frac{(\Sigma y)^2}{n} = 0.489500$$

Sums of squares removed by successive orders of the regression, from Eqs. (9.64) and Table 9.33, for $n = 8$, are given in Table 9.35.

It appears that the third-degree term may not remove a significant amount of the deviation. This statement can be tested by the F test.

Total $\Sigma' y^2$ = 0.489500

Total sum of squares removed by all three terms = 0.489246

Residual $\Sigma' \hat{y}^2$ with $8 - 4 = 4$ degrees of freedom = 0.000254

Variance of estimate (error) for third-degree equation

$$s^2(\hat{y}) = \frac{\Sigma'\hat{y}^2}{4} = 0.000064$$

Contribution of third-degree term with
1 degree of freedom (from Table 8.35) = 0.004525

Testing this sum of squares against the error mean square, we get

$$F = \frac{4{,}525}{64} = 70.7^{**}$$

$$F_{0.01,1,4} = 21.2$$

i.e., the third-degree term is significant at the 0.01 level.

TABLE 9.35

j	y_j	First degree h_{1j}	Second degree h_{2j}	Third degree h_{3j}
1	0.846	× (−7) = −5.922	× (7) = 5.922	× (−7) = −5.922
2	0.573	× (−5) = −2.865	× (1) = 0.573	× (5) = 2.865
3	0.401	× (−3) = −1.203	× (−3) = −1.203	× (7) = 2.807
4	0.288	× (−1) = −0.288	× (−5) = −1.440	× (3) = 0.864
5	0.209	× (1) = 0.209	× (−5) = −1.045	× (−3) = −0.627
6	0.153	× (3) = 0.459	× (−3) = −0.459	× (−7) = −1.071
7	0.111	× (5) = 0.555	× (1) = 0.111	× (−5) = −0.555
8	0.078	× (7) = 0.546	× (7) = 0.546	× (7) = 0.546

$n = 8$, $\Sigma hy = -8.509$, $= 3.005$, $= -1.093$

$(\Sigma hy)^2 = 72.403081$, $= 9.030025$, $= 1.194649$

Σh^2 (from Table 8.33) $= 168$, $= 168$, $= 264$

$$\Sigma'c^2 = \frac{(\Sigma hy)^2}{\Sigma h^2} = 0.430971 \qquad = 0.053750 \qquad = 0.004525$$

Settling for a second-degree equation for the sake of simplicity, since it will account for 99 per cent of $\Sigma'y^2$, the constants for the orthogonal form are calculated from Eq. (9.63) and the λ values from Table 9.33, as follows:

$$B_0 = \bar{y} = 0.332375$$

$$B_1 = \frac{\Sigma h_1 y}{\Sigma h_1^2} \lambda_1 = \left(\frac{-8.509}{168}\right) 2 = -0.101298$$

$$B_2 = \frac{\Sigma h_2 y}{\Sigma h_2^2} \lambda_2 = \left(\frac{3.005}{168}\right) 1 = 0.017887$$

The orthogonal-form equation therefore is

$$\hat{y} = 0.332375 - 0.101298\xi_1 + 0.017887\xi_2$$

To put this in terms of x, we substitute the values of ξ from Eqs. (9.65) to give

$$\hat{y} = 0.332375 - 0.101298(x - 4.5) + 0.017887[(x + 4.5)^2 - 63\tfrac{3}{12}]$$

or

$$\hat{y} = 1.056521 - 0.262281x + 0.017887x^2$$

In terms of the original pressure data, which were divided by 200 to give x in terms of simple integers, $x = P/200$, the final equation becomes

$$\hat{y} = 1.0565 - (1.311)(10^{-3})P + (4.472)(10^{-7})P^2$$

which is identical with the solution obtained by multiple regression in example 9.14.

9.3.3 Calculation of the Orthogonal Polynomials. Fisher[1] has published a simple stepwise procedure for calculating the orthogonal

TABLE 9.36

x	0 order y	1st degree	2d degree	3d degree
1	0.846	0.846	0.846	0.846
2	0.573	1.419	2.265	3.111
3	0.401	1.820	4.085	7.196
4	0.288	2.108	6.193	13.389
5	0.209	2.317	8.510	21.899
6	0.153	2.470	10.980	32.879
7	0.111	2.581	13.561	46.440
8	0.078	2.659	16.220	62.660
$S =$	2.659	16.220	62.660	188.420
$a =$	0.332375	0.450555	0.522167	0.570969
$b =$	0.332375	-0.118180	0.025044	-0.004130
$B =$	0.332375	-0.101297	0.017888	-0.002753
$\Sigma'c^2 =$	8.837851	0.430965	0.053757	0.004561

polynomials without reference to the tables of values. The method is best illustrated by means of a numerical example, and the coded data of example 9.15 will be used. Table 9.36 shows the calculation which is discussed in detail in the following pages.

[1] Fisher, *op. cit.*

Step 1. The data are tabulated according to increasing x values, and for each degree of the solution there is a corresponding column in which each term is the cumulative sum of the corresponding terms of the previous column. For example, the fifth term in the column for the solution of the second-degree polynomial is equal to the sum of the first five terms in the column for the solution of the first-degree polynomial.

Step 2. The sum of each column is determined. This sum must equal the last term in the next column as a check. The sums are designated S_0, S_1, S_2, etc.

Step 3. The first of two auxiliary constants, designated a, is calculated from the following formulas:

$$a_0 = \frac{(1)(S_0)}{(n)}$$

$$a_1 = \frac{(1)(2)(S_1)}{(n)(n+1)}$$

$$a_2 = \frac{(1)(2)(3)(S_2)}{(n)(n+1)(n+2)} \qquad (9.67)$$

$$\vdots$$

$$a_r = \frac{(1)(2)\cdots(r+1)S_r}{(n)(n+1)\cdots(n+r)}$$

Step 4. The second constant, designated b, is calculated from the following formulas:

$$b_0 = a_0$$
$$b_1 = a_0 - a_1$$
$$b_2 = a_0 - 3a_1 + 2a_2$$
$$b_3 = a_0 - 6a_1 + 10a_2 - 5a_3$$
$$b_4 = a_0 - 10a_1 + 30a_2 - 35a_3 + 14a_4 \qquad (9.68)$$
$$b_5 = a_0 - 15a_1 + 70a_2 - 140a_3 + 126a_4 - 42a_5$$

$$b_r = a_0 - \left[\frac{(r)(r+1)c_0}{(1)(2)}\right]a_1 - \left[\frac{(r-1)(r+2)c_1}{(2)(3)}\right]a_2$$
$$- \left[\frac{(r-2)(r+3)c_2}{(3)(4)}\right]a_3 - \cdots$$

where c_0 is the coefficient of a_0, c_1 is the coefficient of a_1, etc.

Step 5. The coefficients of the orthogonal polynomials B are calculated from the second constant as follows:

$$B_0 = b_0$$

$$B_1 = \frac{6b_1}{n-1}$$

$$B_2 = \frac{30b_2}{(n-1)(n-2)}$$

$$B_3 = \frac{104b_3}{(n-1)(n-2)(n-3)}$$

$$B_4 = \frac{630b_4}{(n-1)(n-2)(n-3)(n-4)}$$

$$B_5 = \frac{2772b_5}{(n-1)(n-2)(n-3)(n-4)(n-5)}$$

$$B_r = \frac{(2r+1)!b_r}{(r!)^2(n-1)(n-2)\cdots(n-r)}$$

(9.69)

Step 6. The sums of squares of deviation removed by each degree of the polynomial are calculated from the coefficients by the formulas given below. (The zero-degree term corresponds to $(\Sigma y)^2/n$, which when subtracted from Σy^2 gives $\Sigma' y^2$, or the sum of squares of deviation from the mean.)

$$\Sigma' c_0^2 = nB_0^2$$

$$\Sigma' c_1^2 = \frac{(n)(n^2-1)B_1^2}{12}$$

$$\Sigma' c_2^2 = \frac{(n)(n^2-1)(n^2-4)B_2^2}{180}$$

$$\Sigma' c_3^2 = \frac{(n)(n^2-1)(n^2-4)(n^2-9)B_3^2}{2{,}800}$$

$$\Sigma' c_4^2 = \frac{(n)(n^2-1)(n^2-4)(n^2-9)(n^2-16)B_4^2}{44{,}100}$$

$$\Sigma' c_5^2 = \frac{(n)(n^2-1)(n^2-4)(n^2-9)(n^2-16)(n^2-25)B_5^2}{698{,}544}$$

$$\Sigma' c_r^2 = \frac{(r!)^4(n)(n^2-1)\cdots(n^2-r^2)B_r^2}{(2r)!(2r+1)!}$$

(9.70)

Comparison of the numerical values in Table 9.36 with the results of example 9.15 will demonstrate that except for rounding off differences, the two procedures give identical answers. The calculation of the regression equation in terms of the integer x values and the ultimate decoding to the original data values are done in the same manner detailed in Sec. 9.3.2.

9.4 Correlation and Causality

No discussion of correlation for engineers first being introduced to the subject should be left without a warning that a significant correlation does not mean that changes in one variable cause the changes in the other. To cite an obvious example: if one were to correlate drying rate against mol fraction of nitrogen in an inlet air stream at varying relative humidities, there would be a positive correlation of increasing drying rate with increasing nitrogen concentration. The "cause" of the increased drying would be the decrease in moisture content, not the increase in nitrogen content, although the correlation with nitrogen might be very significant.

Part of the present-day discussion about the relation between cigarette smoking and lung cancer is a result of a very positive correlation between the increase in both these variables. One group suggests that the positive correlation indicates a causal relation. Another group points out that the increase in lung cancer can also be correlated with an increase in gasoline consumption (more noxious gases), an increase in electrical output (more ozone formation), an increase in urban living (more exposure to dust and industrial fumes), etc. The outcome of this discussion will rest on some well-planned experiments. The point, of course, is that the correlation by itself only demonstrates a mathematical relation between the variables.

9.5 Optimization

The subject of experimental attainment of optimum conditions was introduced by G. E. P. Box and K. B. Wilson in 1951.[1] The field has been expanded considerably since the original introduction, principally due to contributions by G. E. P. Box and J. S. Hunter. Some of the articles dealing with this subject are referenced below.[2-4] Only a brief introduction to optimization is given in this section. A thorough presentation of the subject is material for a complete text. Reference

---

[1] G. E. P. Box and K. B. Wilson, *Journal of the Royal Statistical Society*, ser. b, vol. 13, p. 1, 1951.

[2] G. E. P. Box and J. S. Hunter, *Biometrics*, vol. 10, p. 16, 1954.

[3] G. E. P. Box and J. S. Hunter, *Annals of Mathematical Statistics*, vol. 28, p. 195, 1957.

[4] G. E. P. Box and J. S. Hunter, *Technometrics*, vol. 1, p. 97, 1959.

CORRELATION—REGRESSION

to the original Box–Wilson paper, to the other articles mentioned, and to Davies' "Design and Analysis of Industrial Experiments" will supplement the material presented here.

If the response (the results of an experiment) can be expressed in terms of the main effects,

$$Y = b_0 + b_1A + b_2B + b_3C + \cdots \qquad (9.71)$$

Y can be looked upon as the plane of the response variable, and the b terms are the slope of the plane in the direction of each factor variable.

If the response is a function of the factors to a degree higher than one,

$$Y = b_0 + b_1A + b_2B + b_3C + b_{11}A^2 + b_{22}B^2 + b_{12}AB + \cdots$$
$$(9.72)$$

Y is a curved surface with the slopes in the direction of each variable varying from point to point.

The discussion of optimization in this section deals only with situations of the first type. On the assumption that the response surface is a plane, the slopes, or the change of each factor to produce maximum rate of change in response, is calculated. From an initial set of experiments, the direction for maximum rate of change in the response is determined. Experimentation is continued in the direction indicated. When a change in the response indicates the original direction of maximum rate of change is no longer applicable, a second set of experiments is run and a new set of slopes is calculated. If the results of the first set of experiments indicate that a second degree or higher-response surface is involved, the procedure to be followed is still well within the compass of an engineer trained in statistics. However, the procedure for each case will depend on the degree of the response surface, and the detailed explanation is beyond the planned scope of this section. In the large majority of cases, if the levels of the factors involved in the experimental program do not cover too large a range, the approximation of a plane to the response surface function will be satisfactory. Just what range of factor levels will result in a planar response surface is a matter of judgment on the part of the engineer-experimenter.

The discussion here of optimization of first degree response functions follows the nomenclature and analysis methods developed for fractional 2^n factorials in Chap. 8. Familiarity with the material in that chapter is essential for an understanding of what follows.

9.5.1 Linear Interpretation of Factorial Designs. In a full 2^n factorial, if only the main effects were significant, these could be interpreted as the linear response of the dependent variable to the change in level of the independent variables. The mean effect of a factor divided by the difference between the two levels of the factor would be a measure of a change in the response variable per unit change in the factor variable. The slope of the response depends on the difference in the levels of the factor tested. With a planar response function, the slope of the response surface depends on the difference in levels of all of the factors involved. In a multifactor experiment, the direction of steepest incline for the response surface is proportional to the product of the main effects and the difference between the levels of the effects. This is illustrated in example 9.16, which follows.

The first step in endeavoring to locate the point of optimum response is to determine the direction of change from the preliminary experimental conditions. The slopes indicated by the main effects are applicable at the center of the matrix of experimental conditions, the midpoints between the levels of each factor. The direction for further experimentation is from the center of the experiment matrix in the directions indicated by the signs of the slopes calculated from the mean effects of each factor, and the distance to be moved is proportional to the product of the slope and the magnitude of the difference between the factor levels. If the preliminary experiment shows that interaction effects are equal to or greater than the main effects, then the assumption of linearity is not tenable and the procedure described here does not apply.

TABLE 9.37 Calculation of Path of Steepest Ascent from a 2^3 Factorial

				General term
Factors.............	A	B	C	I
Upper level..........	a	b	c	i
Lower level..........	a'	b'	c'	i'
Mean effect..........	AM	BM	CM	IM
Matrix center........	$\dfrac{a+a'}{2}$	$\dfrac{b+b'}{2}$	$\dfrac{c+c'}{2}$	$\dfrac{i+i'}{2}$
Path of steepest ascent.	$(a-a')AM$	$(b-b')BM$	$(c-c')CM$	$(i-i')IM$
Point of further experimentation..........	$\dfrac{a+a'}{2}+\Delta a$	$\dfrac{b+b'}{2}+\Delta b$	$\dfrac{c+c'}{2}+\Delta c$	$\dfrac{i+i'}{2}+\Delta i$

One of the deltas is arbitrarily set, say Δa, and the others are determined by the relations:

$$\Delta b = \frac{\Delta a(b-b')(BM)}{(a-a')(AM)}$$

$$\Delta c = \frac{\Delta a(c-c')(CM)}{(a-a')(AM)}$$

$$\Delta i = \frac{\Delta a(i-i')(IM)}{(a-a')(AM)}$$

Expected response = mean response + $\Sigma(IM)\dfrac{\Delta i}{i-i'}$

If the assumption of linearity does apply and further experimentation along the path of maximum change results in significant change in the response, a second matrix of experiments should be performed further along the path of ascent to confirm or to modify the slopes for further progress along the path of optimization.

Table 9.37 summarizes the calculation of the path of steepest ascent for a factorial design with three factors. The plan includes a general term for the ith factor, and the calculation can be expanded for any number of factors.

example 9.16 An experiment is planned to determine the optimum conditions of temperature, pressure, and time for a particular chemical reaction. A complete 2^3 factorial experiment is run at the following conditions to establish the direction for further experimentation.

Temperature: 200° and 250°
Pressure: 100 and 400 psi
Time: 10 and 30 min.

In the following tabulation of results, t designates the upper level of temperature, p the upper level of pressure, and h the longer time. The results, as per cent conversion, are tabulated in the standard order for calculation, and the calculation of the effects is also shown.

Table 9.38 indicates that the overall mean response was 58 per cent conversion, and only the main effects were significant. The mean temperature effect was 6 conversion per cent units, the mean pressure effect was 4 and the mean time effect was 3 per cent units. The dif-

TABLE 9.38

Experiments	Responses			Total effects	Mean effects
(1)	51	108	224	460	58
t	57	116	236	24	6
p	55	114	12	16	4
tp	61	122	12	0	0
h	54	6	8	12	3
th	60	6	8	0	0
ph	58	6	0	0	0
tph	64	6	0	0	0

ference in temperature levels was 50°, in pressure levels, 300 psi, and in time units, 20 min. Since the sign of each of the effects was positive, the location of a point in the direction of maximum increase in per cent conversion is measured from the center of the experiment matrix and in the direction of each factor variable proportional to $(6)(50°)$: $(4)(300 \text{ psi}):(3)(20 \text{ min.})$. The center of the experiment matrix is at $225°$, 250 psi, and 20 min. If it is decided to move $60°$ from this point to the $285°$ level, the pressure would be increased $(60)(^{120}\!/_{300}) = 240$ psi, and the time would be increased $(60)(^{60}\!/_{300}) = 12$ min. The experimental point at which the response surface is to be tested is $285°$, 490 psi, and 32 min. The predicted response at the new experiment level is

$$58 + (^{6}\!/_{50})(60) + (^{4}\!/_{300})(240) + (^{3}\!/_{20})(12) = 70.2$$

After the path of steepest ascent is established, and an experiment has been run at some point on this path, the response obtained is compared with the predicted response. If the agreement is reasonably good compared to some error estimate, further experimentation along the same path can be continued. If the response is significantly different from the prediction, it may indicate that the point selected was too far along the path from the previous point or from the center of the original matrix. Depending on the experimental conditions, either another factorial can be run with the new point at the center to establish a new direction for further experimentation, or a single experiment can be run at a lesser distance from the previous location. When a maximum along the original path appears to be achieved, then a new matrix is called for to set a new direction.

9.5.2 Interpretation of Fractional Factorial Designs. If four or five factors are involved, the complete factorial might involve more than a practical number of experiments. A 2^5 factorial would require 32 experiments. As pointed out in Chap. 8, by careful selection of the experimental conditions it is possible with only a fraction of the total experiments required for the complete factorial to determine the main effects by aliasing them with the higher order interactions which are usually not significant. The same procedure can be used in the preliminary experiments run to determine the direction of experimentation to obtain the optimum response. The eight experiments required for a complete three factor, two-level factorial can be used to determine the change required in four, five, or under ideal conditions, even in seven experimental variables to obtain the maximum change in the response variable.

The procedure for establishing the experiment design and for determining the aliases is the same as that described in Chap. 8 for fractional factorial experiments. An $n - p$ factorial design is set up and the p factors not included in the complete 2^{n-p} factorial are aliased with one of the higher-order interactions to form a generating contrast. In selecting the higher-order interactions to be aliased with the p factors not included in the initial design, care must be taken to avoid aliasing any main effects with first-order interactions that may be thought to be significant. For example, if two factors, D and E, are said to be added to a complete 2^3 factorial involving factors A, B, and C, D might be aliased with the ABC interaction and E aliased with the BC interaction, giving $D = ABC$ and $E = BC$ for the generating contrasts. $ABCD$ and BCE plus their product, ADE, would make up the defining contrast. Listing only the first-order interactions, A is aliased with DE, B with CE, C with BE, D with AE, and E with BC and AD. The first-order interactions that are not aliased with any main effects are AB, AC, BD, and CD. If any first-order interaction is suspected of being significant, it must be arranged so to be associated with the symbols for one of the interactions free from an alias with a main effect in order to determine the slope of the response surface in the direction of each main effect without its being obscured by an interaction effect.

In using the fractional factorial designs for response surface evaluation, it is also important to know the sign of the calculated effect. Since the direction of movement from the center of the original experiment matrix is indicated by the slopes (main effects/range of factor levels), it is essential to know the sign of the slopes. The most certain way of determining the signs to be associated with any particular effect is to employ the sign table as illustrated in Chap. 8, Tables 8.4 and 8.5. When the upper level of the added factor is associated with the plus signs in the sign table for the aliased interaction, the sign of the calculated effect and hence the slope will indicate the change in response when going from the lower level to the higher level of the factor involved. When the upper level of the added factor is associated with the negative signs in the table of signs for the aliased interaction, the effect and the slope is that obtained when going from the upper level to the lower level of the factor variable. Table 9.39 is a partial sign table for a 2^3 factorial involving factors R, S, and T, with factors U and V aliased with interactions RST and RS

TABLE 9.39 Sign Table for 2^3 Factorial with Two Additional Factors Aliased with Interactions RST and RS

$V = RS$	R	S	T	$U = RST$	$V = RS$	$-V = RS$
$I(v)$	−	−	−	−	+	I
$r(u)$	+	−	−	+	−	$r(uv)$
$s(u)$	−	+	−	+	−	$s(uv)$
$rs(v)$	+	+	−	−	+	rs
$t(uv)$	−	−	+	+	+	$t(u)$
rt	+	−	+	−	−	$rt(v)$
st	−	+	+	−	−	$st(v)$
$rst(uv)$	+	+	+	+	+	$rst(u)$

respectively. The first column on the left shows the experiments to be run when $V = RS$, and the last column on the right shows the experiments to be run when $-V = RS$.

The calculation of the factor effects is carried out in the same manner as described in Chap. 8. The slope of the response surface in the path of maximum ascent at the center of the experiment matrix in the direction of each factor in units of response per unit of the factor is equal to the effect divided by the range between the levels of the factor. The vector (the direction and the distance) indicating the path of steepest ascent is measured from the center of the experiment matrix and is proportional to the product of each main effect and the range between the levels of the factor. The direction is indicated by the sign of the effect. These points are illustrated in example 9.17 which uses data from the original paper on optimization by Box and Wilson.[1]

example 9.17 A 2^{5-2} fractional factorial was employed to determine the path of steepest ascent for five factors with eight experiments. The design involved a chemical reaction and the variables with the upper and lower levels were as follows:

A Concentration of one component............ 93% and 90%
B Ratio of two components.................. 4.5 and 4.0
C Amount of solvent....................... 250 and 200 cc
D Time..................................... 2 and 1 hr
E Ratio of two other components............ 3.5 and 3.0

TABLE 9.40

Experiment	A	B	C	$D = ABC$	$-E = AB$
(1)	−	−	−	−	+
$a(de)$	+	−	−	+	−
$b(de)$	−	+	−	+	−
ab	+	+	−	−	+
$c(d)$	−	−	+	+	+
$ac(e)$	+	−	+	−	−
$bc(e)$	−	+	+	−	−
$abc(d)$	+	+	+	+	+

The only interaction that was thought possibly to be significant was the AC factor, so that D was aliased with the ABC interaction and E was aliased with the $-AB$ interaction, leaving the AC and BC interactions unaliased with any main effects. The E factor was aliased

[1] G. E. P. Box and K. B. Wilson, On the Experimental Attainment of Optimum Conditions, *Journal of Royal Statis. Soc.*, vol. 13, no. 1, p. 1, 1951.

CORRELATION—REGRESSION

with the $-AB$ interaction so that the experiment at the lower level of all factors, experiment (1), would be included in the design. The actual experiment design is shown in Table 9.40.

The results of the experiments and the calculation of the mean effects are shown in Table 9.41.

TABLE 9.41 Results and Calculation of Mean Effects

Experiment	Experimental results	Calculation (See example 8.1)			Mean effects		
(1)	34.4	86.0	162.3	387.6	Overall mean	387.6/8 =	48.45
$a(de)$	51.6	76.3	225.3	47.8	Mean A effect	47.8/4 =	11.95
$b(de)$	31.2	116.5	31.1	-17.4	Mean B effect	$-17.4/4$ =	-4.35
ab	45.1	108.8	16.7	-3.2	Mean E effect	3.2/4 =	0.80
$c(d)$	54.1	17.2	-9.7	63.0	Mean C effect	63.0/4 =	15.75
$ac(e)$	62.4	13.9	-7.9	-14.4	Mean AC effect	$-14.4/4$ =	-3.60
$bc(e)$	50.2	8.3	-3.3	2.0	Mean BC effect	2.0/4 =	0.50
$abc(d)$	58.6	8.4	0.1	3.4	Mean D effect	3.4/4 =	0.85

The direction of further experimentation is calculated from the mean effects and the ranges of the factor levels. This is illustrated in Table 9.42.

TABLE 9.42 Calculation of Direction of Steepest Ascent

Factor	Mean effect (1)	Range of factor levels (2)	Product of (1)(2) = (3)	Center point	Change: [proportional to (3)]	New experimental point
A	11.95	3.0	35.85	91.5	$+2.28$	93.78
B	-4.35	0.5	-2.175	4.25	-0.17	4.08
C	15.75	50.0	787.5	225.00	$+50.00$	275.00
D	0.85	1.0	0.85	1.50	$+0.05$	1.55
E	0.80	0.5	0.40	3.25	$+0.02$	3.27

The actual experiment, as described in the reference, gave a yield of 80 at the new experiment point, compared with the mean value of 48.4 obtained for the first set of results. The path of steepest ascent was extended another 20 units for factor C, and proportionally for the other factors to set the conditions for a third experiment. At this point the yield was 79.4, indicating that a level area of the response surface had been reached. A second 2^{5-2} factorial was carried out with the last experimental point as the center of the design matrix to determine the direction for further experimentation.

9.5.3 Summary. Although the procedure to be followed in any experimental program must be adapted to the specific conditions for the factors and responses involved, the following general summary for a two-level factorial study of response surface may be helpful to set the conditions from which to make the necessary modifications.

1. Select the factors which are considered most likely to affect the response to be studied.

2. The maximum number of experiments should be planned that can be carried out within a homogeneous area of experimentation: time, material, operators, reactors, etc.

3. The upper and lower levels of the factors should be selected with several points in mind:

   a. The range should include values of practical interest.

   b. The range should be small enough so that linearity of response can be reasonably expected.

   c. The range should be large enough so that a significant change in response can be expected.

   d. The ranges of the different factors should be selected so that, as far as can be anticipated from previous experience, the response will change about the same for each factor range change.

4. When fractional factorial designs are employed, first order interactions suspected of having significance should be free of aliases with main effects.

5. After completion of the experiments, the mean effects are calculated by the method outlined in Sec. 8.2.2 and illustrated in examples 8.1, 8.2, 9.16, and 9.17.

6. The path of steepest ascent is measured from the center of the design matrix and in a direction proportional to the product of the mean effects and the range of the factor levels.

7. Additional experiment(s) are carried out along the path indicated by the results of Step 6. The magnitude of the steps is dependent on the judgement of the experimenter. The first experiment along the path indicated by the factorial experiment is usually not far removed from boundary of the original design matrix.

8. If the direction of steepest ascent results in an improvement of the response variable, this path is followed until a maximum is found.

9. If the direction of steepest ascent does not result in an improvement of the response variable, the first step may have been too far removed from the center of the original design matrix. Additional experiments should be run a shorter distance from the design matrix center, but still along the path indicated by the original design results.

10. When a maximum response is indicated along the original path, a new factorial experiment is carried out with this point of maximum as the design center, and the procedure repeated along a new path.

CORRELATION—REGRESSION

11. If from the first factorial design some factors are found to have very small effects, this may be due to
   a. an actual independence of the response to the factor involved,
   b. too small a range of the factor levels, and
   c. selection of a point near the maximum response area for this factor.

12. Even when some factors are found to have very small effects, the path of steepest ascent, including small changes in these factors, should be followed. However, when a second factorial is planned at some point of apparent stability of response, the levels and ranges of the insignificant factors should be changed to try to locate an area when these factors will have a significant effect.

13. If the mean effects from the first factorial design are not significantly larger than the interactions that can be independently determined, the results indicate a nonlinearity of response that is beyond the scope of the procedure outlined here.

9.6 Problems

1. Code the data of example 9.2 by dividing y by 10 and subtracting 70 $[y' = (y/10) - 70]$ and x by subtracting 30, and calculate the least-squares line. $y' = -18.919 + 1.184x'$.

2. Calculate the least-squares line and the correlation coefficient for the following data. What fraction of the sum of squares of deviation is accounted for by the correlation? 0.977.

$X =$	1	2	3	4	5
$Y =$	20	104	168	208	324

3. Calculate the best straight line through the following data points:

$x =$	2.00	2.50	3.00	3.50	4.00	4.50
$y =$	4.17	5.60	6.03	7.03	9.42	9.40

$y = -0.2288 + 2.2063x$.

4. Obtain the 95 per cent confidence limits for the slope for Prob. 3.

5. For the best straight line through the data of Prob. 3, is the predicted value of y at $x = 0$ significantly different from zero at the 95 per cent level?

6. From the following replicate determinations of y at different levels of x, establish whether a linear correlation of y upon x accounts for all but the error variation in y.

$x =$	1.00	1.50	2.20	3.00	4.10	5.20	7.00
$y =$	2.57	4.00	5.10	7.20	11.10	12.25	15.10
	2.60	3.95	5.01	7.00	10.90	13.00	16.00
		4.05	5.20	6.95		12.50	15.75
				7.25			15.80

chapter ten
SEQUENTIAL ANALYSIS

The major portion of the statistics discussed thus far has dealt with drawing some conclusions from a fixed amount of data. These conclusions might be that there is some significant effect, some significant difference, some significant relation; or that the data do not show any significant effect. The point we are making is that the conclusions are drawn from the total data. This is something like a football match or basketball game. The winner is decided after the whole game is played. Or if there is no winner, the game is called a tie. No more data for that particular game are taken. No matter how far ahead one team may get in the beginning, the winner is not decided until all the data are in.

Statistics has available another type of test for specific cases in which each piece of data is considered as received and judgment is weighed with each new observation. As soon as a significant result is obtained, no more data are needed and the experiment can be stopped. This is sequential analysis. It is something like an old-time boxing match which continued until a significant difference between the contestants was established.

The theory of sequential analysis has been developed mostly by the late Professor Wald of Columbia University and is covered by his book[1]

[1] Abraham Wald, "Sequential Analysis," John Wiley & Sons, Inc., New York, 1947.

under this title. In the present chapter we shall merely present the method of sequential analysis as applied to several specific problems. For a more theoretical discussion, the reader is referred to Wald's book or that of Mood.[1]

10.1 General Discussion

In the usual statistical test, a hypothesis is made about the population from which a sample is taken. A statistic is calculated from the sample, and on the basis of the statistic, the hypothesis is either accepted or rejected. When the hypothesis is rejected, it is with a known probability, established by the significance level of the test, of a false rejection. If the hypothesis is accepted, it is usually on the grounds of having insufficient evidence for its rejection, and not with an established probability of a false acceptance. In some cases, as discussed in Sec. 6.5, the probability of false acceptance of the hypothesis, the type II error, can be controlled by the size of the sample. In sequential testing, a hypothesis is established and limits are set for both types of errors. The data are then accumulated one at a time, and as each new observation is obtained the information from all the data is checked against the hypothesis and one of three possibilities follows: the hypothesis is rejected with the preset α chance of error, the hypothesis is accepted with the preset β chance of error, or no decision is made and another datum point is obtained. Data are collected until the hypothesis is either accepted or rejected.

Although at first reading it may appear that this procedure will require more data than if the size of the experiment is set in advance and no judgment made until all the data are collected, it can be shown that on the average less than a half to a third as much data are required for similar results if the sequential analysis can be applied. In most engineering work, problems do not readily lend themselves to sequential analysis and this method is more applicable to inspection sampling. However, because of its interesting formulation, the method is presented here.

In general, the procedure involves plotting some cumulative function of the data as ordinate against the number of observations as abscissa. Two parallel guide lines are drawn, depending on the particular test involved, and the probability levels decided upon. This arrangement is shown schematically in Fig. 10.1. The two parallel lines divide the area into three sections, one for rejection of the hypothesis, one for acceptance of the hypothesis, and the portion between the lines for no decision and continued testing. When the plot of the cumulated function of the data crosses either of the parallel guide lines, the indicated decision is made.

---
[1] A. M. Mood, "Introduction to the Theory of Statistics," McGraw-Hill Book Company, New York, 1950.

As long as the plot stays within the parallel lines, no decision is indicated without a greater probability of error than originally established.

It is not essential actually to plot the data. The maximum and minimum values, corresponding to the two guide lines, can be calculated for each consecutive observation, and these values can be tabulated along

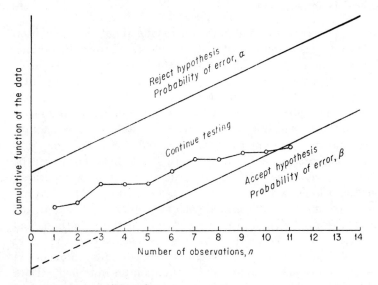

Fig. 10.1 Sequential testing.

with the actual cumulated value from the data. As long as the value from the data remains within the limits set by the two control values, no decision is made. When it exceeds the one or becomes less than the other, the indicated decision to reject or accept the hypothesis is made.

The slope of the guide lines, or the incremental difference between the control values for successive observations, is a function of the hypothesis being tested; and the heights of the guide lines or the magnitude of the control values are established by both the hypothesis being tested and the probability of its false acceptance and false rejection. The following sections demonstrate the formulas for several different applications of sequential testing. These are summarized in Table 10.7 at the end of this chapter.

10.2 Single Test—Binomial Distribution

In a system where a variable can have only one of two possible values (a sample is either good or bad, a student either passes or fails, a run is either successful or not, a coin falls either heads or tails), we can denote

SEQUENTIAL ANALYSIS

the probability of one event p and the alternative probability is $1 - p = q$. The problem is to establish the value of p for the population being sampled within some tolerable limits and with an acceptable chance of error. In a simple sampling problem, suppose we wish to accept material with 1 per cent defectives or less, and we wish to set a 5 per cent limit to the chance of rejecting material that meets this standard. And suppose we wish to reject material that has 5 per cent or more defectives, and we wish to set a 1 per cent limit to the chance of accepting material with this many defectives. We have one hypothesis that $p \leq 0.01$ and set an α value of 0.05 for a false rejection of this hypothesis. We have an alternative hypothesis that $p \geq 0.05$ and set a β value of 0.01 for a false acceptance of the first hypothesis when the second is true.

In general, we set an acceptable limit to $p \leq p_1$ with an α maximum risk of a false rejection; and we set an unacceptable limit to $p \geq p_2$ with a β maximum risk of a false acceptance. For the sequential test, we inspect one item at a time and plot the number of defectives as ordinate against the number of items inspected as abscissa. We draw two control lines on our inspection chart, the upper one:

$$U = h_u + mn \tag{10.1}$$

and a lower one,

$$L = h_l + mn \tag{10.2}$$

where the symbols in these equations are equal to the following quantities:

$$h_u = \frac{\ln \dfrac{1-\beta}{\alpha}}{\ln \dfrac{p_2}{p_1}\dfrac{1-p_1}{1-p_2}} \tag{10.3}$$

$$h_l = \frac{-\ln \dfrac{1-\alpha}{\beta}}{\ln \dfrac{p_2}{p_1}\dfrac{1-p_1}{1-p_2}} \tag{10.4}$$

$$m = \frac{\ln \dfrac{1-p_1}{1-p_2}}{\ln \dfrac{p_2}{p_1}\dfrac{1-p_1}{1-p_2}} \tag{10.5}$$

and n = number of observations.

If

$$g_1 = \ln \frac{p_2}{p_1}; \qquad g_2 = \ln \frac{1 - p_1}{1 - p_2} \qquad (10.6)$$

$$a = \ln \frac{1 - \beta}{\alpha}; \qquad b = \ln \frac{1 - \alpha}{\beta} \qquad (10.7)$$

then

$$h_u = \frac{a}{g_1 + g_2}; \qquad h_l = \frac{-b}{g_1 + g_2}; \qquad m = \frac{g_2}{g_1 + g_2} \qquad (10.8)$$

The terms indicated by a and b are used for all the sequential test procedures, and if this procedure is to be used frequently, it will save considerable time to calculate a table for a and b corresponding to the values of α and β that will be most frequently used. Table 10.1 gives the values for a and b corresponding to α and β values of 0.01, 0.025, and 0.05.

It is obvious from Eqs. (10.1) and (10.2) that these are two parallel lines of slope m and intercepts h_u and h_l. When the plot of number of defectives against sample size crosses either of these two lines, we can make a decision with the selected risk of error. If the plot crosses the lower line, we can accept the material as having p_1 or less fraction defective, with a β risk of error; or if the plot crosses the upper line, we can reject the

TABLE 10.1

β		α		
		0.01	0.025	0.05
0.01	$a =$	4.5951	3.6789	2.9857
0.01	$b =$	4.5951	4.5799	4.5539
0.025	$a =$	4.5799	3.6636	2.9704
0.025	$b =$	3.6789	3.6636	3.6376
0.05	$a =$	4.5539	3.6376	2.9445
0.05	$b =$	2.9857	2.9704	2.9445

material as having p_2 or greater fraction defective with an α chance of error. As long as the plot remains between these lines, no decision is made.

Note that in Eqs. (10.3), (10.4), and (10.5), the numerators and denominators are in terms of natural logarithms in all cases. The values would therefore not be affected if common logarithms were employed. How-

SEQUENTIAL ANALYSIS

ever, inasmuch as in some of the later formulations, logarithms are not used in both numerator and denominator and in these cases the natural base is required, we have shown the equations with natural logarithms in all formulations.

Before we illustrate the method, there are three more numbers which will be of value in using the sequential method. It is possible to calculate estimates of the number of observations that will be required on the basis of the control limits and significance levels established. If it appears that the number of observations will be excessive, the limits can be changed before the test is undertaken. The formulas for three estimates of the number of observations, one based on the assumption that $p = p_1$, one on the assumption that $p = p_2$, and the third on the assumption that p equals the probability corresponding to the slope m of the two control lines, are as follows:

$$\bar{n}_1 = \frac{(1-\alpha)h_l + \alpha h_u}{p_1 - m}, \text{ estimate of } n \text{ if } p = p_1 \qquad (10.9)$$

$$\bar{n}_2 = \frac{(1-\beta)h_u + \beta h_l}{p_2 - m}, \text{ estimate of } n \text{ if } p = p_2 \qquad (10.10)$$

$$\bar{n}_m = \frac{-h_l h_u}{m(1-m)}, \text{ estimate of } n \text{ if } p = m \qquad (10.11)$$

example 10.1. We shall use the conditions of the introductory paragraph to illustrate the procedure. In an inspection program the desired quality is 1 per cent defectives and we set a maximum of 5 per cent chance of rejecting material that has 1 per cent or less defectives. The poorest quality we can tolerate is 5 per cent defectives, and we set a maximum of 1 per cent chance of shipping material with 5 per cent or more defectives. Therefore

$$p_1 = 0.01, \quad \alpha = 0.05$$
$$p_2 = 0.05, \quad \beta = 0.01$$

$$a = \ln \frac{1-\beta}{\alpha} = 2.9857 \qquad \text{(Table 10.1)}$$

$$b = \ln \frac{1-\alpha}{\beta} = 4.5539 \qquad \text{(Table 10.1)}$$

$$g_1 = \ln \frac{p_2}{p_1} = 1.6904$$

$$g_2 = \ln \frac{1-p_1}{1-p_2} = 0.0411$$

$$h_u = \frac{a}{g_1 + g_2} = 1.8089$$

$$h_l = \frac{-b}{g_1 + g_2} = -2.7590$$

$$m = \frac{g_2}{g_1 + g_2} = 0.0249$$

The average number of samples will be:

If $p = p_1$,

$$\bar{n}_1 = \frac{(1 - \alpha)h_l + \alpha h_u}{p_1 - m} = 170$$

If $p = p_2$,

$$\bar{n}_2 = \frac{(1 - \beta)h_u + \beta h_l}{p_2 - m} = 70$$

If $p = m$,

$$\bar{n}_m = \frac{-h_l h_u}{m(1 - m)} = 205$$

For purposes of the problem, we shall assume that the 80th, the 128th, and the 184th item inspected were defective of 234 items tested. A plot of these inspections is shown in Fig. 10.2. At the 234th observa-

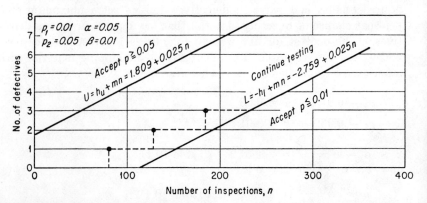

Fig. 10.2 Data from example 10.1.

tion, the data cross the lower control line and the material represented by the sample can be accepted as equal to or less than 0.01 defective with a maximum chance of error of 0.01 that there are actually more than 0.05 defectives.

10.3 The Double-dichotomy, or Odds-ratio, Test

A dichotomy is a division into two parts. By a double dichotomy is meant the division of two things into two parts each. In the sequential test described in the last section, we were measuring a single dichotomy against a theoretical or desired division, i.e., the observed fraction defectives against a specification quality. The situation often arises when two sets of measurements, each of which can be divided into two groups, are to be compared. Does one shipment have a lower percentage of defectives than another? Does one aiming device score a larger percentage of hits than another? When all the data are available, the 2 × 2 χ^2 test of Sec. 5.5.2 can be applied. In many cases the data may be very costly or time-consuming to collect and it may be important to come to a decision with a minimum amount of data. Two parallel production programs might be compared sequentially while they are in operation and a decision made before an entire test program could be completed. Or shipments from two sources might be tested by sequential comparison with less inspection than if the comparison was made between samples of set size from both shipments. To apply this particular type of sequential analysis it is necessary that the inspection be by dichotomy; good or bad, hit or miss, effective or defective, etc.

If we designate the fraction of successes in the first process as p_1, since this is equivalent to the probability of a success in a single observation, and the fraction of successes in the second process as p_2, and q_1 and q_2 as the corresponding fraction of failures in the two processes, where $q = 1 - p$; then p_1/q_1 is the reciprocal of the odds for a success in a single observation from the first process, and p_2/q_2 is the reciprocal of the odds for a success in a single observation from the second process. These ratios will be designated o_1 and o_2. A larger value of o_1 indicates the first process is better; a larger value of o_2 indicates the second process is better. o_2/o_1 is a measure of the relative superiority of process 2 over process 1. o_2/o_1 is the odds ratio, designated R. It is the odds ratio that is tested in the double-dichotomy sequential test.

As with the previous test, two limits to the odds ratio may be set: one, R_1, less than which the first process may be accepted as superior, i.e., a higher ratio of successes to failure; and a second odds ratio, R_2, beyond which the second process may be accepted as superior. Two hypotheses are established: one, that the second process is not superior, i.e., $R \leq R_1$; and a second hypothesis, that the second process is superior, i.e., $R \geq R_2$. If the maximum risks are established, α of falsely rejecting the first hypothesis when it is true, and β of falsely accepting it when the second hypothesis is true, we have all the terms necessary for a sequential test of the two dichotomies. Note that p in this explanation is the fraction of successes, o is the ratio, $\overset{.}{p/q}$, and R is the ratio of the second process probabilities to the first. If the odds ratio were inverted, or if p were taken as the fraction of failures, the same test would apply but with the terms suitably reversed.

The upper and lower control lines are identical with Eq. (10.1) and (10.2):

$$U = h_u + mn \qquad (10.1)$$
$$L = h_l + mn \qquad (10.2)$$

h_u, h_l, and m for the double-dichotomy test are as follows:

$$h_u = \frac{a}{\ln(R_2/R_1)} \qquad (10.12)$$

$$h_l = \frac{-b}{\ln(R_2/R_1)} \qquad (10.13)$$

$$m = \frac{\ln \dfrac{1+R_2}{1+R_1}}{\ln(R_2/R_1)} \qquad (10.14)$$

a and b have already been defined in Eq. (10.7):

$$a = \ln\frac{1-\beta}{\alpha} \qquad b = \ln\frac{1-\alpha}{\beta} \qquad (10.7)$$

n in this case is the number of pairs of data, one from each process, in which there is a difference, i.e., a success in one and a failure in the other. Pairs which show no difference are not counted. In the problem as set up, the number of pairs favorable to the second process is plotted as ordinate against the number of pairs favorable to either method. When the plot crosses the upper control line, then R, the ratio of successes to failures in the second process to the ratio of successes to failures in the first process, may be accepted as exceeding the established value R_2 with an α chance of error. When the plot crosses the lower control line, the alternative assumption, that R is equal to or less than R_1, may be accepted with a maximum chance of error equal to β.

The estimate of the amount of inspection required to establish either hypothesis can be obtained from the following formulas for the mean number of favorable pairs required:

If $R = R_1$,

$$\bar{n}_1 = \frac{(1-\alpha)h_l + \alpha h_u}{R_1/(1+R_1) - m} \qquad (10.15)$$

SEQUENTIAL ANALYSIS

If $R = R_2$,

$$\bar{n}_2 = \frac{(1-\beta)h_n + \beta h_l}{R_2/(1+R_2) - m} \qquad (10.16)$$

If $R = m$,

$$\bar{n}_m = \frac{-h_l h_u}{m(1-m)} \qquad (10.17)$$

It is not necessary actually to plot the data in order to determine when the number of observations falling within a certain category reaches the level necessary to a decision. The values of the control lines corresponding to the different number of total inspections can be tabulated. A tabulation of the data, compared to the tabular values of the control lines, will indicate when one of the control limits is crossed. This method is illustrated in example 10.2.

example 10.2. A plant has a manufacturing process which is reasonably satisfactory, producing about 95 per cent acceptable units depending on the quality of the raw material. A modification in the process is proposed for which the claim is made that it will decrease the fraction of defectives. A process is set up so that the two methods may be run in parallel using the same raw material and inspections run on the dual production line. If the modified process is only about as good as the standard process, no change will be made, so an odds ratio of 1.0 is set for a lower limit. If the modification will increase the acceptable production to 98 per cent when it would ordinarily be 95 per cent, then the change will be made. An odds ratio of $(98/2)/(95/5)$ is set for the upper limit. A maximum error of 5 per cent is set for both cases. Therefore

$H_0;\ R \leq R_1 = 1.0000$

$H_1;\ R \geq R_2 = 2.5790$

$\alpha = \beta = 0.05$

$a = b = 2.9445 \qquad$ (Table 10.1)

$\dfrac{R_2}{R_1} = 2.579; \qquad \dfrac{1+R_2}{1+R_1} = 1.7895$

$h_u = \dfrac{2.9445}{\ln 2.579} = 3.107$

$h_l = \dfrac{-2.9445}{\ln 2.579} = -3.107$

$m = \dfrac{\ln 1.7895}{\ln 2.579} = 0.6145$

An estimate of the average number of favorable pairs that will be required to establish a difference between the methods is:

If $R = R_1$,

$$\bar{n}_1 = \frac{(0.95)(-3.107) + (0.05)(3.107)}{(1.0)/(2.0) - 0.6145} = 24.4$$

If $R = R_2$,

$$\bar{n}_2 = \frac{(0.95)(3.107) + (0.05)(-3.107)}{(2.5790)/(3.5790) - 0.6145} = 26.3$$

If $R = m$,

$$n_m = \frac{(3.107)(3.107)}{(0.6144)(0.3856)} = 40.7$$

The upper and lower limits corresponding to the sequence of integer values of n can be readily obtained by the repeated additions of m to the values of h_u and h_l from Eqs. (10.1) and (10.2):

$$U = h_u + mn$$
$$L = h_l + mn$$

Table 10.2 gives the values of U and L for favorable pairs up to 19.

TABLE 10.2

n	U	L	n	U	L
0	3.107	−3.107	10	9.249	3.035
1	3.721	−2.493	11	9.863	3.649
2	4.335	−1.879	12	10.477	4.263
3	4.950	−1.264	13	11.092	4.878
4	5.564	−0.650	14	11.706	5.492
5	6.178	−0.036	15	12.320	6.106
6	6.792	+0.578	16	12.934	6.720
7	7.406	1.192	17	13.548	7.334
8	8.021	1.807	18	14.163	7.949
9	8.635	2.421	19	14.777	8.563

Inasmuch as the number of pairs favorable to either method must be in whole integers, the limit values of Table 10.2 may be rounded off to whole numbers. In rounding off, the values for the upper limits are rounded to the next higher whole number and the values for the lower limits are rounded to the next lower whole number. Since to

SEQUENTIAL ANALYSIS

be significant the observed number of favorable pairs must either exceed the values of the upper limit or be less than the values of the lower limit, the rounding off must be carried out in the manner described. Table 10.3 shows the rounded-off control limits and the count of favorable pairs in the actual experiment.

TABLE 10.3

Number of favorable pairs	Upper limit	Number of pairs favorable to new method	Lower limit
1	4	1	−3
2	5	2	−2
3	5	2	−2
4	6	3	−1
5	7	4	−1
6	7	5	0
7	8	6	1
8	9	7	1
9	9	8	2
10	10	9	3
11	10	9	3
12	11	10	4
13	12	11	4
14	12	11	5
15	13	12	6
16	13	12	6
17	14	13	7
18	15	14	7
19	15	15	8

With the 19th pair, the number favorable to the new method of production reached the limit set by the odds ratio corresponding to $R_2 = 2.5790$, and we can say with a maximum chance of error equal to 0.05 that the new process is better than the old.

10.4 Test of a Mean against a Standard-value One-sided Test

In the two previous sequential tests, we were dealing with measures that fell into one of two categories, heads or tails, accepts or rejects, etc. The sequential testing can also be applied to continuous measurements of the type which are probably more common in engineering work, where the mean temperature, or reaction rate, or yield, etc., is compared with some standard value. In the sequential tests discussed in this book it is assumed that the standard deviation is known. Sequential tests are available for testing means when the standard deviations are estimated from the data.[1,2] These procedures, however, are not discussed here.

In a process of some kind, involving a measurement of loss, or impurity, we are often interested in demonstrating whether the mean value is equal to or less than some standard value. This is similar to the one-sided t test of Chap. 6. If the measurement is x and the standard value is x_1, we would say the process is acceptable if the mean value was equal to or less than x_1. We could set a maximum deviation d beyond x_1 which would make the process unacceptable, i.e., if the mean was equal to or more than $x_1 + d$. If we set limits to the tolerable risks of error, an α risk of rejecting a mean equal to or less than x_1, and a β risk of accepting a mean equal to or greater than $x_1 + d$, and if we know the standard deviation σ, we have all the necessary terms for establishing a sequential test of the mean. A parallel procedure can be used to test $\bar{x} \geqq x_1$, reversing the roles of α and β.

To use the sequential test, control limits are set by the equations listed; a plot or tabulation is made of Σx (not the mean) against the number of inspections or data points. When Σx exceeds the upper level, we accept $\bar{x} = x_1 + d$. When Σx is less than the lower limit, we accept $\bar{x} = x_1$ with the preset maximum chances of error.

The equations for the upper and lower limits are, as before,

$$U = h_u + mn \qquad (10.1)$$
$$L = h_l + mn \qquad (10.2)$$

The equations for h_u, h_l, and m for the one-sided mean test are:

$$h_u = \frac{a\sigma^2}{d} \qquad (10.18)$$

$$h_l = \frac{-b\sigma^2}{d} \qquad (10.19)$$

$$m = x_1 + \frac{d}{2} \qquad (10.20)$$

a and b are as previously defined:

$$a = \ln \frac{1-\beta}{\alpha} \qquad b = \ln \frac{1-\alpha}{\beta} \qquad (10.7)$$

---

[1] G. A. Barnard, Statistical Inference, *J. Roy. Statis. Soc.*, Suppl. II (1949).
[2] S. Rushton, Sequential t Test, *Biometrika*, **37**:326 (1950).

SEQUENTIAL ANALYSIS

The mean number of samples required when $\bar{x} = x_1$, when $\bar{x} = x_1 + d$, and when $\bar{x} = m$ are calculated from the following equations. If these values indicate that too many tests will be required, the number can be adjusted by varying the values selected for α, β, and d.

If $x = x_1$,

$$\bar{n}_1 = \frac{(1-\alpha)h_l - \alpha h_u}{x_1 - m} \tag{10.21}$$

If $x = x_1 + d$,

$$\bar{n}_2 = \frac{(1-\beta)h_u - \beta h_l}{(x_1 + d) - m} \tag{10.22}$$

If $x = m$,

$$\bar{n}_m = \frac{-h_l h_u}{\sigma^2} \tag{10.23}$$

example 10.3. In example 6.1 at the beginning of the discussion of the t test data were presented from 11 octane-number determinations to see if the mean differed significantly from 87.5. In a case of this kind it might be more usual to seek to learn whether the mean was significantly less than a standard value of 87.5 and not to care much whether the mean could be greater. On this basis, we shall apply sequential analysis to the data, assuming the results were obtained in the order in which they are given in Table 6.2.

It is necessary to set a lower limit to the mean value beyond which we would have only a small risk of accepting the mean as being 87.5. Let us say, if the mean is as low as 86.5, we would want to assume only a 5 per cent risk of stating the value was 87.5. $x_1 = 86.5$, $d = 1.0$, $\alpha = 0.05$. If we set a 5 per cent risk of rejecting the material when it is actually 87.5 octane number, $\beta = 0.05$. We now need only the standard deviation of the measurements to calculate the control limits for the sequential test. For purposes of this problem, we shall assume that we know the standard deviation to be equal to 0.795, the value estimated from the original data. It is not unusual for a plant to have a reasonably good estimate of the standard deviation of a process or measurement that has been in operation for some period of time.

Solution

$$x_1 = 86.5$$
$$d = 1.0$$

$$\sigma^2 = 0.632$$
$$\alpha = \beta = 0.05$$
$$a = b = 2.9445$$
$$h_u = \frac{(2.9445)(0.632)}{1.0} = 1.861$$
$$h_l = -\frac{(2.9445)(0.632)}{1.0} = -1.861$$
$$m = 86.5 + 0.5 = 87.0$$
$$U = 1.861 + 87.0n$$
$$L = -1.861 + 87.0n$$

If the data are plotted, Σx against n, the control lines corresponding to U and L will be only 3.7 units apart while the value of Σx will be of the order of several hundred, depending on the number of measurements required to establish significance. The same results will be obtained if $\Sigma(x - 80)$, or the summation of the octane numbers coded by some other constant, is plotted against the number of inspections, provided the slope of the control lines, m, is coded in the same manner, that is, $m = (x_1 - 80) + d/2$.

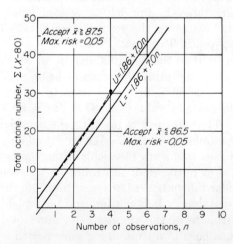

Fig. 10.3 Data from example 10.3.

Figure 10.3 shows the data of example 6.1 plotted against the control limits, coded by subtracting 80 from each value. The sum of the first four terms exceeds the upper limit, and $\bar{x} = 87.5$ can be accepted after the fourth inspection.

Table 10.4 gives the data, their cumulative sum, and the numerical values corresponding to the control limits to illustrate the tabular solution. In this case coding by subtracting 80 is not necessary and the same result is obtained: the cumulative value with the fourth inspection exceeds the limit set for $\bar{x} = 87.5$.

TABLE 10.4

n	Upper limit	x	Σx	Lower limit
1	88.861	88.6	88.6	85.139
2	175.861	86.4	175.0	172.139
3	262.861	87.2	262.2	259.139
4	349.861	88.4	350.6	346.139

10.5 Test of a Mean against a Standard-value Two-sided Test

If a mean is to be tested against a standard value for deviation of either sign, the sequential method may be employed provided the standard deviation is known. The method presented here was developed by the Statistical Research Group at Columbia University.[1]

If the measurements involved are designated x, we may wish to establish whether the mean of the measurements is within some fixed range of a standard value, i.e., within $x_1 \pm d$. We set an α maximum risk of rejecting the mean when it does lie within the range $x_1 \pm d$, and a β maximum risk of accepting the mean when it lies outside that range. The sum of the absolute deviations of x from x_1, $\Sigma|x - x_1|$, is plotted against the number of observations n. When an upper limit is exceeded, $\bar{x} = x_1 \pm d$ may be accepted, and when a lower limit is exceeded, $\bar{x} = x_1$ may be accepted with the preset maximum risks.

The equation for the upper limit is, as before,

$$U = h_n + mn \tag{10.1}$$

where for the two-sided mean test

$$h_u = \frac{(a + 0.693)\sigma^2}{d} \tag{10.24}$$

The equation for the lower limit is, also as before,

$$L = h_l + mn \tag{10.2}$$

[1] "Sequential Analysis of Statistical Data: Applications," Columbia University Press, New York, 1945.

and for the two-sided mean test

$$h_l = -\frac{(b - 0.693)\sigma^2}{d} \tag{10.25}$$

$$m = \frac{d}{2} \tag{10.26}$$

a and b have been defined several times in Eq. (9.7).

example 10.4. Again using the data of example 6.1, suppose that the problem was to establish whether the results, assumed to be obtained sequentially, could have come from a process with a mean octane number of 86.5, or whether the deviation from 86.5 could have been as much as ± 1.0. Using the same variance, 0.632, and setting the α and β limits at 0.05, we have the following formulas for calculating the sequential limits:

$x_1 = 86.5$

$\alpha = \beta = 0.05$

$a = b = 2.945$

$\sigma^2 = 0.632$

$h_u = \dfrac{(2.945 + 0.693)(0.632)}{1.0} = 2.299$

$h_l = -\dfrac{(2.945 - 0.693)(0.632)}{1.0} = -1.423$

$m = \dfrac{1.0}{2} = 0.5$

$U = 2.299 + 0.5n$

$L = -1.423 + 0.5n$

The following table gives the upper and lower limits and the corresponding values of $\Sigma|x - x_1|$ from the data of Table 6.2.

TABLE 10.5

n	Upper limit, U	Lower limit, L	x	$|x - x_1|$	$\Sigma|x - x_1|$
1	2.80	−0.92	88.6	2.1	2.1
2	3.30	−0.42	86.4	0.1	2.2
3	3.80	0.08	87.2	0.7	2.9
4	4.30	0.58	88.4	1.9	4.8

With the fourth value $\Sigma|x - x_1|$ exceeds the upper limit set for $\bar{x} = x_1 \pm d$ and we may state that the mean differs from 86.5 by 1.0 or more.

The same results could obviously be obtained from a graphical solution.

10.6 Test of Standard Deviation against a Given Value

In a case where a new analytical method is being compared to a standard procedure to determine whether the results are as consistent, or in a manufacturing process where the variation in results is more important than the mean (in the manufacture of chain, we are told it is not the mean strength but the strength of the weakest link that controls), we would want to test the standard deviation or variance against some established value. The following formulas provide for sequential-testing a cumulative sum of squares of deviation from the mean to establish with preset risks whether a standard deviation of the population is equal or less than some established value or whether it exceeds this value by some definite amount.

If the standard deviation of the process estimated from the sample is σ, and σ_1 designates an established value, we would be interested in testing whether σ was equal to or less than σ_1, or whether σ exceeded σ_1 by some amount δ. If we designate $\sigma_1 + \delta$ as σ_2, we have two tests:

$$\text{Acceptable: } \sigma \leq \sigma_1 \tag{10.27}$$

$$\text{Unacceptable: } \sigma \geq \sigma_2 \tag{10.28}$$

Let α be the risk of rejecting σ when it is actually equal to or less than σ_1, and let β be the risk of accepting σ when it is actually equal to or greater than σ_2, $\sigma_1 + \delta$. With these values, we can set up the sequential-test limits according to the following formulas. In using these formulas, $\Sigma(x - \mu)^2$ is plotted against n, the number of observations, where μ designates the true mean. If the true mean is not known, $\Sigma(x - \bar{x})^2$ is plotted against $n - 1$ to allow for the bias involved in calculating the sum of squares of deviations from the sample mean instead of the true mean. $\bar{x}$ must be calculated anew with each new datum point.

The equations for the upper and lower limits are, as before,

$$U = h_u + mn \tag{10.1}$$

$$L = h_l + mn \tag{10.2}$$

The equations of h_u, h_l, and m are:

$$h_u = \frac{2a}{1/\sigma_1^2 - 1/\sigma_2^2} \qquad (10.29)$$

$$h_l = \frac{-2b}{1/\sigma_1^2 - 1/\sigma_2^2} \qquad (10.30)$$

$$m = \frac{\ln(\sigma_2^2/\sigma_1^2)}{1/\sigma_1^2 - 1/\sigma_2^2} \qquad (10.31)$$

a and b are as defined previously in Eq. (10.7).

When $\Sigma(x - \mu)^2$ becomes less than the lower limit, σ can be accepted as equal to or less than σ_1, or if $\Sigma(x - \mu)^2$ becomes greater than the upper limit, σ can be accepted as equal to or greater than σ_2, with the present maximum risks of error.

example 10.5 We shall again employ the data of example 6.1, assuming the values occur in the order they are given in Table 6.2.

TABLE 10.6

n	$(x - \mu)$	$(x - \mu)^2$	$\Sigma(x - \mu)^2$	Upper limit	Lower limit
1	1.5	2.25	2.25	9.70	−6.00
2	−0.7	0.49	2.74	11.54	−4.15
3	0.1	0.01	2.75	13.39	−2.31
4	1.3	1.69	4.44	15.24	−0.46
5	0.1	0.01	4.45	17.09	1.39
6	−0.3	0.09	4.54	18.94	3.24
7	−1.0	1.00	5.54	20.79	5.09
8	0.2	0.04	5.58	22.63	6.93

We are willing to accept the process behind these figures if the standard deviation is 1.0 or less, and we do not want to accept the process if the standard deviation is as large as 2.0. We shall assume the mean octane number is 87.1, as it appears in the original example, and we shall set α and β limits of 0.05 for the two risks involved.

The terms necessary for the sequential analysis are:

$$\sigma_1 = 1.0; \qquad \sigma_1^2 = 1.0$$
$$\sigma_2 = 2.0; \qquad \sigma_2^2 = 4.0$$
$$\alpha = \beta = 0.05; \qquad a = b = 2.945$$

$$h_u = \frac{(2)(2.945)}{1 - 0.25} = 7.85$$

$$h_l = \frac{-(2)(2.945)}{1 - 0.25} = -7.85$$

$$m = \frac{\ln 4.00}{1.00 - 0.25} = 1.848$$

$$\mu = 87.1$$

Using the data of Table 6.2 in the order in which they occur, we give in Table 10.6 the values of $(x - \mu)$, $(x - \mu)^2$, $\Sigma(x - \mu)^2$, and upper and lower limits from the foregoing equations. With the eighth value the sum of squares is exceeded by the lower limit and the hypothesis of $\sigma \leq 1.00$ may be accepted.

10.7 Summary

Table 10.7 summarizes the several sequential tests discussed in this chapter and is presented to emphasize the similarity of the formulations.

TABLE 10.7 Summary of Sequential-test Formulas

	Single dichotomy	Double dichotomy	Mean one-sided	Mean two-sided	Variance
Hypotheses:					
H_0;	$p \leq p_1$	$\dfrac{p_2(1-p_1)}{p_1(1-p_2)} \leq R_1$	$\bar{x} \leq x_1$	$\bar{x} = x_1$	$\sigma \leq \sigma_1$
H_1;	$p \geq p_2$	$\dfrac{p_2(1-p_1)}{p_1(1-p_2)} \geq R_2$	$\bar{x} \geq x_1 + d$	$\bar{x} = x_1 \pm d$	$\sigma \geq \sigma_2$
Risks:					
α of rejecting H_0, when...	$p \leq p_1$	$R \leq R_1$	$\bar{x} \leq x_1$	$\bar{x} = x_1$	$\sigma \leq \sigma_1$
β of accepting H_0, when...	$p \geq p_2$	$R \geq R_2$	$\bar{x} \geq x_1 + d$	$\bar{x} = x_1 \pm d$	$\sigma \geq \sigma_2$
Procedure:					
Plot against n............	Σp	Σ pairs favorable to (2)	Σx	$\Sigma \lvert x - x_1 \rvert$	$\Sigma(x - \bar{x})^2$
Upper limit =	$h_u + mn$	$h_u + mn$	$h_u + mn$	$h_u + mn$	$h_u + mn$
Lower limit =	$h_l + mn$	$h_l + mn$	$h_l + mn$	$h_l + mn$	$h_l + mn$
$a =$	$\ln \dfrac{1-\beta}{\alpha}$	$\ln \dfrac{1-\beta}{\alpha}$	$\ln \dfrac{1-\beta}{\alpha}$	$\ln \dfrac{1-\beta}{\alpha}$	$\ln \dfrac{1-\beta}{\alpha}$
$b =$	$\ln \dfrac{1-\alpha}{\beta}$	$\ln \dfrac{1-\alpha}{\beta}$	$\ln \dfrac{1-\alpha}{\beta}$	$\ln \dfrac{1-\alpha}{\beta}$	$\ln \dfrac{1-\alpha}{\beta}$

	Col 1	Col 2	Col 3	Col 4	Col 5
$h_u =$	$\dfrac{a}{g_1+g_2}$	$\dfrac{a}{\ln(R_2/R_1)}$	$\dfrac{a\sigma^2}{d}$	$\dfrac{\sigma^2}{d}(a+0.693)$	$\dfrac{2a}{1/\sigma_1^2 - 1/\sigma_2^2}$
$h_l =$	$\dfrac{-b}{g_1+g_2}$	$\dfrac{-b}{\ln(R_2/R_1)}$	$\dfrac{-b\sigma^2}{d}$	$\dfrac{-\sigma^2}{d}(b-0.693)$	$\dfrac{-2b}{1/\sigma_1^2 - 1/\sigma_2^2}$
$m =$	$\dfrac{g_2}{g_1+g_2}$	$\dfrac{\ln\dfrac{1+R_2}{1+R_1}}{\ln(R_2/R_1)}$	$x_1 + \dfrac{d}{2}$	$\dfrac{d}{2}$	$\dfrac{\ln(\sigma_2^2/\sigma_1^2)}{1/\sigma_1^2 - 1/\sigma_2^2}$
Auxiliary formulas:	$g_1 = \ln\dfrac{p_2}{p_1}$ $g_2 = \ln\dfrac{1-p_1}{1-p_2}$	$0 = \dfrac{p}{1-p} = \dfrac{p}{q}$ $R = \dfrac{0_2}{0_1}$			
Average number of samples:					
$\bar{n}_1 =$	$\dfrac{(1-\alpha)h_l + \alpha h_u}{p_1 - m}$	$\dfrac{(1-\alpha)h_l + \alpha h_u}{\dfrac{R_1}{1+R_1} - m}$	$\dfrac{(1-\alpha)h_l + \alpha h_u}{x_1 - m}$		$\dfrac{(1-\alpha)h_l + \alpha h_u}{\sigma_1^2 - m}$
$\bar{n}_2 =$	$\dfrac{(1-\beta)h_u + \beta h_l}{p_2 - m}$	$\dfrac{(1-\beta)h_u + \beta h_l}{\dfrac{R_2}{1-R_2} - m}$	$\dfrac{(1-\beta)h_u + \beta h_l}{(x_1+d) - m}$		$\dfrac{(1-\beta)h_u + \beta h_l}{\sigma_2^2 - m}$
$\bar{n}_m =$	$\dfrac{-h_l h_u}{m(1-m)}$	$\dfrac{-h_l h_u}{m(1-m)}$	$\dfrac{-h_l h_u}{\sigma^2}$		$\dfrac{-h_l h_u}{2m^2}$

chapter eleven
NONPARAMETRIC STATISTICS

Each of the chapter headings in this book is a subject which, if developed in detail, could command an entire text for itself. Several books do exist which deal specifically with one of the subjects, to which only a chapter is allowed in this work. These remarks are especially applicable to this chapter and to the previous one in which we merely present a number of statistical tests without any attempt at either justification or explanation. Texts specializing in these subjects should be sought out by the student who is interested in more than just the application of the procedures to specific problems. A very complete bibliography of the nonparametric statistics literature was published by Savage in 1953.[1] A more recent book[2] written for the "behavioral sciences" covers a very large number of nonparametric tests. Included in this chapter are those nonparametric procedures that might be of particular interest to engineers.

The parametric statistics involve the calculation of a statistic from a sample and the comparison of this statistic with an assumed population parameter. If the statistic corresponds to a value of the parameter which

---

[1] I. R. Savage, *J. Am. Statis. Assoc.*, **48**:44 (1953).
[2] Sidney Siegel, "Nonparametric Statistics," McGraw-Hill Book Company, New York, 1956.

would occur very infrequently, the sample is assumed not to have come from the same population as the parameter. The nonparametric statistics do not involve any assumption of distribution of the population from which the sample may have come; they are parameter-free. The χ^2 test is nonparametric when applied in the form of a contingency test where the internal homogeneity of the data is tested.

Also included in this chapter are some tests for discarding data which deviate extremely from the rest of the sample.

11.1 Ranking Tests

Ranking denotes numbering observations in order according to some criterion, qualitative, quantitative, or subjective: judging a beauty contest, classifying the states in the Union according to the ratio of rural to urban land areas, listing metals in order of their corrosion resistance. Observations are ranked from 1 to n according to the particular classification, the ranking numbers indicating merely the order in the classification and having no necessary relation to any quantitative measure $1 < 2 < 3 < 4 < 5 < \cdots$ or $1 > 2 > 3 > 4 > 5 > \cdots$, where the symbols "less than" and "greater than" may have a quantitative connotation, or they may have only a qualitative designation such as "more desirable than," "less dangerous than," etc. The relations between consecutive ranks do not have to be uniform. If we rank the coins of the United States currency from 1 to 6, the 1-cent coin having rank 1, the 5-cent coin rank 2, etc. (or the \$1 coin rank 1, the 50-cent coin rank 2, etc.), the relation between successive ranks is not the same. Or if we rank the courses of our engineering curriculum, not according to the grades obtained, but according to how much we enjoyed them, there is no quantitative relation between successively ranked courses.

Ranking is used when there is no quantitative measure available, or if the quantitative measure is not necessary to the information being conveyed. Ranking the planets according to their distances from the sun: Mercury = 1, Venus = 2, Earth = 3, ..., Neptune = 8, Pluto = 9 (1956), might give information more applicable to a particular situation than if the actual distances were tabulated. Ranking is also used when the information of interest is whether two different classifications of the same observations may be related without requiring an actual quantitative relation. We might rank the baseball teams according to their standing at the end of the season (or at any other time) and also rank them according to the number of left-handed players to see if there was any correlation between the two methods of ranking. A third reason for ranking data is that it is quick and simple. It involves no calculations and does convey a certain amount of information.

When ties occur in ranking, the mean value of the ranking numbers is applied to each of the tied observations. For example, if two observations are tied for fourth rank, they are both ranked as 4.5 and the next observation is rank number 6. If three observations are tied for seventh rank, they are all ranked as 8.0, and the next observation after the tie is rank 10.

11.1.1 Spearman Rank Correlation. A quick and simple test to determine whether correlation exists between two classifications of the same observations is the Spearman rank-correlation test, often referred to simply as the rank correlation, although there are others. The observations are ranked according to one classification, and the same observations are then ranked according to a second classification. You might rank a number of concerns in a particular field according to size, i.e., capitalization or production or sales; and then you might rank the same concerns according to dividend yield or price to earnings ratio, to see if there was a correlation between size and yield. You might rank magazines in a particular field first according to amount of advertising and second according to amount of circulation to see if there is a correlation between the two. The Spearman rank-correlation coefficient is a function of the sum of the squares of the differences of the two rankings for each observation and the number of observations.

$$r_s = 1 - \frac{6\Sigma d_i^2}{n(n^2 - 1)} \qquad (11.1)$$

where r_s = Spearman rank-correlation coefficient
d_i = difference between the rankings of the ith observation
n = number of observations

It is obvious that if there is perfect positive correlation, all the differences will be zero and the correlation coefficient will be unity. What is not so obvious, but is equally true, is that if there is perfect negative correlation, the low-ranking observation in one classification is the high-ranking observation in the other, the term $6\Sigma d^2$ will be equal to $2n(n^2 - 1)$, and the correlation coefficient will be equal to -1.0. Table 11.1 gives the maximum values of r_s that can be expected at the indicated probability levels when there is no correlation between the rankings. The probabilities enumerated are the one-sided values, i.e., the probability of r_s being larger than the tabulated figures, or the probability of r_s being less than the negative of the tabulated values, but not both. The sign to be used with the table depends on whether a negative or positive correlation is hypothesized. If neither type of correlation is specified before the value of r_s is calculated, the probabilities to be used with Table 11.1 are 0.10 and 0.02.

For example, if with nine different magazines we hypothesize that there is positive correlation between circulation and quantity of cigarette advertising and find a rank correlation of 0.679, we can state that a value of this magnitude could occur by chance with less than 0.05 probability but more

TABLE 11.1 Maximum Values of r_s Expected by Chance at the Indicated Probability Levels†

N	0.05	0.01	N	0.05	0.01
4	1.000		16	0.425	0.601
5	0.900	1.000	18	0.399	0.564
6	0.829	0.943	20	0.377	0.534
7	0.714	0.893	22	0.359	0.508
8	0.643	0.833	24	0.343	0.485
9	0.600	0.783	26	0.329	0.465
10	0.564	0.746	28	0.317	0.448
12	0.506	0.712	30	0.306	0.432

than 0.01 probability if there were no *positive* correlation, and therefore accept with the same chances of error that there was positive correlation. If we had not set our hypothesis for positive correlation (if we had had no previous opinions and might just as well have believed before seeing any data that there might have been negative correlation between cigarette advertising and circulation), but were merely testing to see if there was either positive or negative correlation, the confidence of our statement of correlation with the 0.679 value of r_s would be set at between the 0.10 and 0.02 levels.

example 11.1. The data[1] in Table 11.2 show the fraction of childless women under 44 years of age and the ratio of men to women in 12 New England and North Central states in 1950. We shall test whether an increase in the fraction of childless women correlates with a decrease in the ratio of men to women.

The states are ranked in columns 4 and 5 of Table 11.2 according to increase in the fraction of childless women and according to decrease in the ratio of men to women in the 12 states. Columns 6 and 7 give the d and d^2 values. The calculation of r_s is at the foot of the table.

† Sidney Siegel, "Nonparametric Statistics," McGraw-Hill Book Company, New York, 1956.
[1] Statistical Abstract of the United States, 1955, Tables 20, 42.

TABLE 11.2

State	Data				Rankings	
	Fraction childless women	Ratio men to women	Childless women	Men to women	d	d^2
Maine..	0.412	0.988	2	2.5	0.5	0.25
N.H....	0.418	0.969	4.5	10	5.5	30.25
Vt......	0.413	0.988	3	2.5	0.5	0.25
Mass. ..	0.479	0.938	12	12	0	0
R.I.....	0.470	0.973	11	6	5	25
Conn. ..	0.455	0.970	9	8.5	0.5	0.25
N.Y.....	0.465	0.954	10	11	1	1
N.J.....	0.449	0.972	8	7	1	1
Pa......	0.448	0.970	7	8.5	1.5	2.25
Ohio....	0.418	0.978	4.5	5	0.5	0.25
Ind.....	0.400	0.991	1	1	0	0
Ill......	0.434	0.983	6	4	2	4
Total.	.....	.....	....	....	...	64.50

$$r_s = 1 - \frac{6\Sigma d^2}{n(n^2 - 1)} = 1 - \frac{387}{(143)(12)} = 1 - 0.226 = 0.774$$

The value of 0.774 for r_s exceeds the 0.01 value of Table 11.1 for 12 observations. We may therefore say that there is less than 0.01 chance of having observed this value of the correlation coefficient if there were no correlation between the two variables.

Note that the same result would be obtained, with a negative value for r_s, if either of the rankings were in the opposite order. The student may verify this for himself. (The slight difference in the absolute value of the coefficient by the two methods is due to the presence of the tie ranks which introduce a very small error.)

11.1.2 Wilcoxon Paired-replicate Rank Test. The t test provides a method for determining whether the mean difference between a series of duplicate observations is significantly different from zero. The following test devised by Wilcoxon[1] offers a quick method of finding a significant difference between pairs of data, although it does not give a quantitative estimate of the difference. In the paired-replicate rank test, the differ-

---

[1] Frank Wilcoxon, "Some Rapid Approximate Statistical Procedures," American Cyanamid Co., New York, 1949.

ences between pairs are ranked without reference to the sign of the difference. If pairs have zero differences, they are not included in the test. Rank ties are handled in the same manner as previously described still without reference to sign, so that a difference of -2 and of $+2$ would be

TABLE 11.3 Minimum Value of Smaller Sums of Ranks of Either Sign at Indicated Probability Level for Replicates That Do Not Differ†

No. of replicates, n	Probability		
	0.05	0.02	0.01
6	0		
7	2	0	
8	4	2	0
9	6	3	2
10	8	5	3
11	11	7	5
12	14	10	7
13	17	13	10
14	21	16	13
15	25	20	16
16	30	24	20
17	35	28	23
18	40	33	28
19	46	38	32
20	52	43	38
21	59	49	43
22	66	56	49
23	73	62	55
24	81	69	61
25	89	77	68

† Frank Wilcoxon, "Some Rapid Approximate Statistical Procedures," American Cyanamid Co., New York, 1949.

considered a tie and both would be given the rank corresponding to the mean of the next two rank numbers to be used. After the ranking is completed, signs corresponding to the signs of the differences are applied to the rank numbers, and the smaller sum of the ranks of the same sign is used for test purposes.

If there was no significant difference between the pairs, we would expect the sum of the ranks with positive and negative signs to be about the same, equal to half the sum of the n ranking numbers. Table 11.3 from Wilcoxon gives the minimum sum of rank of either sign that might be expected at the indicated probability levels for different numbers of pairs if there were no difference between pairs. If the *smaller* sum obtained from ranking differences is *less* than the tabulated value, it indicates a significant difference between pairs with the corresponding chance of error.

The probability values in Table 11.3 apply to a two-sided test for determining a significant difference between pairs of either sign. If the sign of the smaller rank total is set in advance, i.e., if the test is to establish whether method A is significantly larger than method B (or smaller, but not both), then the probability levels correspond to the one-sided test and are half the values shown in Table 11.3.

example 11.2. The following hypothetical data represent a series of duplicate octane-number determinations on several different gasolines run by two different methods. We wish to make a quick test to determine whether the methods give significantly different results. The two octane numbers are given in the first two columns of Table 11.4.

TABLE 11.4

Octane numbers		Difference, $A - B$	Rank
Method A	Method B		
82	74	8	8
64	60	4	5
87	84	3	4
72	74	−2	−2.5
70	68	2	2.5
80	75	5	6
92	93	−1	−1
72	66	6	7

The differences, the first result minus the second, are in the third column. The rank of the absolute value of the differences, with signs corresponding to the sign of the difference, is given in column 4.

The smaller sum of ranks of like sign is -3.5. Referring to Table 11.3, we see that for eight replicates 4 is the minimum sum at the 0.05 level and 2 is the minimum sum at the 0.02 level. Therefore we may say, with less than a 0.05 but not less than 0.02 chance of error, that there is significant difference between the test methods.

11.1.3 Wilcoxon-White Rank Test of Unpaired Measurements. The previous test provided a quick method for comparing several sets of paired observations. The present ranking test is for determining a difference between two sets of data, not paired. Wilcoxon[1] developed the

[1] *Ibid.*

TABLE 11.5 Minimum Value of Rank Sum for Groups That Do Not Differ†
0.05 level

Larger group size, n_2	Smaller group size, n_1													
	2	3	4	5	6	7	8	9	10	11	12	13	14	15
4			10											
5		6	11	15										
6		7	12	18	26									
7		7	13	20	27	36								
8	3	8	14	21	29	38	49							
9	3	8	15	22	31	40	51	63						
10	3	9	15	23	32	42	53	65	78					
11	4	9	16	24	34	44	55	68	81	96				
12	4	10	17	26	35	46	58	71	85	99	115			
13	4	10	18	27	37	48	60	73	88	103	119	137		
14	4	11	19	28	38	50	63	76	91	106	123	141	160	
15	4	11	20	29	40	52	65	79	94	110	127	145	164	185
16	4	12	21	31	42	54	67	82	97	114	131	150	169	
17	5	12	21	32	43	56	70	84	100	117	135	154		
18	5	13	22	33	45	58	72	87	103	121	139			
19	5	13	23	34	46	60	74	90	107	124				
20	5	14	24	35	48	62	77	93	110					
21	6	14	25	37	50	64	79	95						
22	6	15	26	38	51	66	82							
23	6	15	27	39	53	68								
24	6	16	28	40	55									
25	6	16	28	42										
26	7	17	29											
27	7	17												
28	7													

† Reprinted with permission of the publishers and the author, Colin White, from *Biometrics*, **8**:33 (1952).

TABLE 11.6 Minimum Value of Rank Sum for Groups That Do Not Differ†
0.01 level

Larger group size, n_2	Smaller group size, n_1													
	2	3	4	5	6	7	8	9	10	11	12	13	14	15
5				15										
6			10	16	23									
7			10	17	24	32								
8			11	17	25	34	43							
9		6	11	18	26	35	45	56						
10		6	12	19	27	37	47	58	71					
11		6	12	20	28	38	49	61	74	87				
12		7	13	21	30	40	51	63	76	90	106			
13		7	14	22	31	41	53	65	79	93	109	125		
14		7	14	22	32	43	54	67	81	96	112	129	147	
15			8	15	23	33	44	56	70	84	99	115	133	151	171
16			8	15	24	34	46	58	72	86	102	119	137	155
17			8	16	25	36	47	60	74	89	105	122	140	
18			8	16	26	37	49	62	76	92	108	125		
19	3		9	17	27	38	50	64	78	94	111			
20	3		9	18	28	39	52	66	81	97				
21	3		9	18	29	40	53	68	83					
22	3		10	19	29	42	55	70						
23	3		10	19	30	43	57							
24	3		10	20	31	44								
25	3		11	20	32									
26	3		11	21										
27	4		11											
28	4													

† Reprinted with permission of the publishers and the author, Colin White, from *Biometrics*, **8**:33 (1952).

method for sets containing the same number of observations, and White[1] extended the method to groups of unequal sizes.

The procedure is to rank the data from the two sets of observations as if they were one array. After the ranking is completed, with ties assigned the mean rank regardless of whether they occur in one set or two, the sum of the ranks for each set is obtained. The smaller sum is compared with the values given in Tables 11.5 and 11.6. The tabulated values are

[1] Colin White, *Biometrics*, **8**:33 (1952).

NONPARAMETRIC STATISTICS 373

the minimum that can be expected at the indicated probability levels from groups that do not differ. If smaller values are obtained, a significant difference can be attributed to the two groups with a maximum chance of error set by the probability levels of the two tables. As it was pointed out in the beginning of the discussion of ranking, the order of the items to be ranked may be either from the largest to the smallest, or from the smallest to the largest. The Wilcoxon-White test for unpaired measurements requires the ranking to be done so as to produce the least possible sum. If the test indicates a sum less than the minimum nonsignificant value given in the tables, it is not necessary to repeat the calculation with the ranking in the reverse order. However, if the first ranking does not produce a significantly small sum, it is necessary to repeat the calculation in the reverse order to confirm that there is not a significant difference between the two sets of data. This condition is illustrated in example 11.3. It is not necessary actually to rerank the data and to perform the actual summation of the new ranks. The rank sum in one order is related in a simple way to the rank sum in the reverse order. If there are n items in a group and the total number of ranked items is N, the total in one ranking order is equal to $(N + 1)n$ minus the rank total in the reverse order. There is therefore no necessity for recalculating the rank totals in reverse order when the groups are of equal size because, as the foregoing relationship shows, the total of one group in one rank order will equal the total of the second group in the reverse order. The total of ranks for both groups will be $(N - 1)N/2$.

example 11.3. Suppose we were interested in a quick test to see whether the fraction of childless women was different in the Northern states than in the Southern states (1950 data). Table 11.7 gives the 1950 data[1] for 12 Northern states, repeated from Table 11.2, and additional data for 13 Southern states. In columns 3 and 6 are the ranking numbers for all 25 states.

The smaller sum of ranks is 137. Referring to Table 11.5, we see that for groups of 12 and 13 observations, the lowest nonsignificant sum is 119. Since the smaller sum is not significantly small, the ranking is reversed to check whether the smaller sum is the least possible. With the ranking reversed—the smallest fraction ranked 25 and the largest fraction ranked 1—the totals of the ranks become 124 for the North and 201 for the South ($[12 \times 26] - 188 = 124$, and $[13 \times 26] - 137 = 201$). The lesser sum, although smaller than the first smaller rank total, is still larger than the tabulated value of 119, so that no significant difference can be attributed to the two sets of data. Therefore, unless we are willing to accept more than a 5 per cent

[1] Statistical Abstract of the United States, 1955.

chance of error, we cannot say there is significant difference between these two groups of states.

It appears that the District of Columbia might make the total of the ranks of Southern states too high, and it is questionable whether the District of Columbia may rightfully be included in this category.

TABLE 11.7

	North			South	
State	Fraction childless women	Rank	State	Fraction childless women	Rank
Maine	0.412	8	Del.	0.447	18
N.H.	0.418	11.5	Md.	0.425	14
Vt.	0.413	9	D.C.	0.558	25
Mass.	0.479	24	Va.	0.430	15.5
R.I.	0.470	23	W.Va.	0.403	6
Conn.	0.455	21	N.C.	0.430	15.5
N.Y.	0.465	22	S.C.	0.420	13
N.J.	0.449	20	Ga.	0.402	4.5
Pa.	0.448	19	Fla.	0.415	10
Ohio	0.418	11.5	Ky.	0.401	3
Ind.	0.400	2	Tenn.	0.410	7
Ill.	0.434	17	Ala.	0.402	4.5
			Miss.	0.392	1
Total		188	Total		137

Elimination of the District of Columbia would decrease the sum of the Southern-state ranks to 112, which sum is just significant for two groups of 12 each. So there may be a significant difference, after all, if we are willing to discard the data from the District of Columbia.

A t test of means will not be significant (as the student may verify for himself) inasmuch as the variance within groups is so large that the small difference between means is not significant.

11.2 Sign Test

A simple test for determining a difference between paired observations is to count the number of positive and negative differences. If there were no real difference between two classifications, we would expect the number of positive and negative differences between pairs to be about the same. If the probability of a positive difference or a negative difference is 0.50, the cumulative binomial distribution provides a means of calculating the probability of observing less than any specific number of differences of one sign for various pairs of observations. Dixon and Mood[1] have published tables, which are condensed here, giving the maximum significant number for different probability levels of the less frequent sign. In other words, if the less frequent sign occurs a number of times equal to or less than the tabulated values, there is evidence at the cor-

TABLE 11.8 Maximum Significant Number of Less Frequent Sign in n Observations When $P(+) = 0.50$

n	Probability level			
	0.25	0.10	0.05	0.01
8	1	1	0	0
9	2	1	1	0
10	2	1	1	0
12	3	2	2	1
14	4	3	2	1
16	5	4	3	2
18	6	5	4	3
20	6	5	5	3
25	9	7	7	5
30	11	10	9	7
35	13	12	11	9
40	15	14	13	11
45	18	16	15	13
50	20	18	17	15
55	22	20	19	17
60	25	23	21	19
75	32	29	28	25
100	43	41	39	36

responding probability level of a difference between the classifications. The maximum significant values are given in Table 11.8.

The test is applied by observing the sign of the difference between pairs of data and ignoring the magnitude of the difference. If any pairs are

[1] W. J. Dixon and A. M. Mood, *J. Am. Statis. Assoc.*, **41**:557 (1946).

exactly tied, i.e., there is no difference, these are disregarded in the sign test. After the signs of all the differences have been noted, the number of the less frequent sign is obtained and compared with Table 11.8. If it is equal or less than the tabulated value for the number of pairs involved, there is evidence that there is a difference between the classifications of the observations forming the pairs.

Since the magnitude of the differences is not used in the sign test, it is obvious that it will be less sensitive than the paired-replicate rank test which takes into account both the sign and the magnitude of the differences. For instance, in example 11.2, there were eight pairs of replicates, of which two have negative differences and six positive. From Table 11.8, we can see that we could not say from the sign test alone that there was a difference between the two methods with even as low as a 0.25 probability of error. However, the sign test is exceedingly simple to apply: if two analysts differ 9 times out of 10, there is less than 0.05 chance that there is no difference between them—that familiarity with the values of Table 11.8 should be useful.

11.3 The Range

The difference between the largest and smallest values of a group of observations is the range. The range, like the variance, is a measure of the spread of the observations, and for comparable groups of data, a comparison of ranges will indicate differences in the internal consistency of the groups. Since the range is calculated from only two of the data points in a group, it is obviously not as sensitive as the variance which involves all the data. However, for small groups of data, the range is almost as efficient at the variance. ("One statistic is more efficient than another if it varies less from sample to sample for a fixed population. . . ."[1]) Snedecor[2] lists the efficiency of the range for samples of different sizes for estimating the standard deviation. For samples smaller than 8, the efficiency is more than 90 per cent; for samples from 8 to 10, the efficiency is greater than 85 per cent; and for sample sizes 11 to 20, the efficiency is at least 70 per cent. Because of its relatively high efficiency, and because it is extremely easy to calculate, the range is used in inspection and quality-control work with small samples where the variation is an important factor.

A considerable amount of literature has been accumulating in the past

---

[1] C. Eisenhart (eds.), "Techniques of Statistical Analysis," McGraw-Hill Book Company, New York, 1947, p. 227.

[2] George W. Snedecor, "Statistical Methods," 5th ed., Iowa State College Press, Ames, Iowa, 1956.

several years dealing with the use of the range for a number of statistical tests. Some of the tests are presented in this section.[1]

11.3.1 Estimating the Standard Deviation from the Range.

Factors are available in quality-control manuals for establishing the expected value and control limits for ranges when the standard deviation of the population is known. The factor for predicting the range from the standard deviation is universally referred to as d_2. The reciprocal of this value provides a factor for estimating the standard deviation from the range.

$$\sigma = R \frac{1}{d_2} \qquad (11.2)$$

where R is the range, and $1/d_2$ is a conversion factor depending on the sample size. Table 11.9 lists the values of $1/d_2$ for various sample sizes.

TABLE 11.9 Factors for Estimating the Standard Deviation from the Range

n	$1/d_2$	n	$1/d_2$
2	0.886	9	0.337
3	0.591	10	0.325
4	0.486	12	0.307
5	0.430	14	0.294
6	0.395	16	0.283
7	0.370	18	0.275
8	0.351	20	0.268

In example 6.1 the 11 observations of an octane-number determination were given. The standard deviation estimated from the data was 0.795. The range of the 11 observations was $88.6 - 86.1 = 2.5$. Interpolation from Table 11.9 gives a factor for estimating the standard deviation from 11 observations of 0.316. The standard deviation estimated from the range is $(2.5)(0.316) = 0.790$.

11.3.2 Tests Using the Range.

Inasmuch as the range provides a reasonable estimate of the standard deviation, it can be used in the denominator of a t-type test for setting a confidence range to means; for determining a significant difference between a mean and some standard value; or for checking the difference between two means. The tests are set up in a manner exactly paralleling the t tests. Lord[2] had proposed the symbol u instead of t for these tests.

---

[1] For more extensive discussions, the student is referred to E. Lord, *Biometrika*, **34**:41 (1947); John W. Tukey, *Trans. N.Y. Acad. Sci.*, **16**:88 (1953); J. E. Jackson and E. L. Ross, *J. Am. Statis. Assoc.*, **50**:417 (1955).

[2] E. Lord, *op. cit.*

For testing whether a mean is significantly different from some established value, the following formulation is used:

$$u = \frac{|\bar{x} - m|}{R} \tag{11.3}$$

The maximum values of u that can be expected with various probabilities for different-size samples drawn from a population with a mean

TABLE 11.10 Critical Values of u for Various Sample Sizes, n, and Probability Levels, α†

$$\left[u = \frac{|x - m|}{R} \right]$$

n \ α	0.10	0.05	0.02	0.01
2	3.157	6.351	15.910	31.828
3	0.885	1.304	2.111	3.008
4	0.529	0.717	1.023	1.316
5	0.388	0.507	0.685	0.843
6	0.312	0.399	0.523	0.628
7	0.263	0.333	0.429	0.507
8	0.230	0.288	0.366	0.429
9	0.205	0.255	0.322	0.374
10	0.186	0.230	0.288	0.333
11	0.170	0.210	0.262	0.302
12	0.158	0.194	0.241	0.277
13	0.147	0.181	0.224	0.256
14	0.138	0.170	0.209	0.239
15	0.131	0.160	0.197	0.224
16	0.124	0.151	0.186	0.212
17	0.118	0.144	0.177	0.201
18	0.113	0.137	0.168	0.191
19	0.108	0.131	0.161	0.182
20	0.104	0.126	0.154	0.175

† Condensed from Tables 9 and 10 from E. Lord, *Biometrika*, **34**:41 (1947).

of m are given in Table 11.10. The test is handled in the same way as the t test. The value of u, calculated from the sample mean and sample range and the hypothesized value of the true mean, is compared with the critical values given in Table 11.10. If u is greater than the tabulated value for the particular sample size, the hypothesis that the mean estimated by the sample is equal to m may be rejected with the indicated probability of a false rejection.

example 11.4. Again using the data of example 6.1: 11 octane numbers were obtained with a mean of 87.1. The question was raised as to whether these could have come from a material with a true octane number of 87.5. The highest and lowest observed values were 88.6 and 86.1, giving a range of 2.5. u is calculated as follows:

$$u = \frac{87.5 - 87.1}{2.5} = 0.160 \tag{11.4}$$

Table 11.10 indicates that with 11 observations a u value of this magnitude can be expected somewhat more than 10 per cent of the time when the hypothesis being tested is true. If we reject the hypothesis and say the mean is not 87.5, we take more than 10 per cent chance of being in error.

u and the range can be used to set confidence ranges for the true mean in the same way that t and the standard deviation were used in Eq. (6.15). If Eq. (11.3) is transposed and m replaced by μ to designate the true mean rather than an expected value, we obtain

$$\mu = \bar{x} \pm (u)R \tag{11.5}$$

Using the numerical values from example 11.4, we can set a 90 per cent confidence range for the true mean of

$\mu = 87.1 \pm (0.210)(2.5)$
$87.6 \geq \mu \geq 86.6$

or a 99 per cent confidence range of

$\mu = 87.1 \pm (0.302)(2.5)$
$87.9 \geq \mu \geq 86.3$

The range may also be used to establish a significant difference between two means in the same way that t and the pooled estimate of standard deviation were used. In this case the formulation is

$$u_{\bar{R}} = \frac{|\bar{x}_1 - \bar{x}_2|}{\bar{R}} \tag{11.6}$$

$\bar{R}$ is the mean range of the two samples. The two samples must be of the same size to use the test as given here.

The critical values, i.e., the maximum values of $u_{\bar{R}}$ that can be expected with various probabilities for samples of equal size drawn from populations with the same mean, are given in Table 11.11. The hypothesis for this test is that the two means estimated by the samples are equal. If $u_{\bar{R}}$ calculated from the two sample means and the mean range of the two samples

TABLE 11.11 Critical Values of $u_{\bar{R}}$ for Various Equal Sample Sizes, n, and Probability Levels, α†

$$\left[u_{\bar{R}} = \frac{|\bar{x}_1 - \bar{x}_2|}{\bar{R}} \right]$$

n \ α	0.10	0.05	0.02	0.01
2	2.322	3.427	5.553	7.916
3	0.974	1.272	1.715	2.093
4	0.644	0.813	1.047	1.237
5	0.493	0.613	0.772	0.896
6	0.405	0.499	0.621	0.714
7	0.347	0.426	0.525	0.600
8	0.306	0.373	0.459	0.521
9	0.275	0.334	0.409	0.464
10	0.250	0.304	0.371	0.419
11	0.233	0.280	0.340	0.384
12	0.214	0.260	0.315	0.355
13	0.201	0.243	0.294	0.331
14	0.189	0.228	0.276	0.311
15	0.179	0.216	0.261	0.293
16	0.170	0.205	0.247	0.278
17	0.162	0.195	0.236	0.264
18	0.155	0.187	0.225	0.252
19	0.149	0.179	0.216	0.242
20	0.143	0.172	0.207	0.232

† Condensed from Tables 9 and 10 from E. Lord, *Biometrika*, 34:41 (1947).

is greater than the tabulated value at the proper level of n, which is the number of observations in *each* sample, the hypothesis of equal means can be rejected with the indicated risk of a false rejection.

example 11.5. In example 6.5 the results of two sets of five replicate flow determinations on some powdered metal were presented and tested by the t test to determine whether a significant difference existed in the mean flow times of the two powders.

TABLE 11.12 Summary of Data
from Example 6.6

	Test 1	Test 2
No. of runs............	5	5
Mean time, sec.......	29.96	30.84
Range R.............	1.0	0.4
Mean range $\bar{R}$.......	0.7	

$$u \text{ test: } u_{\bar{R}} = \frac{30.84 - 29.96}{0.7} = 1.257**$$

The data are summarized in Table 11.12.

This value of $u_{\bar{R}}$ exceeds the value for five observations at the 0.01 significance level, and we may therefore reject the hypothesis of equal means for the two tests with less than 0.01 chance of being wrong.

The simplicity of the significance tests using the range compared to those using the estimated standard deviation with apparently similar results might bring up the question as to why more time has not been given to this statistic. The range is an important statistic in quality-control work where groups of small numbers of observations are involved. With samples of five to eight observations, the range is about as efficient, 89 to 96 per cent, as the sample standard deviation. However, the efficiency falls off rapidly with increase in sample size. With samples greater than 20 it is obvious that the variation of the data is being measured with less than 10 per cent of the observations when the range is used, while the standard deviation is rapidly approaching the population value as indicated by the approach of the t values to those of the standard deviate of the normal distribution.

For quick estimates from small quantities of data, the range is an effective statistic to use. In the examples used for illustration in this book, small quantities of data were used so that the arithmetic would be easy to follow. In these cases the range is as effective as the standard deviation. It is far more common in engineering practice to have larger quantities of data for statistical tests, and the range would ordinarily not be used when the number of observations was greater than 10.

11.4 The Corner Test

Olmstead and Tukey[1] have developed a simple and rapid test for checking association between two variables. The engineer often plots an accu-

---

[1] P. S. Olmstead and J. W. Tukey, *Ann. Math. Statis.*, **18**:495 (1947).

mulation of data and wonders whether there is evidence of an association. Although we attempted an eloquent argument for calculating the correlation coefficient in Chap. 9, it is possible to determine (statistically) the presence of an association between two variables without any numerical calculations, other than counting, if sufficient data are available. This method is best described by an illustration.

The data are plotted on rectangular coordinate paper. See Fig. 11.1 wherein a group of carbon analyses from a furnace melt and from a ladle are plotted in the usual manner.

Medians for the x coordinate and for the y coordinate are drawn through the data dividing the figure into four quadrants. The upper-right and lower-left quadrants are designated positive, and the upper-left and lower-right are designated negative. (A median is a value which is exceeded by half the data and which exceeds half the data. For an even number of observations, with no ties, the medians will not intersect any of the data. The procedure with an odd number of observations and when ties are encountered will be discussed in just a moment.)

The significance of the association of the variables is established by the outlying data points. A line parallel to and above the y median is drawn so that it excludes from the main body of data as many points as possible all on one side of the x median. A similar parallel is drawn below the y

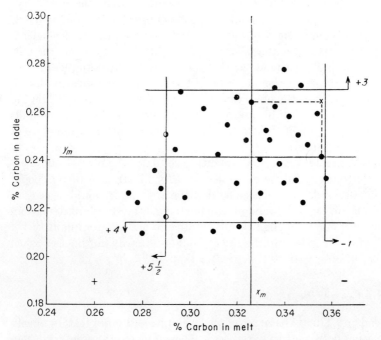

Fig. 11.1 The Corner-test method—carbon in ladle vs. carbon in melt.

median. Two vertical lines parallel to the x median are drawn on the right and left sides, both of which exclude as many data points as possible all on one side of the y median (not necessarily the same side for the two lines).

The points excluded by each of the guide lines are counted separately and assigned the sign of the quadrant in which they fall. If a point falls beyond two of the guide lines, it is counted twice. The algebraic sum of all the outlying points is obtained. Table 11.13 from Olmstead and

TABLE 11.13 Critical Values of Corner Association Test

Significance level	Quadrant sum
0.10	9
0.05	11
0.02	13
0.01	14–15
0.005	15–17
0.002	17–19
0.001	18–21

Tukey[1] gives the critical values of this total corresponding to different probability levels. If the total from the data is equal to or greater than the values in Table 11.13, the hypothesis of no association between the variables may be rejected at the indicated significance levels. A positive sign to the significant total indicates a positive association, and a negative sign indicates a negative association. The quadrant sum from Fig. 11.1 is $+3, -1, +5\frac{1}{2}, +4 = 11\frac{1}{2}$; indicating a significant correlation at the 0.05 level.

In general, the test is independent of the number of data points involved. The range of critical values of the quadrant sums for the lower probability levels is for samples of different sizes, the lower values applying to larger samples.

If the data consist of an odd number of points, the medians will pass through one or two of them. The intersected points need be especially considered only if they are involved in the final counting. In this case the two points are replaced by a single point which has the x, y coordinates corresponding to the x coordinate of the point intersected by the y median, and the y coordinate of the point intersected by the x median. An example of this replacement is shown in the upper-right quadrant of Fig. 11.1.

In cases of ties, again they need only be considered when they affect the final counting. However, if the data tend to have a large number of ties, the method is not applicable. If a median goes through a number of ties, these may be distributed on alternate sides of the median, the location

[1] *Ibid.*

of the first point being decided by the toss of a coin. If a guide line intersects tied data on alternate sides of a median line, the recommended procedure is to include in the quadrant count the number of tied points according to the following formula:

$$\text{Number} = \frac{\text{number favorable to continuing the count}}{1 + \text{number favorable to discontinuing the count}}$$
(11.7)

An example of a tied pair is shown in the left side of Fig. 11.1. Here, of the tied pair, one is favorable to continuing the count and one is unfavorable, so that the number of points added to the quadrant sum from the tied pair is $\frac{1}{2}$.

11.5 Discarding Data

A question close to probably every engineer's heart is, when may data be discarded as "obviously" not consistent with other measurements? Despite the best-laid plans of directors and engineers, air will get in the gas sample, or a thermocouple will be connected backward, or part of the product will be spilled before being weighed. Besides cases when mistakes occur, there are others when the measurements include the effect of variables (perhaps intermittent in operation) which are not germane to the problem being studied. A system being studied for catalyst activity might suddenly fluctuate more than usual because of variation of cooling water temperature which had previously been constant. An increased variation in an operation might be caused by unknown changes in flow due to foreign matter in an orifice metered line.

Observations can be expected to fall within some range of the mean, the deviations from the mean being set by the frequency and the total variance of the process, when the variance is composed of a number of small effects acting at random. If a new effect is added, or a bias (gross error) occurs, deviations from the mean that would be expected to occur with extremely small frequency under normal circumstances will be observed. When such deviations occur they may be discarded on the grounds (1) that they represent the effect of variables that are not under study, or (2) that they are the result of absolute errors and do not represent the actual observations, or (3) even if assumption 1 or 2 is not true, the large deviation can be considered the result of a combination of factors that is atypical and not representative of usual operating conditions. If a sample (measurement or observation) is drawn from a population (process or experiment) with a mean μ and a standard deviation σ, and the population is normally distributed, we have learned that there is only 1 chance in 20 of observing a deviation from the mean as large as $\pm 1.96\sigma$. There is only

1 chance in 100 of a deviation as large as $\pm 2.58\sigma$, 1 chance in 1,000 of $\pm 3.219\sigma$, 1 chance in 1 million of $\pm 4.89\sigma$. Under this ideal situation, it is reasonable to discard data which deviate from the mean by 5σ, or 4σ ($P = 0.000063$, 1 chance in 158,000). In discarding the observation with the large deviation, we might say either that we were probably in error in assuming the sample came from the specific population since the occurrence of so large a deviation is highly unlikely; or we might say that we are seeking a sample to be representative of the population, and a sample with this large a deviation is not representative. On either basis, we are justified in discarding the sample.

The usual situation with industrial or experimental data is somewhat different. Here we have a group of measurements with an apparent outlying member. We need some basis for deciding whether the outlier is a reasonable result based on the other observations or whether it may be discarded as being an improbable member of the group. Before we give some rules for discarding data, we want to emphasize that all data contain some information, if it is only about the fallibility of the data taker or the limitations of the data analyzer. Rather than say that the outlying data are "wrong," let us say we cannot use these data for the problem at hand.

Several methods have been suggested for judging outlying observations. We shall give tables for two: one from Nair,[1] using the estimated variance, and one from Dixon,[2] based on the range. Both of these methods apply to rejecting a single outlying point from a small group of data.

The method of Nair gives the maximum deviations from the sample mean that can be expected for single values in samples of different size when the standard deviation is known or has been estimated from samples of another size. Table 11.14 gives values of the extreme deviates (from the sample mean) at various probability levels. $\bar{x}$ is the sample mean, and x_m is the maximum or minimum value of x that can be expected in a sample of size n at the indicated probability level. s_ν is the estimated standard deviation, estimated from some other source with ν degrees of freedom. If a deviate that is larger than the tabulated value is observed in a sample, this single observation may be discarded as being significantly different from the remainder of the sample, with the corresponding risk of error.

example 11.6. A modification is proposed to a partial-combustion process for producing acetylene in order to increase the yield, which is normally about 7.0 per cent. A trial run with 20 determinations (19 degrees of freedom) gives a standard deviation for the process and analysis of 0.40. Five runs with the modified process give the follow-

[1] K. R. Nair, *Biometrika*, **39**:190 (1952).
[2] W. J. Dixon, *Ann. Math. Statis.*, **22**:68 (1951).

TABLE 11.14 Percentage Points of the Extreme Deviate[†]

$$\frac{|x_m - \bar{x}|}{s_\nu}$$

f \ n	10% level							5% level						
	3	4	5	6	7	8	9	3	4	5	6	7	8	9
10	1.68	1.93	2.11	2.25	2.36	2.46	2.54	2.02	2.29	2.49	2.63	2.75	2.85	2.93
11	1.66	1.91	2.08	2.21	2.32	2.42	2.49	1.99	2.26	2.44	2.58	2.70	2.79	2.87
12	1.65	1.89	2.05	2.19	2.29	2.38	2.46	1.97	2.22	2.40	2.54	2.65	2.75	2.83
13	1.63	1.87	2.04	2.16	2.27	2.36	2.43	1.95	2.20	2.38	2.51	2.62	2.71	2.79
14	1.62	1.85	2.02	2.14	2.25	2.33	2.41	1.93	2.18	2.35	2.48	2.59	2.68	2.76
15	1.61	1.84	2.00	2.13	2.23	2.31	2.39	1.92	2.16	2.33	2.46	2.56	2.65	2.73
16	1.61	1.83	1.99	2.12	2.22	2.30	2.37	1.90	2.14	2.31	2.44	2.54	2.63	2.70
17	1.60	1.82	1.98	2.10	2.20	2.28	2.35	1.89	2.13	2.30	2.42	2.52	2.61	2.68
18	1.59	1.82	1.97	2.09	2.19	2.27	2.34	1.88	2.12	2.28	2.41	2.51	2.59	2.66
19	1.59	1.81	1.97	2.09	2.18	2.26	2.33	1.87	2.11	2.27	2.39	2.49	2.58	2.65
20	1.58	1.80	1.96	2.08	2.17	2.25	2.32	1.87	2.10	2.26	2.38	2.48	2.56	2.63
24	1.57	1.78	1.94	2.05	2.15	2.22	2.29	1.84	2.07	2.23	2.35	2.44	2.52	2.59
30	1.55	1.77	1.92	2.03	2.12	2.20	2.26	1.82	2.04	2.20	2.31	2.40	2.48	2.55
40	1.54	1.75	1.89	2.01	2.09	2.17	2.23	1.80	2.02	2.17	2.28	2.37	2.44	2.51
60	1.52	1.73	1.87	1.98	2.07	2.14	2.20	1.78	1.99	2.14	2.25	2.33	2.41	2.47
120	1.51	1.71	1.85	1.96	2.05	2.12	2.18	1.76	1.97	2.11	2.21	2.30	2.37	2.43
∞	1.50	1.70	1.83	1.94	2.02	2.09	2.15	1.74	1.94	2.08	2.18	2.27	2.33	2.39
	2.5% level							1% level						
10	2.36	2.64	2.84	2.99	3.10	3.20	3.28	2.76	3.05	3.25	3.39	3.50	3.59	3.67
11	2.31	2.59	2.78	2.93	3.04	3.14	3.22	2.71	3.00	3.19	3.33	3.44	3.53	3.61
12	2.28	2.55	2.74	2.88	2.99	3.08	3.16	2.67	2.95	3.14	3.28	3.39	3.48	3.55
13	2.25	2.52	2.70	2.84	2.95	3.04	3.12	2.63	2.91	3.10	3.24	3.34	3.43	3.51
14	2.23	2.49	2.67	2.80	2.91	3.00	3.08	2.60	2.87	3.06	3.20	3.30	3.39	3.47
15	2.20	2.46	2.64	2.77	2.88	2.97	3.04	2.57	2.84	3.02	3.16	3.27	3.35	3.43
16	2.19	2.44	2.62	2.75	2.85	2.94	3.01	2.55	2.81	3.00	3.13	3.24	3.32	3.39
17	2.17	2.42	2.60	2.73	2.83	2.92	2.99	2.52	2.79	2.97	3.10	3.21	3.29	3.36
18	2.16	2.41	2.58	2.71	2.81	2.89	2.97	2.50	2.77	2.95	3.08	3.18	3.27	3.34
19	2.15	2.39	2.56	2.69	2.79	2.87	2.95	2.49	2.75	2.92	3.06	3.16	3.24	3.31
20	2.14	2.38	2.55	2.67	2.77	2.86	2.93	2.47	2.73	2.91	3.04	3.14	3.22	3.29
24	2.10	2.34	2.50	2.62	2.72	2.80	2.87	2.43	2.68	2.85	2.97	3.07	3.15	3.22
30	2.07	2.30	2.46	2.58	2.67	2.75	2.82	2.38	2.62	2.79	2.91	3.01	3.08	3.15
40	2.04	2.27	2.42	2.53	2.62	2.70	2.76	2.34	2.57	2.73	2.85	2.94	3.02	3.08
60	2.01	2.23	2.38	2.49	2.58	2.65	2.71	2.30	2.52	2.68	2.79	2.88	2.95	3.01
120	1.98	2.20	2.34	2.45	2.53	2.60	2.66	2.25	2.48	2.62	2.73	2.82	2.89	2.95
∞	1.95	2.16	2.30	2.41	2.49	2.56	2.61	2.22	2.43	2.57	2.68	2.76	2.83	2.88
	0.5% level							0.1% level						
10	3.03	3.32	3.52	3.65	3.76	3.85	3.92	3.54	3.84	4.04	4.17	4.28	4.35	4.40
11	2.98	3.27	3.46	3.60	3.71	3.79	3.86	3.49	3.80	3.99	4.12	4.23	4.30	4.36
12	2.93	3.22	3.41	3.55	3.66	3.74	3.81	3.45	3.75	3.94	4.07	4.19	4.26	4.31
13	2.89	3.18	3.37	3.50	3.61	3.70	3.77	3.41	3.71	3.90	4.03	4.14	4.22	4.28
14	2.86	3.14	3.33	3.46	3.57	3.66	3.73	3.38	3.67	3.86	4.00	4.10	4.18	4.24
15	2.83	3.11	3.29	3.43	3.53	3.62	3.69	3.35	3.64	3.83	3.96	4.06	4.15	4.21
16	2.80	3.08	3.26	3.39	3.50	3.58	3.66	3.32	3.51	3.80	3.93	4.03	4.12	4.18
17	2.77	3.05	3.23	3.36	3.47	3.55	3.62	3.29	3.58	3.77	3.90	4.00	4.09	4.15
18	2.75	3.02	3.21	3.34	3.44	3.53	3.60	3.27	3.55	3.74	3.88	3.98	4.06	4.12
19	2.73	3.00	3.18	3.31	3.42	3.50	3.57	3.25	3.53	3.72	3.85	3.95	4.03	4.10
20	2.72	2.98	3.16	3.29	3.39	3.48	3.55	3.23	3.51	3.70	3.83	3.93	4.01	4.08
24	2.66	2.92	3.09	3.22	3.32	3.40	3.47	3.16	3.44	3.62	3.75	3.85	3.93	4.00
30	2.60	2.86	3.02	3.15	3.25	3.32	3.39	3.08	3.36	3.53	3.66	3.76	3.84	3.90
40	2.55	2.79	2.96	3.08	3.17	3.25	3.31	3.01	3.27	3.44	3.57	3.66	3.74	3.81
60	2.50	2.73	2.89	3.01	3.10	3.17	3.23	2.93	3.19	3.35	3.47	3.56	3.64	3.70
120	2.45	2.67	2.83	2.94	3.02	3.09	3.15	2.85	3.10	3.26	3.37	3.46	3.53	3.59
∞	2.40	2.62	2.76	2.87	2.95	3.02	3.07	2.78	3.01	3.17	3.28	3.36	3.43	3.48

[†] K. R. Nair, *Biometrika*, Vol. 39 (1952).

ing percentages of yield:

8.4, 8.2, 8.5, 7.3, 8.1

We wish to test whether the low yield of 7.3 per cent might reasonably be discarded.

The mean of the five results is 8.1 per cent. The deviation of the low value from the mean, 8.1 − 7.3, is 0.8. In units of the standard deviation, this deviate is 0.8/0.4 = 2.0. Entering Table 11.14 with a deviate of 2.0 for a sample of $n = 5$ and a standard deviation estimated with 19 degrees of freedom, we see that the observed value exceeds the 10 per cent extreme deviate but is less than the 5 per cent value; i.e., we may discard the 7.3 per cent value with less than 10 per cent but more than 5 per cent risk of error.

The method of Dixon[1] provides a test for extreme values when no estimation of the standard deviation is available. If the observations in the sample are ranked, the individual values can be identified x_1, x_2, x_3, . . . , x_{n-1}, x_n. It is immaterial whether the ranking proceeds from high values to low or from low values to high. The Dixon extreme-value test gives the maximum ratio of differences between extreme-ranking observations to be expected at various probability levels and for different sample sizes. Table 11.15 gives the test ratios and maximum expected values. For samples less than about eight observations, the ratio of the difference between the extreme and the next-to-extreme value to the total range is compared with the tabulated values for the same sample size. If the observed ratio exceeds the tabulated maximum expected ratio, the extreme value may be rejected with the risk of error set by the probability level. For samples between about 9 and 14, test the ratio of the difference between the first and third ranking observations to the difference between the first and next to last. For samples of 15 or more, use the ratio of the difference between the first and third ranking observations to the difference between the first- and the second-from-last observation.

In example 11.6 the difference between the first and second values is 0.8, and the difference between the first and last is 1.2, giving a ratio of 0.8/1.2 = 0.667. For a sample with five observations, this value lies between the 0.05 and 0.01 probability levels, so that by this range ratio test we could reject the low value of 7.3 per cent with less than 0.05 chance of error.

Note that a different conclusion would be drawn from the two tests. This difference between the results of the two tests raises the obvious question of which one to use. This is an academic question and would not be raised with actual data. The two tests for extreme deviates

[1] *Ibid.*

TABLE 11.15 Maximum Ratio of Extreme Ranking Observations[†]

Recommended for sample size	Rank difference ratio	Sample size, n	Maximum ratio		
			Probability level		
			0.10	0.05	0.01
$n < 8$	$\dfrac{x_2 - x_1}{x_n - x_1}$	3	0.886	0.941	0.988
		4	0.679	0.765	0.889
		5	0.557	0.642	0.780
		6	0.482	0.560	0.698
		7	0.434	0.507	0.637
$8 < n < 15$	$\dfrac{x_3 - x_1}{x_{n-1} - x_1}$	8	0.650	0.710	0.829
		9	0.594	0.657	0.776
		10	0.551	0.612	0.726
		11	0.517	0.576	0.679
		12	0.490	0.546	0.642
		13	0.467	0.521	0.615
		14	0.448	0.501	0.593
$n \geq 15$	$\dfrac{x_3 - x_1}{x_{n-2} - x_1}$	15	0.472	0.525	0.616
		16	0.454	0.507	0.595
		17	0.438	0.490	0.577
		18	0.424	0.475	0.561
		19	0.412	0.462	0.547
		20	0.401	0.450	0.535

[†] W. J. Dixon, *Ann. Math. Statis.*, **22**:68 (1951).

involve two different hypotheses. One sets the maximum deviation that might be expected in a sample drawn from a normal population when an independent estimate of the standard deviation of the population is available. The other sets the maximum difference between extreme observations that might be expected in a sample drawn from a normal population when there is no estimate of the standard deviation available. The test to be used in a specific case would depend on whether an estimate of the standard deviation were available from some previous source.

In textbook problems, the data provide all the evidence on which judgments are made. In the practical case, it is often necessary to take into account information about the population (system) that comes from other sources than the actual measurements of the data. How the engineer-statistician divides his judgments, or how the engineer and statistician come to an agreement, is often a nice question. We would only add to Polonius' advice: "to thine own self be true . . ." and take more data.

chapter twelve
COMPUTER CALCULATIONS

All, or practically all, the statistical tests and procedures discussed in the preceding chapters of this book can be done with prearranged programs in the various computers that are available to engineers. As with all engineering problems, it is necessary to size the equipment to the problem. You would not plant your crocuses with a 40 cu ft power shovel, and you would not measure blast furnace output on a gram balance. So you would not calculate the mean of half a dozen samples with a "giant" computer, and you would not (unless absolutely necessary) do a multiple logarithmic regression analysis with a 1940 desk calculator. The purpose of this chapter is to discuss some of the facilities available for the different types of statistical problem.

12.1 Desk Calculators

Calculations with small quantities of data—means, t tests, standard deviations, 2^n factorials, and even linear regressions—can be readily done on the numerous desk calculators that are available. For almost all of the statistical tests, the calculations involve Σx, Σx^2, Σy, Σy^2, Σxy, and some algebraic manipulation of these values. A number of the desk calculators will perform the summation of single entries and the summation of products (and/or squares) in one operation. In other words, by entering a series of x values or a series of x and y values, Σx, and Σxy, or Σx, and Σx^2 can be obtained at the same time. If the data involve only one or two significant digits, Σx, Σx^2, Σy, Σy^2, and $2\Sigma xy$ can be obtained in one operation. Several of the mechanical desk calculators also have automatic square root operations so that, for statistical calculations, the standard deviation and correlation coefficients, as well as other functions, can be calculated by the simple entry of the number for which the root is desired.

At this writing (1968 to 1969) there are some eight or twelve different desk calculators available in the United States that are particularly suitable for engineering statistical calculations. The author's current favorite is the Wang,[1] which is an electronic calculator that not only has all of the features mentioned in the preceding paragraph, but is also capable of the direct conversion of data to logarithms and exponential values. This feature permits the direct correlation both of data as entered and/or logarithms of the data, which is common practice when semilog or log-log plots of the data indicate exponential relationships between the variables.

A number of the popular desk calculators have larger models which provide programming facilities for routine calculations that are identical from calculation to calculation except for the data involved. The methods of handling repetitive operations differ among machines. One method includes facilities for punching a paper tape when the first set of operations is performed. This tape can then be reused for similar operations with the entry of different sets of data. Another method incorporates facilities for storing programs within the calculator so that when the operation is first entered, it is stored and can be used again as often as desired for similar calculations. The Wang calculator has a "programmer" which is plugged into the electric circuit between the desk calculator and the power source. Programs are handpunched on standard IBM cards and placed in the "programmer." Data are entered in the keyboard of the regular calculator and the mathematical operations are carried out from the "instructions" punched into the program card.

All of these procedures are very adequate for repetitive routine calculations of the various statistical tests discussed in this book. Manufacturers of the desk calculators that are equipped with programming facilities are prepared to supply programs for the more common statistical procedures. It is the author's experience, however, that except perhaps for the more complicated programs such as multiple regression analysis, programs written by the user for his specific needs are more satisfactory than the general programs written to satisfy all users. Programming for

[1] The Wang Laboratories, Massachusetts.

the desk calculators is very easily learned, but the methods are specific for each type of calculator so they can not be discussed here. Some discussion of programming for digital computers is included in the next section.

Single statistical calculations can be efficiently carried out on any of the desk calculators available. The author's preference has been stated. Different models have different features that make them more applicable to different circumstances. Often the choice of a particular model is not made on its applicability to statistical calculations alone. For statistical calculations the summation of squares and products and the square-root calculation are particularly valuable. A counting feature to note the number of entries is also of value in entering tabulated data. The value of programming facilities for a desk calculator depends entirely on the amount of repetitive calculations that are carried out. And at some point, when the number of repetitive calculations becomes quite large, a digital computer becomes more applicable. For multiple component regression analyses, for polynomial regression analyses, and for multicomponent factorial analyses, the digital computer is particularly suited. The use of digital computers in statistical analyses is discussed briefly in the next section.

12.2 Digital Computer Calculations

The discussion in these final sections assumes some familiarity on the part of the reader with the Fortran language of computer programming. If you are not familiar with Fortran, stay with the text anyway. If you continue in the field of engineering statistics, Fortran will catch up with you. The principal digital computers designed for engineering use have "statistical packages," i.e., programs, available for routine statistical calculations. While it will be necessary to understand the basis for the statistical calculation, more and more of the actual arithmetic is being performed by the computers and the current language of the computers is Fortran.

One principal source for statistical programs is the BMD, Biomedical Computer Programs published by the University of California. Individual computer manufacturers supply statistical packages for use in their computers. The following list enumerates some of the programs that are available.

Simple descriptive statistics:
    Arithmetic mean
    Geometric mean
    Harmonic mean
    Median

Mode
Range
Mean deviation
Standard deviation
Maximum and minimum values
Chi square tests
   2×2 and $n \times m$ contingency tables
   Proportions
   Goodness-of-fit to Poisson and normal distributions
Distribution: fitting, plotting, and calculating
   Binomial
   Poisson
   Normal
   Hypergeometric
Analysis of Variance
   General factorial up to 8 factors at 5 levels
   Balanced blocks (Latin squares)
   Nested factorials
   Complete and fractional 2^n factorials
Regression analysis
   Simple linear regression
   Multiple factor regression
   Polynomial regression
Nonparametric analyses
   Rank correlation
   Range tests

All of these programs are extremely efficient for doing exactly what they are programmed to do. Their shortcoming, if it can be called that, is that they do exactly and only exactly what they are programmed to do. For example, the t tests in the statistical package of one large computer work only for the comparison of means of two sets of data if the number of observations in both sets is the same. In order to test the mean of a single set of observations against some preassigned value, it is necessary to enter a "dummy" variable inasmuch as the program has been written to deal with two or more sets of data, and cannot handle only one.

Some of the preprogrammed regression analyses provide for the entering of several sets of data and then permit the calculation of correlation between different pairs (for a two-variable calculation), or among several variables (for a multiple regression calculation). The programs usually permit the manipulation—deletion, transforming by logarithms or by exponentiation, multiplying by a constant, etc.—after a preliminary regression is run. However, if a log-log or a semilog correlation is desired initially, it is not possible to obtain it. The program provides for the transformation of data that has been used in a previous regression analysis. It is necessary to let the program calculate a simple linear regression before it will calculate a log-log or a semilog regression. Also, the regression programs do not provide for the cases involving the restriction of forcing the regression through the origin of the coordinates, the 0., 0. point.

The computer calculations are extremely fast and accurate, and their contribution to engineering and statistical calculations is firmly established. The only point of these apparently critical statements is to suggest that the engineer is often better advised to provide his own program or to modify existing programs to his particular needs, rather than to adjust his data to fit the "prepackaged" programs that are available. It is also important to know exactly what the program will do and how it handles restrictions, if any, of the statistical test involved. The chi square programs in one statistical package, for example, do not take into special consideration problems involving categories with five or less observations, and observations of this magnitude tend to bias chi square too high.

12.3 Some Statistical Computation Programs

The programming of routine statistical calculations is relatively simple inasmuch as the analytical expressions for the statistics involved are well defined. It is merely necessary to program the same calculation routine that would be carried out by hand or by desk computation. The following sample programs are written in Fortran. The variable names are selected to be self-identifying to make the programs easier to follow. The variable names follow the symbols and nomenclature used in the earlier chapters of this book. In addition, some explanatory statements are included at the right of the programs. These explanatory notes are not part of the actual computer program, but are merely descriptive. The equation numbers in the explanatory notes refer to equations in this text where the particular statistical functions are discussed.

Data input are given as "READ 0, . . ." statements. The input statements will vary among computers and will vary for different types of data input. No output statements are included in the programs inasmuch as these will differ among different computer systems, and each engineer will use his own format for output. The calculated statistics are identified in each program by the explanatory statements, and each user can select the items he desires to have printed in the output. No defining statements, dimension statements, or end statements are included.

Considerable latitude exists in programming, and the following are principally descriptive and not necessarily the "best" programs for the particular calculation. The purpose of these programs in the form they are presented is to show the steps involved in the calculation with a minimum of explanatory notes. The programs that follow are some that are not readily available in the standard statistics packages with which the author is familiar.

12.3.1 *t* Test for Difference between Two Means

READ 0, N1, N2	Size of the two samples
READ 0, (X1(I), I = 1, N1)	First set of data
READ 0, (X2(I), I = 1, N2)	Second set of data
SUMX1 = 0.	
SUMX2 = 0.	
SUMX12 = 0.	
SUMX22 = 0.	
DO 1 I = 1, N1	
SUMX1 = SUMX1 + X1(I)	Σx_1
1 SUMX12 = SUMX12 + X1(I) * X1(I)	Σx_1^2
DO 2 I = 1, N2	
SUMX2 = SUMX2 + X2(I)	Σx_2
2 SUMX22 = SUMX22 + X2(I) * X2(I)	Σx_2^2
X1MEAN = SUMX1/FLOAT(N1)	$\bar{x}_1$
X2MEAN = SUMX2/FLOAT(N2)	$\bar{x}_2$
SPX12 = SUMX12 − X1MEAN * SUMX1	$\Sigma' x_1^2$ [Eq. (4.9)]
SPX22 = SUMX22 − X2MEAN * SUMX2	$\Sigma' x_2^2$ [Eq. (4.9)]
SP = SQRT((SPX12 + SPX22)/(FLOAT(N1 + N2 − 2)))	Pooled standard deviation, $\bar{s}$, [Eq. (6.25)]
T = ABS(X1MEAN − X2MEAN)/(SP * SQRT(1./FLOAT(N1) + 1./FLOAT(N2)))	*t* value [Eq. (6.24)]
NU = N1 + N2 − 2	Degrees of freedom

12.3.2 Regression through Origin, $Y = B * X$

This program follows the discussion of Sec. 9.1.8, and the nomenclature of Table 9.18.

READ 0, N	Number of observations
READ 0, (X(I), I = 1, N)	Independent variable values

```
      READ 0, (Y(I), I = 1, N)                  Dependent vari-
                                                  able values
      SUMY = 0.
      SUMY2 = 0.
      SUMX2 = 0.
      SUMXY = 0.
      DO 1 I = 1, N
      SUMY = SUMY + Y(I)                        Σy
      SUMY2 = SUMY2 + Y(I) * Y(I)               Σy²
      SUMX2 = SUMX2 + X(I) * X(I)               Σx²
   1  SUMXY = SUMXY + X(I) * Y(I)               Σxy
      B = SUMXY/SUMX2                           Regression coeffi-
                                                  cient
      C = B * SUMXY                             Sum of squares
                                                  due to regres-
                                                  sion
      RESID = SUMY2 - C                         Residual sum of
                                                  squares
      SPY2 = SUMY2 - SUMY ** 2/FLOAT(N)         Total sum of
                                                  squares [Eq.
                                                  (4.9)]
      IF(RESID) 2, 2, 3                         These statements
   2  CORR = 1.                                  provide for the
      GO TO 4                                    possibility of
   3  IF (SPY2 - RESID) 5, 5, 6                  correlation
   5  CORR = 0.                                  coefficients of
      GO TO 4                                    exactly 1. or 0.
   6  CORR = SQRT((SPY2 - RESID)/SPY2)          Correlation
                                                  coefficient
                                                  [Eq. (9.13)]
   4  NU = N - 1                                Degrees of free-
                                                  dom
```

12.3.3 Log-log Regression Calculation, $Y = A * X ** B$

If the usual linear regression program is used for a log-log regression calculation, after first transforming the variables to their log values, the results obtained will be for the equation $\log(Y) = \log(A) + B * \log(X)$, and the sums of squares and correlation coefficient will be based on deviations from the mean of the $\log(Y)$ values. The values of interest are a comparison of the sums of squares of deviations from the log-log regression line and the sums of squares of deviations of the Y values from their mean. The following program calculates the log-log regression line and then calculates the correlation coefficient based on the comparison of the sums of squares of deviations from the log-log line and original Y values.

READ 0, N	Number of observations
READ 0, (X(I), I = 1, N)	Independent variable values
READ 0, (Y(I), I = 1, N)	Dependent variable values
SUMXL = 0.	
SUMXL2 = 0.	
SUMY = 0.	
SUMYL = 0.	
SUMY2 = 0.	
SXLYL = 0.	
DO 1 I = 1, N	
XL = ALOG(X(I))	
YL = ALOG(Y(I))	
SUMXL = SUMXL + XL	$\Sigma \log(x)$
SUMXL2 = SUMXL2 + XL * XL	$\Sigma (\log(x))^2$
SUMY = SUMY + Y(I)	Σy
SUMYL = SUMYL + YL	$\Sigma \log(y)$
SUMY2 = SUMY2 + Y(I) * Y(I)	Σy^2
1 SXLYL = SXLYL + XL * YL	$\Sigma \log(x).\log(y)$
SPXL2 = SUMXL2 − SUMXL * SUMXL/FLOAT(N)	$\Sigma'(\log(x))^2$
SPY2 = SUMY2 − SUMY ** 2/FLOAT(N)	$\Sigma' y^2$
SPXLYL = SXLYL − SUMXL * SUMYL/FLOAT(N)	$\Sigma'(\log(x).\log(y))$
B = SPXLYL/SPXL2	Regression coefficient [(Eq. (9.4)]
LA = SUMYL/FLOAT(N) − B * SUMXL/FLOAT(N)	Intercept of regression in log form (Eq. (9.2)]
A = EXP(LA)	Coefficient A of regression in exponential form: $Y = AX^B$
RESID = 0.	
DO 2 I = 1, N	
2 RESID = RESID + (Y(I) − A * X(I) ** B) ** 2	Residual sum of squares
IF(SPY2 − RESID) 3, 3, 4	These statements provide for the possibility of correlation coefficient of exactly 0. or 1.
3 CORR = 0.	
GO TO 5	
4 IF(RESID) 6, 6, 7	
6 CORR = 1.	
GO TO 5	

COMPUTER CALCULATIONS

7   CORR = SQRT((SPY2 − RESID)/SPY2)   Correlation coefficient [Eq. (9.13)]

5   NU = N − 2   Degrees of freedom

12.3.4 Semi-log Regression Calculation, $Y = A * EXP(B * X)$

A similar problem exists for the routine calculation of a semilog regression as for the log-log regression. If the logarithms of the dependent variable are used for the regression analysis, the sums of squares and the correlation coefficient will be calculated with reference to the mean of the logarithms rather than the mean of the original observations. The following program calculates the semilog regression, but calculates the correlation coefficient based on a comparison with the deviations from the mean of the original observations.

READ 0, N	Number of observations
READ 0, (X(I), I = 1, N)	Independent variable values
READ 0, (Y(I), I = 1, N)	Dependent variable values
SUMX = 0.	
SUMX2 = 0.	
SUMY = 0.	
SUMY2 = 0.	
SUMYL = 0.	
SUMXYL = 0.	
DO 1 I = 1, N	
YL = ALOG(Y(I))	
SUMX = SUMX + X(I)	Σx
SUMX2 = SUMX2 + X(I) * X(I)	Σx^2
SUMY = SUMY + Y(I)	Σy
SUMY2 = SUMY2 + Y(I) * Y(I)	Σy^2
SUMYL = SUMYL + YL	$\Sigma \log(y)$
1 SUMXYL = SUMXYL + X(I) * YL	$\Sigma x.\log(y)$
SPX2 = SUMX2 − SUMX ** 2/FLOAT(N)	$\Sigma' x^2$
SPY2 = SUMY2 − SUMY ** 2/FLOAT(N)	$\Sigma' y^2$
SPXYL = SUMXYL − SUMX * SUMYL/FLOAT(N)	$\Sigma' x.\log(y)$
B = SPXYL/SPX2	Regression coefficient [Eq. (9.4)]
LA = SUMYL/FLOAT(N) − B * SUMX/FLOAT(N)	Intercept of regression in semilog form [Eq. (9.2)]

	A = EXP(LA)	Coefficient A of regression in exponential form: $Y = A.\text{EXP}(BX)$
	RESID = 0.	
	DO 2 I = 1, N	
2	RESID = RESID + (Y(I) − A * EXP(B * X(I))) ** 2	Residual sums of squares
	IF (RESID) 3, 3, 4	
3	CORR = 1.	These statements
	GO TO 5	provide for the
4	IF (SPY2 − RESID) 6, 6, 7	possibility of
6	CORR = 0.	correlation
	GO TO 5	coefficient of exactly 1. or 0.
7	CORR = SQRT((SPY2 − RESID)/SPY2)	Correlation coefficient [Eq. (9.13)]
5	NU = N − 2	Degrees of freedom

12.3.5 Multiple Regression through Origin with Two Independent Variables, $Y = B_1 X_1 + B_2 X_2$

Many multiple variable regression analysis programs are available. The singular feature of the following program is the restriction of the regression through the origin. The calculations conform to those given in Table 9.28, with the modification that the relations in the following program include the 0., 0. restriction while the equations in Table 9.28 apply to the general case.

READ 0, N	Number of observations
READ 0, (X1(I), I = 1, N)	First independent variable
READ 0, (X2(I), I = 1, N)	Second independent variable
READ 0, (Y(I), I = 1, N)	Dependent variable
SUMX1 = 0.	
SUMX12 = 0.	
SUMX2 = 0.	
SUMX22 = 0.	
SX1X2 = 0.	
SUMY = 0.	

```
        SUMY2 = 0.
        SX1Y = 0.
        SX2Y = 0.
        DO 1 I = 1, N
        SUMX1 = SUMX1 + X1(I)                          $\Sigma x_1$
        SUMX12 = SUMX12 + X1(I) * X1(I)                $\Sigma x_1^2$
        SUMX2 = SUMX2 + X2(I)                          $\Sigma x_2$
        SUMX22 = SUMX22 + X2(I) * X2(I)                $\Sigma x_2^2$
        SX1X2 = SX1X2 + X1(I) * X2(I)                  $\Sigma x_1.x_2$
        SUMY = SUMY + Y(I)                             $\Sigma y$
        SUMY2 = SUMY2 + Y(I) * Y(I)                    $\Sigma y^2$
        SX1Y = SX1Y + X1(I) * Y(I)                     $\Sigma x_1.y$
1       SX2Y = SX2Y + X2(I) * Y(I)                     $\Sigma x_2.y$
        DENOM = SUMX12 * SUMX22 − SX1X2 ** 2
        B1 = (SX1Y * SUMX22 − SX2Y * SX1X2)/DENOM      First regression
                                                       coefficient
        B2 = (SX2Y * SUMX12 − SX1Y * SX1X2)/DENOM      Second regres-
                                                       sion coefficient
        SPY2 = SUMY2 − SUMY * SUMY/FLOAT(N)            $\Sigma' y^2$ Total sum of
                                                       squares
        RESID = SUMY2 − B1 * SX1Y − B2 * SX2Y          Residual sum of
                                                       squares
        IF (RESID) 2, 2, 3                             These statements
2       CORR = 1.                                      provide for the
        GO TO 4                                        possibility of
3       IF (SPY2 − RESID) 5, 5, 6                      the correlation
5       CORR = 0.                                      coefficient
        GO TO 4                                        being exactly
                                                       1. or 0.
6       CORR = SQRT((SPY2 − RESID)/SPY2)               Correlation
                                                       coefficient
4       NU = N − 2                                     Degrees of
                                                       freedom
```

TABLES

Table	Text no.	Title	Text reference, page
I	3.12	Area under the Normal Curve, $F(z)$	55
II	5.1	χ^2	84
III	6.1	t	110
IV	7.8	$F_{0.05}$	160
V	7.9	$F_{0.01}$	160

TABLE I Area under the Normal Curve, $F(z)$

$$F(z) = \int_{-\infty}^{z} \frac{1}{\sqrt{2\pi}} e^{-z^2/2} \, dz$$

z	.00	.01	.02	.03	.04	.05	.06	.07	.08	.09
.0	.5000	.5040	.5080	.5120	.5160	.5199	.5239	.5279	.5319	.5359
.1	.5398	.5438	.5478	.5517	.5557	.5596	.5636	.5675	.5714	.5753
.2	.5793	.5832	.5871	.5910	.5948	.5987	.6026	.6064	.6103	.6141
.3	.6179	.6217	.6255	.6293	.6331	.6368	.6406	.6443	.6480	.6517
.4	.6554	.6591	.6628	.6664	.6700	.6736	.6772	.6808	.6844	.6879
.5	.6915	.6950	.6985	.7019	.7054	.7088	.7123	.7157	.7190	.7224
.6	.7257	.7291	.7324	.7357	.7389	.7422	.7454	.7486	.7517	.7549
.7	.7580	.7611	.7642	.7673	.7704	.7734	.7764	.7794	.7823	.7852
.8	.7881	.7910	.7939	.7967	.7995	.8023	.8051	.8078	.8106	.8133
.9	.8159	.8186	.8212	.8238	.8264	.8289	.8315	.8340	.8365	.8389
1.0	.8413	.8438	.8461	.8485	.8508	.8531	.8554	.8577	.8599	.8621
1.1	.8643	.8665	.8686	.8708	.8729	.8749	.8770	.8790	.8810	.8830
1.2	.8849	.8869	.8888	.8907	.8925	.8944	.8962	.8980	.8997	.9015
1.3	.9032	.9049	.9066	.9082	.9099	.9115	.9131	.9147	.9162	.9177
1.4	.9192	.9207	.9222	.9236	.9251	.9265	.9279	.9292	.9306	.9319
1.5	.9332	.9345	.9357	.9370	.9382	.9394	.9406	.9418	.9429	.9441
1.6	.9452	.9463	.9474	.9484	.9495	.9505	.9515	.9525	.9535	.9545
1.7	.9554	.9564	.9573	.9582	.9591	.9599	.9608	.9616	.9625	.9633
1.8	.9641	.9649	.9656	.9664	.9671	.9678	.9686	.9693	.9699	.9706
1.9	.9713	.9719	.9726	.9732	.9738	.9744	.9750	.9756	.9761	.9767
2.0	.9772	.9778	.9783	.9788	.9793	.9798	.9803	.9808	.9812	.9817
2.1	.9821	.9826	.9830	.9834	.9838	.9842	.9846	.9850	.9854	.9857
2.2	.9861	.9864	.9868	.9871	.9875	.9878	.9881	.9884	.9887	.9890
2.3	.9893	.9896	.9898	.9901	.9904	.9906	.9909	.9911	.9913	.9916
2.4	.9918	.9920	.9922	.9925	.9927	.9929	.9931	.9932	.9934	.9936
2.5	.9938	.9940	.9941	.9943	.9945	.9946	.9948	.9949	.9951	.9952
2.6	.9953	.9955	.9956	.9957	.9959	.9960	.9961	.9962	.9963	.9964
2.7	.9965	.9966	.9967	.9968	.9969	.9970	.9971	.9972	.9973	.9974
2.8	.9974	.9975	.9976	.9977	.9977	.9978	.9979	.9979	.9980	.9981
2.9	.9981	.9982	.9982	.9983	.9984	.9984	.9985	.9985	.9986	.9986
3.0	.9987	.9987	.9987	.9988	.9988	.9989	.9989	.9989	.9990	.9990
3.1	.9990	.9991	.9991	.9991	.9992	.9992	.9992	.9992	.9993	.9993
3.2	.9993	.9993	.9994	.9994	.9994	.9994	.9994	.9995	.9995	.9995
3.3	.9995	.9995	.9995	.9996	.9996	.9996	.9996	.9996	.9996	.9997
3.4	.9997	.9997	.9997	.9997	.9997	.9997	.9997	.9997	.9997	.9998

Even percentage points of the normal distribution

$F(z)$	.75	.90	.95	.975	.99	.995	.999	.9995	.99995	.999995
$\alpha = 2[1 - F(z)]$	.50	.20	.10	.05	.02	.01	.002	.001	.0001	.00001
z	0.674	1.282	1.645	1.960	2.326	2.576	3.090	3.291	3.891	4.417

TABLE II χ^2†

Probability of a larger value of χ^2

D.F.	.99	.98	.95	.90	.80	.70	.50	.30	.20	.10	.05	.02	.01	.001
1	.0³157	.0³628	.00393	.0158	.0642	.148	.455	1.074	1.642	2.706	3.841	5.412	6.635	10.827
2	.0201	.0404	.103	.211	.446	.713	1.386	2.408	3.219	4.605	5.991	7.824	9.210	13.815
3	.115	.185	.352	.584	1.005	1.424	2.366	3.665	4.642	6.251	7.815	9.837	11.345	16.268
4	.297	.429	.711	1.064	1.649	2.195	3.357	4.878	5.989	7.779	9.488	11.668	13.277	18.465
5	.554	.752	1.145	1.610	2.343	3.000	4.351	6.064	7.289	9.236	11.070	13.388	15.086	20.517
6	.872	1.134	1.635	2.204	3.070	3.828	5.348	7.231	8.558	10.645	12.592	15.033	16.812	22.457
7	1.239	1.564	2.167	2.833	3.822	4.671	6.346	8.383	9.803	12.017	14.067	16.622	18.475	24.322
8	1.646	2.032	2.733	3.490	4.594	5.527	7.344	9.524	11.030	13.362	15.507	18.168	20.090	26.125
9	2.088	2.532	3.325	4.168	5.380	6.393	8.343	10.656	12.242	14.684	16.919	19.679	21.666	27.877
10	2.558	3.059	3.940	4.865	6.179	7.267	9.342	11.781	13.442	15.987	18.307	21.161	23.209	29.588
11	3.053	3.609	4.575	5.578	6.989	8.148	10.341	12.899	14.631	17.275	19.675	22.618	24.725	31.264
12	3.571	4.178	5.226	6.304	7.807	9.034	11.340	14.011	15.812	18.549	21.026	24.054	26.217	32.909
13	4.107	4.765	5.892	7.042	8.634	9.926	12.340	15.119	16.985	19.812	22.362	25.472	27.688	34.528
14	4.660	5.368	6.571	7.790	9.467	10.821	13.339	16.222	18.151	21.064	23.685	26.873	29.141	36.123
15	5.229	5.985	7.261	8.547	10.307	11.721	14.339	17.322	19.313	22.307	24.996	28.259	30.578	37.697
16	5.812	6.614	7.962	9.312	11.152	12.624	15.338	18.418	20.465	23.542	26.296	29.633	32.000	39.252
17	6.408	7.255	8.672	10.085	12.002	13.531	16.338	19.511	21.615	24.769	27.587	30.995	33.409	40.790
18	7.015	7.906	9.390	10.865	12.857	14.440	17.338	20.601	22.760	25.989	28.869	32.346	34.805	42.312
19	7.633	8.567	10.117	11.651	13.716	15.352	18.338	21.689	23.900	27.204	30.144	33.687	36.191	43.820
20	8.260	9.237	10.851	12.443	14.578	16.266	19.337	22.775	25.038	28.412	31.410	35.020	37.566	45.315
21	8.897	9.915	11.591	13.240	15.445	17.182	20.337	23.858	26.171	29.615	32.671	36.343	38.932	46.797
22	9.542	10.600	12.338	14.041	16.314	18.101	21.337	24.939	27.301	30.813	33.924	37.659	40.289	48.268
23	10.196	11.293	13.091	14.848	17.187	19.021	22.337	26.018	28.429	32.007	35.172	38.968	41.638	49.728
24	10.856	11.992	13.848	15.659	18.062	19.943	23.337	27.096	29.553	33.196	36.415	40.270	42.980	51.179
25	11.524	12.697	14.611	16.473	18.940	20.867	24.337	28.172	30.675	34.382	37.652	41.566	44.314	52.620
26	12.198	13.409	15.379	17.292	19.820	21.792	25.336	29.246	31.795	35.563	38.885	42.856	45.642	54.052
27	12.879	14.125	16.151	18.114	20.703	22.719	26.336	30.319	32.912	36.741	40.113	44.140	46.963	55.476
28	13.565	14.847	16.928	18.939	21.588	23.647	27.336	31.391	34.027	37.916	41.337	45.419	48.278	56.893
29	14.256	15.574	17.708	19.768	22.475	24.577	28.336	32.461	35.139	39.087	42.557	46.693	49.588	58.302
30	14.953	16.306	18.493	20.599	23.364	25.508	29.336	33.530	36.250	40.256	43.773	47.962	50.892	59.703

† Reprinted from Table IV of R. A. Fisher and Frank Yates, "Statistical Tables for Biological, Agricultural and Medical Research," Oliver & Boyd, Ltd., Edinburgh and London, 1953, by permission of the authors and publishers.

TABLE III t†

α, probability of a larger absolute value of t

D.F	.9	.8	.7	.6	.5	.4	.3	.2	.1	.05	.02	.01	.001
1	.158	.325	.510	.727	1.000	1.376	1.963	3.078	6.314	12.706	31.821	63.657	636.619
2	.142	.289	.445	.617	.816	1.061	1.386	1.886	2.920	4.303	6.965	9.925	31.598
3	.137	.277	.424	.584	.765	.978	1.250	1.638	2.353	3.182	4.541	5.841	12.941
4	.134	.271	.414	.569	.741	.941	1.190	1.533	2.132	2.776	3.747	4.604	8.610
5	.132	.267	.408	.559	.727	.920	1.156	1.476	2.015	2.571	3.365	4.032	6.859
6	.131	.265	.404	.553	.718	.906	1.134	1.440	1.943	2.447	3.143	3.707	5.959
7	.130	.263	.402	.549	.711	.896	1.119	1.415	1.895	2.365	2.998	3.499	5.405
8	.130	.262	.399	.546	.706	.889	1.108	1.397	1.860	2.306	2.896	3.355	5.041
9	.129	.261	.398	.543	.703	.883	1.100	1.383	1.833	2.262	2.821	3.250	4.781
10	.129	.260	.397	.542	.700	.879	1.093	1.372	1.812	2.228	2.764	3.169	4.578
11	.129	.260	.396	.540	.697	.876	1.088	1.363	1.796	2.201	2.718	3.106	4.437
12	.128	.259	.395	.539	.695	.873	1.083	1.356	1.782	2.179	2.681	3.055	4.318
13	.128	.259	.394	.538	.694	.870	1.079	1.350	1.771	2.160	2.650	3.012	4.221
14	.128	.258	.393	.537	.692	.868	1.076	1.345	1.761	2.145	2.624	2.977	4.140
15	.128	.258	.393	.536	.691	.866	1.074	1.341	1.753	2.131	2.602	2.947	4.073
16	.128	.258	.392	.535	.690	.865	1.071	1.337	1.746	2.120	2.583	2.921	4.015
17	.128	.257	.392	.534	.689	.863	1.069	1.333	1.740	2.110	2.567	2.898	3.965
18	.127	.257	.392	.534	.688	.862	1.067	1.330	1.734	2.101	2.552	2.878	3.922
19	.127	.257	.391	.533	.688	.861	1.066	1.328	1.729	2.093	2.539	2.861	3.883
20	.127	.257	.391	.533	.687	.860	1.064	1.325	1.725	2.086	2.528	2.845	3.850
21	.127	.257	.391	.532	.686	.859	1.063	1.323	1.721	2.080	2.518	2.831	3.819
22	.127	.256	.390	.532	.686	.858	1.061	1.321	1.717	2.074	2.508	2.819	3.792
23	.127	.256	.390	.532	.685	.858	1.060	1.319	1.714	2.069	2.500	2.807	3.767
24	.127	.256	.390	.531	.685	.857	1.059	1.318	1.711	2.064	2.492	2.797	3.745
25	.127	.256	.390	.531	.684	.856	1.058	1.316	1.708	2.060	2.485	2.787	3.725
26	.127	.256	.390	.531	.684	.856	1.058	1.315	1.706	2.056	2.479	2.779	3.707
27	.127	.256	.389	.531	.684	.855	1.057	1.314	1.703	2.052	2.473	2.771	3.690
28	.127	.256	.389	.530	.683	.855	1.056	1.313	1.701	2.048	2.467	2.763	3.674
29	.127	.256	.389	.530	.683	.854	1.055	1.311	1.699	2.045	2.462	2.756	3.659
30	.127	.256	.389	.530	.683	.854	1.055	1.310	1.697	2.042	2.457	2.750	3.646
40	.126	.255	.388	.529	.681	.851	1.050	1.303	1.684	2.021	2.423	2.704	3.551
60	.126	.254	.387	.527	.679	.848	1.046	1.296	1.671	2.000	2.390	2.660	3.460
120	.126	.254	.386	.526	.677	.845	1.041	1.289	1.658	1.980	2.358	2.617	3.373
∞	.126	.253	.385	.524	.674	.842	1.036	1.282	1.645	1.960	2.326	2.576	3.291

† Reprinted from Table III of R. A. Fisher and Frank Yates, "Statistical Tables for Biological, Agricultural and Medical Research," Oliver & Boyd, Ltd., Edinburgh and London, 1953, by permission of the authors and publishers.

TABLE IV $F_{0.05}$†
0.05 probability of a larger value of F

ν_2 = degrees of freedom for numerator
ν_1 = degrees of freedom for denominator

ν_1 \ ν_2	1	2	3	4	5	6	7	8	9	10	11	12	14	16	20	24	30	40	50	75	100	200	500	∞
1	161	200	216	225	230	234	237	239	241	242	243	244	245	246	248	249	250	251	252	253	253	254	254	254
2	18.51	19.00	19.16	19.25	19.30	19.33	19.36	19.37	19.38	19.39	19.40	19.41	19.42	19.42	19.44	19.45	19.46	19.47	19.47	19.48	19.49	19.49	19.50	19.50
3	10.13	9.55	9.28	9.12	9.01	8.94	8.88	8.84	8.81	8.78	8.76	8.74	8.71	8.69	8.66	8.64	8.62	8.60	8.58	8.57	8.56	8.54	8.54	8.53
4	7.71	6.94	6.59	6.39	6.26	6.16	6.09	6.04	6.00	5.96	5.93	5.91	5.87	5.84	5.80	5.77	5.74	5.71	5.70	5.68	5.66	5.65	5.64	5.63
5	6.61	5.79	5.41	5.19	5.05	4.95	4.88	4.82	4.78	4.74	4.70	4.68	4.64	4.60	4.56	4.53	4.50	4.46	4.44	4.42	4.40	4.38	4.37	4.36
6	5.99	5.14	4.76	4.53	4.39	4.28	4.21	4.15	4.10	4.06	4.03	4.00	3.96	3.92	3.87	3.84	3.81	3.77	3.75	3.72	3.71	3.69	3.68	3.67
7	5.59	4.74	4.35	4.12	3.97	3.87	3.79	3.73	3.68	3.63	3.60	3.57	3.52	3.49	3.44	3.41	3.38	3.34	3.32	3.29	3.28	3.25	3.24	3.23
8	5.32	4.46	4.07	3.84	3.69	3.58	3.50	3.44	3.39	3.34	3.31	3.28	3.23	3.20	3.15	3.12	3.08	3.05	3.03	3.00	2.98	2.96	2.94	2.93
9	5.12	4.26	3.86	3.63	3.48	3.37	3.29	3.23	3.18	3.13	3.10	3.07	3.02	2.98	2.93	2.90	2.86	2.82	2.80	2.77	2.76	2.73	2.72	2.71
10	4.96	4.10	3.71	3.48	3.33	3.22	3.14	3.07	3.02	2.97	2.94	2.91	2.86	2.82	2.77	2.74	2.70	2.67	2.64	2.61	2.59	2.56	2.55	2.54
11	4.84	3.98	3.59	3.36	3.20	3.09	3.01	2.95	2.90	2.86	2.82	2.79	2.74	2.70	2.65	2.61	2.57	2.53	2.50	2.47	2.45	2.42	2.41	2.40
12	4.75	3.88	3.49	3.26	3.11	3.00	2.92	2.85	2.80	2.76	2.72	2.69	2.64	2.60	2.54	2.50	2.46	2.42	2.40	2.36	2.35	2.32	2.31	2.30
13	4.67	3.80	3.41	3.18	3.02	2.92	2.84	2.77	2.72	2.67	2.63	2.60	2.55	2.51	2.46	2.42	2.38	2.34	2.32	2.28	2.26	2.24	2.22	2.21
14	4.60	3.74	3.34	3.11	2.96	2.85	2.77	2.70	2.65	2.60	2.56	2.53	2.48	2.44	2.39	2.35	2.31	2.27	2.24	2.21	2.19	2.16	2.14	2.13
15	4.54	3.68	3.29	3.06	2.90	2.79	2.70	2.64	2.59	2.55	2.51	2.48	2.43	2.39	2.33	2.29	2.25	2.21	2.18	2.15	2.12	2.10	2.08	2.07
16	4.49	3.63	3.24	3.01	2.85	2.74	2.66	2.59	2.54	2.49	2.45	2.42	2.37	2.33	2.28	2.24	2.20	2.16	2.13	2.09	2.07	2.04	2.02	2.01
17	4.45	3.59	3.20	2.96	2.81	2.70	2.62	2.55	2.50	2.45	2.41	2.38	2.33	2.29	2.23	2.19	2.15	2.11	2.08	2.04	2.02	1.99	1.97	1.96
18	4.41	3.55	3.16	2.93	2.77	2.66	2.58	2.51	2.46	2.41	2.37	2.34	2.29	2.25	2.19	2.15	2.11	2.07	2.04	2.00	1.98	1.95	1.93	1.92
19	4.38	3.52	3.13	2.90	2.74	2.63	2.55	2.48	2.43	2.38	2.34	2.31	2.26	2.21	2.15	2.11	2.07	2.02	2.00	1.96	1.94	1.91	1.90	1.88
20	4.35	3.49	3.10	2.87	2.71	2.60	2.52	2.45	2.40	2.35	2.31	2.28	2.23	2.18	2.12	2.08	2.04	1.99	1.96	1.92	1.90	1.87	1.85	1.84
21	4.32	3.47	3.07	2.84	2.68	2.57	2.49	2.42	2.37	2.32	2.28	2.25	2.20	2.15	2.09	2.05	2.00	1.96	1.93	1.89	1.87	1.84	1.82	1.81
22	4.30	3.44	3.05	2.82	2.66	2.55	2.47	2.40	2.35	2.30	2.26	2.23	2.18	2.13	2.07	2.03	1.98	1.93	1.91	1.87	1.84	1.81	1.80	1.78
23	4.28	3.42	3.03	2.80	2.64	2.53	2.45	2.38	2.32	2.28	2.24	2.20	2.14	2.10	2.04	2.00	1.96	1.91	1.88	1.84	1.82	1.79	1.77	1.76
24	4.26	3.40	3.01	2.78	2.62	2.51	2.43	2.36	2.30	2.26	2.22	2.18	2.13	2.09	2.02	1.98	1.94	1.89	1.86	1.82	1.80	1.76	1.74	1.73
25	4.24	3.38	2.99	2.76	2.60	2.49	2.41	2.34	2.28	2.24	2.20	2.16	2.11	2.06	2.00	1.96	1.92	1.87	1.84	1.80	1.77	1.74	1.72	1.71

TABLES

26	4.22	3.37	2.98	2.74	2.59	2.47	2.39	2.32	2.27	2.22	2.18	2.15	2.10	2.05	1.99	1.95	1.90	1.85	1.82	1.78	1.76	1.72	1.70	1.69
27	4.21	3.35	2.96	2.73	2.57	2.46	2.37	2.30	2.25	2.20	2.16	2.13	2.08	2.03	1.97	1.93	1.88	1.84	1.80	1.76	1.74	1.71	1.68	1.67
28	4.20	3.34	2.95	2.71	2.56	2.44	2.36	2.29	2.24	2.19	2.15	2.12	2.06	2.02	1.96	1.91	1.87	1.81	1.78	1.75	1.72	1.69	1.67	1.65
29	4.18	3.33	2.93	2.70	2.54	2.43	2.35	2.28	2.22	2.18	2.14	2.10	2.05	2.00	1.94	1.90	1.85	1.80	1.77	1.73	1.71	1.68	1.65	1.64
30	4.17	3.32	2.92	2.69	2.53	2.42	2.34	2.27	2.21	2.16	2.12	2.09	2.04	1.99	1.93	1.89	1.84	1.79	1.76	1.72	1.69	1.66	1.64	1.62
32	4.15	3.30	2.90	2.67	2.51	2.40	2.32	2.25	2.19	2.14	2.10	2.07	2.02	1.97	1.91	1.86	1.82	1.76	1.74	1.69	1.67	1.64	1.61	1.59
34	4.13	3.28	2.88	2.65	2.49	2.38	2.30	2.23	2.17	2.12	2.08	2.05	2.00	1.95	1.89	1.84	1.80	1.74	1.71	1.67	1.64	1.61	1.59	1.57
36	4.11	3.26	2.86	2.63	2.48	2.36	2.28	2.21	2.15	2.10	2.06	2.03	1.98	1.93	1.87	1.82	1.78	1.72	1.69	1.65	1.62	1.59	1.56	1.55
38	4.10	3.25	2.85	2.62	2.46	2.35	2.26	2.19	2.14	2.09	2.05	2.02	1.96	1.92	1.85	1.80	1.76	1.71	1.67	1.63	1.60	1.57	1.54	1.53
40	4.08	3.23	2.84	2.61	2.45	2.34	2.25	2.18	2.12	2.07	2.04	2.00	1.95	1.90	1.84	1.79	1.74	1.69	1.66	1.61	1.59	1.55	1.53	1.51
42	4.07	3.22	2.83	2.59	2.44	2.32	2.24	2.17	2.11	2.06	2.02	1.99	1.94	1.89	1.82	1.78	1.73	1.68	1.64	1.60	1.57	1.54	1.51	1.49
44	4.06	3.21	2.82	2.58	2.43	2.31	2.23	2.16	2.10	2.05	2.01	1.98	1.92	1.88	1.81	1.76	1.72	1.66	1.63	1.58	1.56	1.52	1.50	1.48
46	4.05	3.20	2.81	2.57	2.42	2.30	2.22	2.14	2.09	2.04	2.00	1.97	1.91	1.87	1.80	1.75	1.71	1.65	1.62	1.57	1.54	1.51	1.48	1.46
48	4.04	3.19	2.80	2.56	2.41	2.30	2.21	2.14	2.08	2.03	1.99	1.96	1.90	1.86	1.79	1.74	1.70	1.64	1.61	1.56	1.53	1.50	1.47	1.45
50	4.03	3.18	2.79	2.56	2.40	2.29	2.20	2.13	2.07	2.02	1.98	1.95	1.90	1.85	1.78	1.74	1.69	1.63	1.60	1.55	1.52	1.48	1.46	1.44
55	4.02	3.17	2.78	2.54	2.38	2.27	2.18	2.11	2.05	2.00	1.97	1.93	1.88	1.83	1.76	1.72	1.67	1.61	1.58	1.52	1.50	1.46	1.43	1.41
60	4.00	3.15	2.76	2.52	2.37	2.25	2.17	2.10	2.04	1.99	1.95	1.92	1.86	1.81	1.75	1.70	1.65	1.59	1.56	1.50	1.48	1.44	1.41	1.39
65	3.99	3.14	2.75	2.51	2.36	2.24	2.15	2.08	2.02	1.98	1.94	1.90	1.85	1.80	1.73	1.68	1.63	1.57	1.54	1.49	1.46	1.42	1.39	1.37
70	3.98	3.13	2.74	2.50	2.35	2.23	2.14	2.07	2.01	1.97	1.93	1.89	1.84	1.79	1.72	1.67	1.62	1.56	1.53	1.47	1.45	1.40	1.37	1.35
80	3.96	3.11	2.72	2.48	2.33	2.21	2.12	2.05	1.99	1.95	1.91	1.88	1.82	1.77	1.70	1.65	1.60	1.54	1.51	1.45	1.42	1.38	1.35	1.32
100	3.94	3.09	2.70	2.46	2.30	2.19	2.10	2.03	1.97	1.92	1.88	1.85	1.79	1.75	1.68	1.63	1.57	1.51	1.48	1.42	1.39	1.34	1.30	1.28
125	3.92	3.07	2.68	2.44	2.29	2.17	2.08	2.01	1.95	1.90	1.86	1.83	1.77	1.72	1.65	1.60	1.55	1.49	1.45	1.39	1.36	1.31	1.27	1.25
150	3.91	3.06	2.67	2.43	2.27	2.16	2.07	2.00	1.94	1.89	1.85	1.82	1.76	1.71	1.64	1.59	1.54	1.47	1.44	1.37	1.34	1.29	1.25	1.22
200	3.89	3.04	2.65	2.41	2.26	2.14	2.05	1.98	1.92	1.87	1.83	1.80	1.74	1.69	1.62	1.57	1.52	1.45	1.42	1.35	1.32	1.26	1.22	1.19
400	3.86	3.02	2.62	2.39	2.23	2.12	2.03	1.96	1.90	1.85	1.81	1.78	1.72	1.67	1.60	1.54	1.49	1.42	1.38	1.32	1.28	1.22	1.16	1.13
1000	3.85	3.00	2.61	2.38	2.22	2.10	2.02	1.95	1.89	1.84	1.80	1.76	1.70	1.65	1.58	1.53	1.47	1.41	1.36	1.30	1.26	1.19	1.13	1.08
∞	3.84	2.99	2.60	2.37	2.21	2.09	2.01	1.94	1.88	1.83	1.79	1.75	1.69	1.64	1.57	1.52	1.46	1.40	1.35	1.28	1.24	1.17	1.11	1.00

† Reprinted from Table 10.5.3 in George W. Snedecor, "Statistical Methods," 5th ed., Iowa State College Press, Ames, Iowa, 1956, by permission of the author and publisher.

TABLE V $F_{0.01}$[†]

0.01 probability of a larger value of F

ν_2 = degrees of freedom for numerator
ν_1 = degrees of freedom for denominator

ν_1 \ ν_2	1	2	3	4	5	6	7	8	9	10	11	12	14	16	20	24	30	50	75	100	200	500	∞
1	4,052	4,999	5,403	5,625	5,764	5,859	5,928	5,981	6,022	6,056	6,082	6,106	5,142	6,169	6,208	6,234	6,258	6,302	6,323	6,334	6,352	6,361	6,366
2	98.49	99.00	99.17	99.25	99.30	99.33	99.34	99.36	99.38	99.40	99.41	99.42	99.43	99.44	99.45	99.46	99.47	99.48	99.49	99.49	99.49	99.50	99.50
3	34.12	30.82	29.46	28.71	28.24	27.91	27.67	27.49	27.34	27.23	27.13	27.05	26.92	26.83	26.69	26.60	26.50	26.41	26.27	26.23	26.18	26.14	26.12
4	21.20	18.00	16.69	15.98	15.52	15.21	14.98	14.80	14.66	14.54	14.45	14.37	14.24	14.15	14.02	13.93	13.83	13.69	13.61	13.57	13.52	13.48	13.46
5	16.26	13.27	12.06	11.39	10.97	10.67	10.45	10.27	10.15	10.05	9.96	9.89	9.77	9.68	9.55	9.47	9.38	9.24	9.17	9.13	9.07	9.04	9.02
6	13.74	10.92	9.78	9.15	8.75	8.47	8.26	8.10	7.98	7.87	7.79	7.72	7.60	7.52	7.39	7.31	7.23	7.09	7.02	6.99	6.94	6.90	6.88
7	12.25	9.55	8.45	7.85	7.46	7.19	7.00	6.84	6.71	6.62	6.54	6.47	6.35	6.27	6.15	6.07	5.98	5.85	5.78	5.75	5.70	5.67	5.65
8	11.26	8.65	7.59	7.01	6.63	6.37	6.19	6.03	5.91	5.82	5.74	5.67	5.56	5.48	5.36	5.28	5.20	5.06	5.00	4.96	4.91	4.88	4.86
9	10.56	8.02	6.99	6.42	6.06	5.80	5.62	5.47	5.35	5.26	5.18	5.11	5.00	4.92	4.80	4.73	4.64	4.51	4.45	4.41	4.36	4.33	4.31
10	10.04	7.56	6.55	5.99	5.64	5.39	5.21	5.06	4.95	4.85	4.78	4.71	4.60	4.52	4.41	4.33	4.25	4.12	4.05	4.01	3.96	3.93	3.91
11	9.65	7.20	6.22	5.67	5.32	5.07	4.88	4.74	4.63	4.54	4.46	4.40	4.29	4.21	4.10	4.02	3.94	3.80	3.74	3.70	3.66	3.62	3.60
12	9.33	6.93	5.95	5.41	5.06	4.82	4.65	4.50	4.39	4.30	4.22	4.16	4.05	3.98	3.86	3.78	3.70	3.56	3.49	3.46	3.41	3.38	3.36
13	9.07	6.70	5.74	5.20	4.86	4.62	4.44	4.30	4.19	4.10	4.02	3.96	3.85	3.78	3.67	3.59	3.51	3.37	3.30	3.27	3.21	3.18	3.16
14	8.86	6.51	5.56	5.03	4.69	4.46	4.28	4.14	4.03	3.94	3.86	3.80	3.70	3.62	3.51	3.43	3.34	3.21	3.14	3.11	3.06	3.02	3.00
15	8.68	6.36	5.42	4.89	4.56	4.32	4.14	4.00	3.89	3.80	3.73	3.67	3.56	3.48	3.36	3.29	3.20	3.07	3.00	2.97	2.92	2.89	2.87
16	8.53	6.23	5.29	4.77	4.44	4.20	4.03	3.89	3.78	3.69	3.61	3.55	3.45	3.37	3.25	3.18	3.10	2.96	2.89	2.86	2.80	2.77	2.75
17	8.40	6.11	5.18	4.67	4.34	4.10	3.93	3.79	3.68	3.59	3.52	3.45	3.35	3.27	3.16	3.08	3.00	2.86	2.70	2.76	2.70	2.67	2.65
18	8.28	6.01	5.09	4.58	4.25	4.01	3.85	3.71	3.60	3.51	3.44	3.37	3.27	3.19	3.07	3.00	2.91	2.78	2.71	2.68	2.62	2.59	2.57
19	8.18	5.93	5.01	4.50	4.17	3.94	3.77	3.63	3.52	3.43	3.36	3.30	3.19	3.12	3.00	2.92	2.84	2.70	2.63	2.60	2.54	2.51	2.49
20	8.10	5.85	4.94	4.43	4.10	3.87	3.71	3.56	3.45	3.37	3.30	3.23	3.13	3.05	2.94	2.86	2.77	2.63	2.56	2.53	2.47	2.44	2.42
21	8.02	5.78	4.87	4.37	4.04	3.81	3.65	3.51	3.40	3.31	3.24	3.17	3.07	2.99	2.88	2.80	2.72	2.58	2.51	2.47	2.42	2.38	2.36
22	7.94	5.72	4.82	4.31	3.99	3.76	3.59	3.45	3.35	3.26	3.18	3.12	3.02	2.94	2.83	2.75	2.67	2.53	2.46	2.42	2.37	2.33	2.31
23	7.88	5.66	4.76	4.26	3.94	3.71	3.54	3.41	3.30	3.21	3.14	3.07	2.97	2.89	2.78	2.70	2.62	2.48	2.41	2.37	2.32	2.28	2.26
24	7.82	5.61	4.72	4.22	3.90	3.67	3.50	3.36	3.25	3.17	3.09	3.03	2.93	2.85	2.74	2.66	2.58	2.44	2.36	2.33	2.27	2.23	2.21
25	7.77	5.57	4.68	4.18	3.86	3.63	3.46	3.32	3.21	3.13	3.05	2.99	2.89	2.81	2.70	2.62	2.54	2.40	2.32	2.29	2.23	2.19	2.17

26	7.72	5.53	4.64	4.14	3.82	3.59	3.42	3.29	3.17	3.09	3.02	2.96	2.86	2.77	2.66	2.58	2.50	2.41	2.36	2.28	2.25	2.19	2.15	2.13
27	7.68	5.49	4.60	4.11	3.79	3.56	3.39	3.26	3.14	3.06	2.98	2.93	2.83	2.74	2.63	2.55	2.47	2.38	2.33	2.25	2.21	2.16	2.12	2.10
28	7.64	5.45	4.57	4.07	3.76	3.53	3.36	3.23	3.11	3.03	2.95	2.90	2.80	2.71	2.60	2.52	2.44	2.35	2.30	2.22	2.18	2.13	2.09	2.06
29	7.60	5.42	4.54	4.04	3.73	3.50	3.33	3.20	3.08	3.00	2.92	2.87	2.77	2.68	2.57	2.49	2.41	2.32	2.27	2.19	2.15	2.10	2.06	2.03
30	7.56	5.39	4.51	4.02	3.70	3.47	3.30	3.17	3.06	2.98	2.90	2.84	2.74	2.66	2.55	2.47	2.38	2.29	2.24	2.16	2.13	2.07	2.03	2.01
32	7.50	5.34	4.46	3.97	3.66	3.42	3.25	3.12	3.01	2.94	2.86	2.80	2.70	2.62	2.51	2.42	2.34	2.25	2.20	2.12	2.08	2.02	1.98	1.96
34	7.44	5.29	4.42	3.93	3.61	3.38	3.21	3.08	2.97	2.89	2.82	2.76	2.66	2.58	2.47	2.38	2.30	2.21	2.15	2.08	2.04	1.98	1.94	1.91
36	7.39	5.25	4.38	3.89	3.58	3.35	3.18	3.04	2.94	2.86	2.78	2.72	2.62	2.54	2.43	2.35	2.26	2.17	2.12	2.04	2.00	1.94	1.90	1.87
38	7.35	5.21	4.34	3.86	3.54	3.32	3.15	3.02	2.91	2.82	2.75	2.69	2.59	2.51	2.40	2.32	2.22	2.14	2.08	2.00	1.97	1.90	1.86	1.84
40	7.31	5.18	4.31	3.83	3.51	3.29	3.12	2.99	2.88	2.80	2.73	2.66	2.56	2.49	2.37	2.29	2.20	2.11	2.05	1.97	1.94	1.88	1.84	1.81
42	7.27	5.15	4.29	3.80	3.49	3.26	3.10	2.96	2.86	2.77	2.70	2.64	2.54	2.46	2.35	2.26	2.17	2.08	2.02	1.94	1.91	1.85	1.80	1.78
44	7.24	5.12	4.26	3.78	3.46	3.24	3.07	2.94	2.84	2.75	2.68	2.62	2.52	2.44	2.32	2.24	2.15	2.06	2.00	1.92	1.88	1.82	1.78	1.75
46	7.21	5.10	4.24	3.76	3.44	3.22	3.05	2.92	2.82	2.73	2.66	2.60	2.50	2.42	2.30	2.22	2.13	2.04	1.98	1.90	1.86	1.80	1.76	1.72
48	7.19	5.08	4.22	3.74	3.42	3.20	3.04	2.90	2.80	2.71	2.64	2.58	2.48	2.40	2.28	2.20	2.11	2.02	1.96	1.88	1.84	1.78	1.73	1.70
50	7.17	5.06	4.20	3.72	3.41	3.18	3.02	2.88	2.78	2.70	2.62	2.56	2.46	2.39	2.26	2.18	2.10	2.00	1.94	1.82	1.82	1.76	1.71	1.68
55	7.12	5.01	4.16	3.68	3.37	3.15	2.98	2.85	2.75	2.66	2.59	2.53	2.43	2.35	2.23	2.15	2.06	1.96	1.90	1.82	1.78	1.71	1.66	1.64
60	7.08	4.98	4.13	3.65	3.34	3.12	2.95	2.82	2.72	2.63	2.56	2.50	2.40	2.32	2.20	2.12	2.03	1.93	1.87	1.79	1.74	1.68	1.63	1.60
65	7.04	4.95	4.10	3.62	3.31	3.09	2.93	2.79	2.70	2.61	2.54	2.47	2.37	2.30	2.18	2.09	2.00	1.90	1.84	1.76	1.71	1.64	1.60	1.56
70	7.01	4.92	4.08	3.60	3.29	3.07	2.91	2.77	2.67	2.59	2.51	2.45	2.35	2.28	2.15	2.07	1.98	1.88	1.82	1.74	1.69	1.62	1.56	1.53
80	6.96	4.88	4.04	3.56	3.25	3.04	2.87	2.74	2.64	2.55	2.48	2.41	2.32	2.24	2.11	2.03	1.94	1.84	1.78	1.70	1.65	1.57	1.52	1.49
100	6.90	4.82	3.98	3.51	3.20	2.99	2.82	2.69	2.59	2.51	2.43	2.36	2.26	2.19	2.06	1.98	1.89	1.79	1.73	1.64	1.59	1.51	1.46	1.43
125	6.84	4.78	3.94	3.47	3.17	2.95	2.79	2.65	2.56	2.47	2.40	2.33	2.23	2.15	2.03	1.94	1.85	1.75	1.68	1.59	1.54	1.46	1.40	1.37
150	6.81	4.75	3.91	3.44	3.14	2.92	2.76	2.62	2.53	2.44	2.37	2.30	2.20	2.12	2.00	1.91	1.83	1.72	1.66	1.56	1.51	1.43	1.37	1.33
200	6.76	4.71	3.88	3.41	3.11	2.90	2.73	2.60	2.50	2.41	2.34	2.28	2.17	2.09	1.97	1.88	1.79	1.69	1.62	1.53	1.48	1.39	1.33	1.28
400	6.70	4.66	3.83	3.36	3.06	2.85	2.69	2.55	2.46	2.37	2.29	2.23	2.12	2.04	1.92	1.84	1.74	1.64	1.57	1.47	1.42	1.32	1.24	1.19
1000	6.66	4.62	3.80	3.34	3.04	2.82	2.66	2.53	2.43	2.34	2.26	2.20	2.09	2.01	1.89	1.81	1.71	1.61	1.54	1.44	1.38	1.28	1.19	1.11
∞	6.64	4.60	3.78	3.32	3.02	2.80	2.64	2.51	2.41	2.32	2.24	2.18	2.07	1.99	1.87	1.79	1.69	1.59	1.52	1.41	1.36	1.25	1.15	1.00

† Reprinted from Table 10.5.3 in George W. Snedecor, "Statistical Methods," 5th ed., Iowa State College Press, Ames, Iowa, 1956, by permission of the author and publisher.

INDEX

α probabilities, 114, 136
A priori probability, 3
Additivity of chi square, 97
Adjusted chi square, 86
Alias in 2^n factorials, 249
Analysis of variance, 149
  degrees of freedom, 171
  F test, 159
  factorial design, 220
  fractional factorials, 248
  Greco-Latin square, 201
  hierarchical groupings, 178
  interaction, 206
  Latin square, 200
  missing data, 204
  Model 1, 176
  Model 2, 184
  Model 3, 186
  multiple balanced blocks, 200
  nested classifications, 192
  regression, 271
  several means, 171
  single classification with subgrouping, 191
  two factors, 196
    with replication, 206
    with subgrouping and replication, 212

Analysis of variance, 2^n factorials, 237
  variance-estimate calculation, 223
Anderson, R. L., 323n.
Area under normal curve, table 56, 403
ASTM Manual on Quality Control of Materials, 76n.
Average, definition of, 64

β probabilities, 114, 136
Balanced blocks, 235
Bancroft, T. A., 323n.
Barnard, G. A., 354n.
Bartlett, M. S., 167n.
Bartlett's test, 167
Bayes' theorem of probability, 11
Bennett, C. A., 323n.
Best estimate: of mean, 64
  of standard deviation, 70
Best fitting line, 261
Bicking, Charles A., 124n.
Binomial confidence intervals, tables, 38, 39
Binomial distribution, 27
  cumulative, 30
Binomial probability paper, 41
Blocks in 2^n factorials, 245

Box, G. E. P., 332n., 338n.
Buist, J. M., 168n.
Burington, R. S., 30n., 48n., 51n.

Central tendency measures, 63
Chi square (χ^2), 82
  additivity of, 97
  adjusted, 86
  approach to normal distribution, 106
  Bartlett chi square test, 167
  calculation of, for more than 30 degrees of freedom, 106
  contingency tables, 90
  definition of, 83
  degrees of freedom, 84
  goodness of fit test, 99
  null hypothesis, 84
  ratio test, 89
  table, 85, 404
Cochran, W. G., 201n., 221n.
Codex Book Co., 43n.
Coding calculations, 72
Coefficient of correlation, 267
Coefficient of regression, 261
Combinations, 16
Comparison of several slopes, 283
Comparison of two slopes, 278
Computer calculations, 389
Conditional probability, 5
Confidence range: of binomial distribution, 35
  of mean, 120
  of regression, 277
  of slope, 273
  of variances, 166
Confounding, 244
Contingency tables with chi square, 90
Continuity correction to chi square, 86
Corner test for correlation, 381
Correlation, 260
  coefficient of, 260, 267, 311
  corner test, 381
  (See also Regression)
Cox, Gertrude, 201n., 221n.
Critical value, 27
Cumulative binomial distribution, 30
Cumulative Poisson distribution, 49
Curvilinear regression, 318

Davies, O. L., 150n., 201n., 221n., 274n.
Degrees of freedom: in analysis of variance, 175
  in chi square test, 84
  definition of, 76
  in t test, of one mean, 116
  of two means, 128

Deletion of a variable, 306
Deviate, standard, 55
Deviation: mean, 75
  standard (see Standard deviation)
Difference between duplicates, 124
Digits: significant, 77
Discarding data, 384
Distribution: binomial, 27
  chi square, 99
  normal, 53
  Poisson, 46
  t, 109
Dixon, W. J., 59n., 375n., 385n., 388n.
Double-dichotomy test, 349
Duplicate data, 124

Eisenhart, C., 115n., 376n.
Empirical probability, 3
Error: propagation of, 150
Error variance, 204
Errors: of Type I, 134
  of Type II, 134
Estimate of missing data, 204
  of standard deviation, 70
  of true mean, 115

F = chi square/degrees of freedom, 166
$F = t^2$, 272
F tables, 162–165, 406–409
F test, 159
Factorial analysis, 220
  confounding in, 244
  fractional, 248
  2^n design, 235
Factorials, 19
  logs of, 21
Fisher, R. A., 85n., 111n., 200n., 267n., 268n., 319n., 323n., 324n., 404n.
Fortran, 393
Fox, R. T., Jr., 276n.
Fractional 2^n factorials, 299
Franklin, N. L., 323n.

Garner, Norman R., 233n.
Gaussian multipliers, 299
Generating contrast, 246
Geometric mean, 66
Goodness-of-fit test, 91
Goulden, C. H., 323n.
Grant, E. L., 20n., 50n., 115n.
Greco-Latin squares, 200

Half replicate of 2^n, 300
"Handbook of Chemistry and Physics," 20n.

INDEX

Harmonic mean, 67
Hastay, M. W., 115n.
Hicks, Charles, R., 232n.
Hierarchies of subgroups, 225
Homogeneity of variances, 167
Hunter, J. S., 332n.
Hydrocarbon Research, Inc., 269n.
Hypergeometric probability, 21
Hypothesis: null (see Null hypothesis)

Independent variables: one, 261
  several, 297
Interaction, 206
Intercept, 261

Jackson, J. E., 377n.
Johnson, J. Enoch, 287n.

Kay, Webster B., 320n.

Lange, N. A., 20n.
Latin squares, 200
Least significant difference between means, 178
Least-squares line, 261
Lesser, Arthur, 198n.
Levenspiel, Octave, 303n.
Line through origin, 293
Linear regression: more than two variables, 297
  with replication, 289
  two variables, 261
Logarithms of factorials, 20
Lord, E., 377n., 378n., 380n.

Massey, F. S., 59n.
Mathematical expectation, 3
May, D. C., 30n., 48n., 51n.
Mean, 63
  confidence range of, 120
  geometric, 66
  harmonic, 69
  significant digits for, 80
  t test of, 115
  variance of, 112
Mean deviation, 75
Mean difference, 125
Means, difference between, 178
  test of several, 171
Median, 65
Metals Powder Association, 128n., 174n.

Missing data, 204
Mixed models, 249
Mode, 64
Model 1 analysis of variance, 176
Model 2 analysis of variance, 184
Model 3 analysis of variance, 186
Molina, E. C., 48n., 50n.
Mood, A. M., 47n., 70n., 83n., 112n., 343n., 375n.
Mosteller, F., 41n.
Multiple balanced blocks, 200
Multiple regression, 297

Nair, K. R., 385n., 386n.
Nested classes of variables, 225
Nonparametric statistics, 364
Normal curve, area under, table, 56, 403
Normal (standard) deviate, 55
Normal distribution, 53
Normal probability paper, 58
Null hypothesis, 113
  chi-square test, 84
  sequential tests, 343
  t test, 113
  variance test, 159
Number of significant digits, 78

Odds, 3
Odds-ratio test, 349
Olmstead, P. S., 381n.
Optimization, 332
Origin: line through, 356
Orthogonal polynomials, 323

Paired replicate test, 368
Parameter, 26
Partial correlation coefficient, 311
Partial regression coefficient, 298
Pascal's triangle, 28
Pearson, Karl, 83n.
Pelipety, M. G., 263n.
Permutation, 13
Perry, R. H., 58n.
Plewes, A. C., 280n.
Poisson distribution, 56
  cumulative, 49
Poker probabilities, 22
Pooled estimate of standard deviation, 127
Power of t test, 135
Precision, 121
Precision limits, 121

Probability, 1
 a priori, 3
 Bayes' theorem, 11
 conditional 7
 definition, 2
 empirical, 3
 laws of, 5
 multiplication, 5
Programs for digital computers, 473
Propagation of variance, 150

Quadrant sum, 335

Range, 75
 tests with, 376
Rank correlation, 366
Ranking, 365
Ratio test with chi square, 89
Regression, 260
 coefficient of, 261
 confidence range of, 273
 curvilinear, 281
 linear: more than two variables, 297
 two variables, 261
 multiple, 297
 through origin, 293
 with replication, 289
 variance test, 271
Ross, E. L., 377n.
Rounding off digits, 79
Rushton, S., 354n.

Salzer, Herbert E., 20
Sample size, 137
 tables, 142, 144
Savage, I. R., 364n.
Schiffer, Don, 94n.
Schlesinger, M. D., 295n.
Sequential analysis, 342
"Sequential Analysis of Statistical Data," 357n.
Shepherd, Martin, 217n.
Sieder, E. N., 305n.
Siegel, Sidney, 364n., 367n.
Sign test, 374
Significance level, 57
 of chi square test, 84
Significant difference, 57
 of chi square test, 84
 of F test, 159
 of t test, 128
Significant digits, rules for, 77
Sillitto, G. P., 142n., 144n.
Size of sample, 137
 tables, 142

Slope, 261
 confidence limits of, 273
Snedecor, George W., 38n., 40n., 108n., 159n., 163n., 165n., 284n., 311n., 376n., 407n., 409n.
Spearman rank correlation, 366
Standard deviate, 55
Standard deviation, 68
 estimate of, 69
 of mean, 112
 from range, 377
Statistic, 26
"Statistical Abstract of the United States," 1n., 367n., 373n.
Statistics: plural, 1
 singular, 1
Stern, S. Alexander, 320n.
Stewart, P. B., 276n.
Sum of squares, 69

t test, 109
 confidence range of, 120
 degrees of freedom, 113
 null hypothesis, 113
 one sided, 118
 summary, 132
 table, 132, 405
 two means, 127
Table of signs, 237
Tables, chi square, 85, 404
 F, 162–165, 406–409
 normal distribution 56, 403
 t, 111, 405
"Tables of Binomial Probability," 30n.
"Tables of Cumulative Binomial Probability," 33n.
Tate, G. E., 305n.
Tippett, L. H. C., 70n.
Triangle, Pascal's, 28
True mean: confidence range of, 120
 estimate of, 115
Tukey, J. W., 41n., 179n., 181n., 377n., 381n.
Two means: difference between, 127
2^n factorials, 235
Type I errors, 114
Type II errors, 114
 limiting, 134

"United States Life Tables," 47n.
Unpaired rank test, 371

Variability measures, 68

INDEX

Variance, 74
　analysis of, 149
　confidence range of, 166
　of correlation, 271
　of a difference, 154
　of estimate, 212
　F test, 159
　of a general function, 154
　propagation of, 150
　of a sum, 150
Variances: homogeneity of, 167
Verneulen, Theodore, 290n.
Volk, W., 218n.

Wald, A., 261n., 342n.
Wallis, W. A., 115n.

Weighted degrees of freedom, 186
Weighted estimate variance, 128
White, Colin, 371n., 372n.
Wholly significant difference between means, 177
Wilcoxon, Frank, 368n., 369n.
Wilcoxon paired replicate test, 368
Wilcoxon-White unpaired rank test, 371
Williams, G. E., 168n.
Wilson, K. B., 338n.
World Almanac, 1n.

Yates, Frank, 85n., 111n., 201n., 268n., 324n., 325n., 404n.
Youden, W. J., 141n.